The Queen of Mathematics

Kluwer Texts in the Mathematical Sciences

VOLUME 8

A Graduate-Level Book Series

The titles published in this series are listed at the end of this volume.

The Queen of Mathematics

An Introduction to Number Theory

W. S. Anglin

Department of Mathematics and Statistics,
McGill University,
Montreal, Quebec, Canada

KLUWER ACADEMIC PUBLISHERS

DORDRECHT / BOSTON / LONDON

A C.I.P. Catalogue record for this book is available from the Library of Congress.

ISBN 0-7923-3287-3

Published by Kluwer Academic Publishers,
P.O. Box 17, 3300 AA Dordrecht, The Netherlands.

Kluwer Academic Publishers incorporates
the publishing programmes of
D. Reidel, Martinus Nijhoff, Dr W. Junk and MTP Press.

Sold and distributed in the U.S.A. and Canada
by Kluwer Academic Publishers,
101 Philip Drive, Norwell, MA 02061, U.S.A.

In all other countries, sold and distributed
by Kluwer Academic Publishers Group,
P.O. Box 322, 3300 AH Dordrecht, The Netherlands.

Printed on acid-free paper

Printed in the Netherlands

Contents

The stars indicate sections which can be skipped on a first reading. Nothing in an unstarred section depends on anything in a starred section (with the exception of some material in Chapter 7). Since Chapter 6 does not depend on Chapter 5, a first reading might consist of the unstarred sections in Chapters 1 to 4, followed by Chapter 6.

Preface

Like other introductions to the Queen of Mathematics, this one includes the usual curtsy to divisibility theory, the bow to congruence, and the little chat with quadratic reciprocity. It also includes rigorous proofs of historically important results such as

- Lagrange's Four Square Theorem,
- the theorem that n is congruent just in case $y^2 = x^3 - n^2x$ has infinitely many rational points,
- Lucas's theorem on square square pyramid numbers,
- the theorem behind Lucas's test for perfect numbers,
- the theorem that a regular n-gon is constructible just in case $\phi(n)$ is a power of 2,
- the fact that the circle cannot be squared,
- the fact that every natural number is the sum of 3 triangular numbers,
- Fermat's polygonal number conjecture,
- Dirichlet's theorem on primes in arithmetic progressions,
- the Prime Number Theorem, and
- Rademacher's partition theorem.

We have tried to make the proofs of these theorems as accessible as possible. We have avoided higher algebra altogether, and we use analysis only where it is absolutely necessary (and only in the starred, or optional, sections).

Unlike other number theory books, *The Queen of Mathematics* follows the order of history, with the chapter on simple continued fractions preceding the chapter on congruence. This order is just as natural as the more usual order, and it reflects the fact that simple continued fractions are an essential component of much current research in number theory.

Unique to *The Queen of Mathematics* are its presentations of

- the topic of palindromic simple continued fractions,
- an elementary solution of Lucas's square pyramid problem,
- Baker's solution for simultaneous Fermat equations,
- an elementary proof of Fermat's polygonal number conjecture, and
- the Lambek-Moser-Wild theorem.

The reader will also find much historical information about who discovered what when.

For much of the book the only prerequisite is pre-calculus mathematics. However, the reader should be warned that the proofs are tightly written, and will not normally be accessible to someone who has not had several undergraduate courses in mathematics. *The Queen of Mathematics* is an introductory textbook, not for the average mathematics student, but for an Honours student or a first year graduate student.

I thank Andonowati, I. Krisna, J. Lambek, I. Rabinovitch, S. Timruang, M. Tong, and D. D. Zhang for their inspiration and encouragement.

W. S. Anglin, 1995

Chapter 1

Propaedeutics

A natural number is one of the numbers $0, 1, 2, 3, \ldots$. Number Theory, as it is traditionally understood, is that branch of mathematics which studies the natural numbers. It includes ordinary arithmetic. For example, figuring out why long division works is a problem in Number Theory. As we shall see, Number Theory goes much further than this.

The *Concise Oxford Dictionary* defines 'propaedeutics' as 'preliminary learning'. In this chapter, we introduce the basic concepts of Number Theory. However, in order that the reader become intimate with the Queen of Mathematics as soon as possible, we also give some results which, although easy to prove, are usually reserved for the last chapters of introductory books.

1.1 Mathematical Induction

About 500 BC, Pythagoras (or his followers) noticed that numbers such as 3, 6, and 10 can be represented by an isosceles right triangle filled with pebbles. For example, 10 can be represented as in Figure 1.1.

If n is a natural number, the n-th triangular number is defined as

$$0 + 1 + 2 + \cdots + (n-1)$$

For example, 0 is the first triangular number, and 10 is the fifth. From

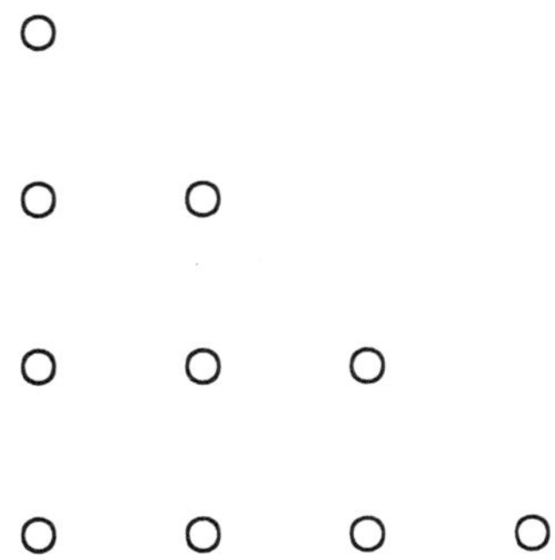

Figure 1.1: Ten as a Triangle

the formula for the area of a right triangle, we see that the n-th triangular number is about

$$\frac{1}{2}\,\text{side} \times \text{side} = \frac{1}{2}n^2$$

In fact,

$$0 + 1 + 2 + \cdots + (n-1) = \frac{1}{2}(n-1)n$$

However, how shall we prove this fact?

A basic property of the natural numbers is that they obey a principle called the Principle of Mathematical Induction (MI). This principle was used by the ancient Greeks, and first stated explicitly by a theologian, Levi ben Gerson (1288–1344), in 1321. The name 'mathematical induction' was introduced by Augustus de Morgan in 1838.

THE PRINCIPLE OF MATHEMATICAL INDUCTION
If (1) something is true of a natural number a, and
if (2) whenever it is true of a natural number b
then it is also true of $b + 1$
then it is true of all natural numbers not less than a.

The 'something' true of a can be any property. For example, it can be the property of 'making the triangular number formula come out true'. (For the benefit of the philosophers, however, we should add that the 'something' is intended not to be a vague property, such as 'is small', or a 'second order' property, such as 'is not nonstandard'.)

The Principle of Mathematical Induction might be called the 'Principle of Upwards Contagion'. Suppose that a natural number a has a contagious disease. Suppose also that whenever a natural number b has this disease, the next higher natural number, $b+1$, catches this disease. Then all the numbers, from a up, are going to be sick. In the case of interest to us here, $a = 1$, and the contagious disease is 'making the triangular number formula come out true'.

The formula

$$0 + 1 + \cdots + (n-1) = \frac{1}{2}(n-1)n$$

works for $n = 1$. Moreover, if it works for n, then we have

$$0 + 1 + 2 + \cdots + (n-1) + n = \frac{1}{2}(n-1)n + n$$

Hence

$$0 + 1 + 2 + \cdots + n = \frac{1}{2}((n+1)-1)(n+1)$$

That is, the formula works for $n+1$. Hence, by mathematical induction, it works for all natural numbers ≥ 1. Our formula for the n-th triangular number is indeed correct.

There are other versions of MI. For example, there is the following.

> Suppose that whenever something is true of all natural numbers less than n then it is also true of n.
> Then it is true of all natural numbers.

Here it is assumed that anything at all is true of all natural numbers less than 0. How can this be? If you claim that all natural numbers less than 0 are pink, I cannot contradict you by pointing to one that is

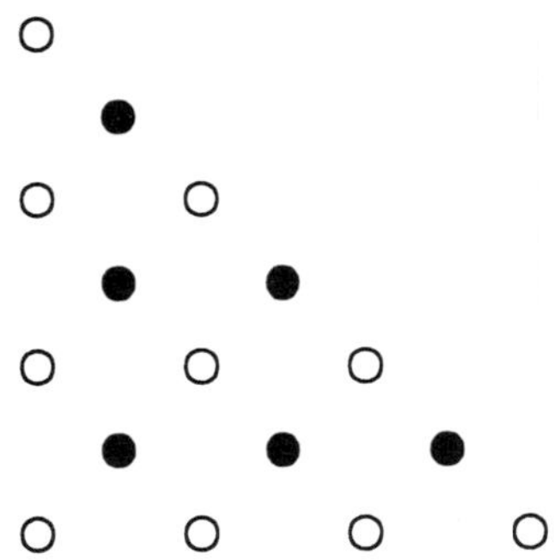

Figure 1.2: Building a Pyramid

not pink— since there is none to point to. Thus I may as well let you have your claim. I shall, however, remark that it is merely 'trivially true'.

Suppose we stack an isosceles right triangle with 6 pebbles on top of the isosceles right triangle with 10 pebbles. The 6 pebbles go in the 6 gaps left by the 10 pebbles, as in Figure 1.2.

On top of that second triangle with 6 pebbles goes an isosceles right triangle with 3 pebbles. Finally, we put 1 pebble on top of the triangle with 3 pebbles. That gives us a complete pyramid with a triangular base. It contains

$$0 + 1 + 3 + 6 + 10 = 20$$

pebbles. We define the n-th *tetrahedral number* as the sum of the first n triangular numbers. For example, 0 is the first tetrahedral number, and 20 is the fifth. Using MI, we can prove that the n-th tetrahedral number is $n(n^2 - 1)/6$.

Certainly, this is true when $n = 1$. Furthermore, if it is true for n, then it follows that the $(n + 1)$-st tetrahedral number is

$$n(n^2 - 1)/6 + n(n + 1)/2 = (n + 1)((n + 1)^2 - 1)/6$$

Hence, by MI, the formula is true for all n.

If we build a pyramid with a square base, we have 1 pebble on top, 4 in the second layer, 9 in the third layer, and so on. The pebbles in any layer (above the base) fit in the holes between the pebbles in the layer beneath it. If the base is a square of side n, then the number of pebbles in the pyramid is

$$1^2 + 2^2 + \cdots + n^2$$

By using MI, one can prove that this equals $n(n+1)(2n+1)/6$. (In the next section we show how to derive such formulas.)

In 1875 Edouard Lucas, who had been a French artillery officer in the Franco-prussian war, challenged the readers of the *Nouvelles Annales de Mathématiques* to prove the following:

> A square pyramid of cannon-balls contains a square number of cannon-balls only when it has 24 cannon-balls along its base.

In other words, the only nontrivial natural number solution of

$$1^2 + 2^2 + \cdots + n^2 = m^2$$

is $n = 24$ and $m = 70$.

Lucas did not live to see his challenge met. The problem was not solved until 1918, when G. N. Watson gave a complicated solution based on a specially extended theory of Jacobian elliptic functions. (See volume 48 of the *Messenger of Mathematics.*) At first, Lucas thought he had a short, completely elementary solution, but no short, completely elementary solution was forthcoming until 1988, when this author found one. The reader can find it in Section 4.4 of this book. There he or she will also find a proof of the fact that the only square tetrahedral numbers are 0, 1, 4, and 19 600.

It was Carl Friedrich Gauss (1777–1855), the 'Prince of Mathematicians', who named Number Theory the 'Queen of Mathematics'.[1] About 1800, Gauss was the first to find a complete proof of the fact that

[1] Sartorius von Waltershausen: Gauss zum Gedächtniss. (Leipzig, 1856), p. 79.

every natural number is a sum of 3 triangular numbers. For example, $7 = 0 + 1 + 6$. We give what is essentially Gauss's own proof of this result in Chapter 6 of this book. We also give Cauchy's generalisation to polygonal numbers.

Exercises 1.1

1. What is the tenth triangular number?
2. Is 41 616 triangular? Why or why not?
3. Prove that the sum of two consecutive triangular numbers is square.
4. Find a square triangular number greater than 1.
5. Express 100 as a sum of 3 triangular numbers.
6. Prove that $1 + 3 + \cdots + (2n - 1) = n^2$.
7. Prove that $1^3 + 2^3 + \cdots + n^3 = (n(n + 1)/2)^2$.
8. Prove that

$$1^3 + 3^3 + 5^3 + \cdots + (2n - 1)^3 = n^2(2n^2 - 1)$$

—a triangular number.
9. Consider the triangle

$$\begin{array}{ccccc} 1 & & & & \\ 3 & 5 & & & \\ 7 & 9 & 11 & & \\ 13 & 15 & 17 & 19 & \text{etc.} \end{array}$$

Nicomachus of Gerasa (Palestine) lived about 100 AD. He was the first to suggest that numbers are (contents of) ideas in the mind of God. He was also the first to note that the sum of the entries in the n-th row of the above triangle is n^3. Prove this.
10. Let $a, b, c,$ and d be natural numbers. Consider the sequence

$$a \quad b \quad ac + bd \quad bc + (ac + bd)d \quad \ldots$$

For example, if $a = b = c = d = 1$, then the sequence is the *Fibonacci sequence*

$$1 \quad 1 \quad 2 \quad 3 \quad 5 \quad 8 \quad \ldots$$

Let w and z be the roots of $x^2 - dx - c$. If $w \neq z$, the n-th term of the sequence is

$$\frac{(b - za)w^{n-1} - (b - wa)z^{n-1}}{w - z}$$

If $w = z$, then the n-th term is

$$(n-1)(d/2)^{n-2}b - (n-2)(d/2)^{n-1}a$$

11. A 'unit fraction' is a fraction of the form $1/c$, where c is an integer greater than 1. Using mathematical induction on a, show that any proper fraction a/b can be written as a sum of distinct unit fractions. (Hint: let 1/q be the largest unit fraction we can subtract from a/b and still get a positive number. Then

$$a/b = 1/q + (qa - b)/bq, \quad qa - b < a, \text{ and } (qa - b)/bq < 1/q.)$$

12. Consider the property 'having all natural numbers not greater than it equal to it'. This is true of 0. For 0 is such that all natural numbers not greater than it are equal to it. Suppose, moreover, that a natural number b has this property. Let c be a natural number not greater than $b+1$. Then $c-1$ is not greater than b. On the 'induction assumption' that b has the given property, it follows that $c-1 = b$. Hence $c = b+1$. Thus $b+1$ is such that all natural numbers not greater than it equal it. Hence, by MI, each natural number is such that all natural numbers not greater than it are equal to it. For example, since 10 is not greater than 20, it follows that $10 = 20$. So find the mistake!

13. Show that 1805/1806 is the largest proper fraction that can be written as a sum of 4 or fewer unit fractions.

14.* This problem is starred because it is quite hard. Prove that there is a proper fraction which cannot be written as a sum of 1000 or fewer unit fractions.

1.2 Bernoulli Numbers *

Let n be a fixed positive integer. Where r is any natural number,

let

$$S(r) = 1^r + 2^r + \cdots + (n-1)^r$$

That is, $S(r)$ is the sum of the r-th powers of the first n natural numbers. For example, $S(0) = n - 1$ and $S(1) = (n-1)n/2 = \frac{1}{2}n^2 - \frac{1}{2}n$.

By the Binomial Theorem,

$$(x+1)^{r+1} - x^{r+1} = \binom{r+1}{1} x^r + \binom{r+1}{2} x^{r-1} + \cdots + 1$$

Substituting $n-1$, $n-2$, ..., 2, 1 for x, we get $n-1$ equations. Adding these equations, we obtain

$$n^{r+1} - 1 = \binom{r+1}{1} S(r) + \binom{r+1}{2} S(r-1) + \cdots + S(0) \qquad (1.1)$$

Thus if we know $S(0)$, $S(1)$, ..., $S(r-1)$, we can compute a formula for $S(r)$. Note that, as can be proved by mathematical induction, $S(r)$ is a polynomial in n of degree $r+1$. Note also that if $r \neq 0$ this polynomial has no constant term.

For any natural number r, let B_r be the coefficient of n in the polynomial equal to $S(r)$. For example, since $S(1) = \frac{1}{2}n^2 - \frac{1}{2}n$, it follows that $B_1 = -\frac{1}{2}$. B_r is called the r-th *Bernoulli number*. As another example, $B_0 = 1$.

The Bernoulli numbers were so named by Abraham De Moivre, in 1730, in recognition of the fact that they were first studied by James Bernoulli (1654–1705). Bernoulli wanted the spiral $r = e^\theta$ engraved on his tombstone, with the inscription 'I shall arise the same, though changed'.

Gathering the coefficients of n in Equation 1.1, we find that

$$0 = \binom{r+1}{1} B_r + \binom{r+1}{2} B_{r-1} + \cdots + \binom{r+1}{r+1} B_0$$

and hence

$$B_r = -\frac{1}{r+1} \sum_{k=2}^{r+1} \binom{r+1}{k} B_{r+1-k}$$

Using this formula, we can calculate the Bernoulli numbers.

n	B_n
0	1
1	$-\frac{1}{2}$
2	$\frac{1}{6}$
3	0
4	$-\frac{1}{30}$
5	0
6	$\frac{1}{42}$

Bernoulli numbers have some fascinating properties.

1. If k is a positive integer, then $B_{2k+1} = 0$.
2. Von Staudt's Theorem. Where k is a positive integer, the denominator of B_{2k} is the product of all primes p such that $p - 1$ is a divisor of $2k$. For example, if $k = 1$, the primes in question are 2 and 3, and, indeed, the denominator of B_2 is their product 6.
3. Euler's Theorem. Let

$$E(t) = 1/1^t + 1/2^t + 1/3^t + \cdots$$

If k is a positive integer,

$$B_{2k} = \pm 2(2k)!E(2k)/(2\pi)^{2k}$$

4. Kummer's Theorem. If p is an odd prime which does not divide evenly into the numerator of any of the numbers $B_2, B_4, \ldots, B_{p-3}$ then there are no positive integers x, y, and z such that $x^p + y^p = z^p$.
5. The Euler-Maclaurin Formula. If k is a positive integer, the coefficient of n^k in the polynomial equal to $S(r)$ is

$$\frac{r!B_{r-k+1}}{(r-k+1)!k!}$$

(Recall that there is no constant term in this polynomial unless $r = 0$.) This can be proved using mathematical induction on n. Bernoulli himself used this formula to calculate the sum of the tenth powers of the natural numbers from 1 to 1000 inclusive. The sum is

$$91,409,924,241,424,243,424,241,924,242,500$$

Exercises 1.2

1. Calculate B_8, B_{10}, B_{12}, and B_{14}.
2. Let $T(r) = (-1)^r + (-2)^r + \cdots + (-(n-1))^r$. Show that

$$-(-(n-1))^{r+1} = \binom{r+1}{1} T(r) + \binom{r+1}{2} T(r-1) + \cdots + T(0)$$

3. Show that if r is odd, then

$$n^{r+1} - 1 - (-(n-1))^{r+1}$$

$$= 2\left(\binom{r+1}{2} S(r-1) + \binom{r+1}{4} S(r-3) + \cdots + S(0)\right)$$

4. Hence, gathering the coefficients of n, if r is odd,

$$\frac{r-1}{2} = \binom{r+1}{2} B_{r-1} + \binom{r+1}{4} B_{r-3} + \cdots + \binom{r+1}{r-1} B_2$$

5. Hence if r is odd and greater than 1, $B_r = 0$.
6. Find a formula for $1^4 + 2^4 + \cdots + n^4$.
7. Show that

$$S(10) = \frac{6n^{11} - 33n^{10} + 55n^9 - 66n^7 + 66n^5 - 33n^3 + 5n}{66}$$

8. Let $f(n)$ be a polynomial with rational coefficients and degree r. Let

$$g(n) = f(1) + f(2) + \cdots + f(n)$$

Prove that $g(n)$ is a polynomial of degree $r+1$. (This result is the foundation of the 'method of differences'. See Chrystal's *Algebra*.)
9. Use Von Staudt's Theorem to show that if p is a prime of the form $3m+1$ then B_{2p} has denominator 6.
10. Assuming Euler's Theorem, show that the absolute value of B_{2k} increases without limit.
11.* Use MI to prove the Euler-Maclaurin Sum Formula.

1.3 Primes

A natural number is a prime if and only if it has exactly two natural number divisors. The first four primes are 2, 3, 5, and 7. Primes are the heart of Number Theory. Almost every question in Number Theory comes down to a question about primes.

As we shall prove, there are infinitely many primes. At the moment (1994), the largest known prime is $2^{848433} - 1$. The first 46 primes are the following.

The First 46 Primes

2	13	31	53	73	101	127	151	179
3	17	37	59	79	103	131	157	181
5	19	41	61	83	107	137	163	191
7	23	43	67	89	109	139	167	193
11	29	47	71	97	113	149	173	197
								199

The pattern of the primes still eludes us. We know that the n-th prime is somewhere in the neighbourhood of $n \log_e n$ but, with the exception of some useless 'artificial' formulas, we do not have any formula giving the n-th prime itself.

One reason that primes are so important is that every natural number greater than 1 has a factorisation into primes, and, disregarding the order of the factors, this factorisation is unique. We prove the uniqueness of the factorisation as follows.

Theorem 1.3.1 *No natural number has more than one prime factorisation.*

Proof: Let n be the smallest natural number, if there is one, which has two factorisations into primes:

$$n = pqr \ldots \text{ and } n = p'q'r' \ldots$$

(with the primes written in nondecreasing order). By n's minimality, $p \neq p'$ and, without loss of generality, we may suppose that $p' < p$.

Since n is not prime, $n \geq p^2$ and hence $n > pp'$. Since $n > n - pp' \geq 1$, it follows that $n - pp'$ has a unique prime factorisation (if it is not equal to 1). By the Distributive Law, p is a factor of $n - pp'$ (and hence $n - pp' \neq 1$) and p' is also a factor of $n - pp'$. Thus

$$pqr \ldots - pp' = pp'Q$$

for some natural number Q. Hence

$$qr \ldots = p'Q + p'$$

Since $qr \ldots < n$, it follows that $qr \ldots$ has a unique prime factorisation. Thus p' is one of the primes q, r, ... But $p' < p \leq q \leq r \ldots$. Contradiction.

The first proof of unique factorisation was given by Gauss in 1801.

The fact that there are infinitely many primes was known to Euclid of Alexandria in 300 BC. He gave the following proof of it.

Theorem 1.3.2 *No finite list of primes is complete.*

Proof: If there are only n primes p_1, p_2, ..., p_n, then let $m = p_1 p_2 \ldots p_n + 1$, and let q be a prime factor of m. Now q is not any of the primes $p_1, \ldots, p_n$ since dividing any of these into m gives remainder 1. So q is not on the list of primes.

If a and b are integers (with a nonzero), we write $a|b$ as an abbreviation for the statement 'a divides evenly into b.' For example, $2|12$ but $2 \not| 11$. If p is a prime and $p|ab$ then the Unique Factorisation Theorem implies that $p|a$ or $p|b$.

We can use primes to develop the theory of greatest common divisors (gcds). The gcd of two integers a and b (not both 0) is the greatest positive integer which divides evenly into both of them. We write this number as $\gcd(a, b)$ or as (a, b). For example, $(12, 15) = 3$ and $(0, 7) = 7$. Note that $(-a, b) = (a, -b) = (a, b)$. Note also that if $(a, b) \neq 1$ then there is a prime which divides both a and b. If there is no such prime then a and b are *relatively prime*.

To find the gcd of two natural numbers a and b, it suffices to find their prime factorisation. Let $p_1, \ldots, p_n$ be the primes which divide into either a or b (or both). Let the prime factorisation of a and b be

$$a = {p_1}^{a_1} \ldots {p_n}^{a_n} \text{ and } b = {p_1}^{b_1} \ldots {p_n}^{b_n}$$

—with a_k and b_k possibly equal to 0. Then

$$(a, b) = {p_1}^{\min(a_1,\, b_1)} \ldots {p_n}^{\min(a_n,\, b_n)}$$

We use the notation (a, b, c) for the greatest common divisor of the three integers a, b, and c.

Exercises 1.3

1. Show that the 47-th prime is 211.
2. Give the prime factorisation of 10 403.
3. Show that a natural number is a square iff all the exponents in its prime factorisation are even.
4. Find a natural number half of which is a square, a third of which is a cube, and a fifth of which is a fifth power.
5. If p is a prime and $p|a^2$ then $p^2|a^2$.
6. If p is a prime factor of both a and $a^2 + b^2$ then $p|b$.
7. If $c|a$ and $c|b$ then $c|(a, b)$.
8. Prove that $(a/(a, b),\ b/(a, b)) = 1$.
9. If $(a, b, c) = 1$ and $a|bc$ then $a = (a, b)(a, c)$.
10. If $(a, b) = 1$ and $ab = c^2$ then a and b are both squares.
11. Prove that $(a + b, b) = (a, b)$.
12. A pair of primes are *twin primes* if they differ by 2. For example, 11 and 13 are twin primes. Find all the twin primes less than 200. (It is not known whether there are infinitely many such pairs.)
13. Prove that there are arbitrarily large gaps between primes. (Hint: consider the sequence $n! + 2, \ldots, n! + n$.)
14. Show that if x is a natural number less than 40 then $x^2 + x + 41$ is prime.
15. In 1675, Jean Prestet proved that if you reduce the fraction a/b, getting m/n in lowest terms, then the least common multiple of a and

b is an. Do the same.
16.* Show that there is no polynomial $f(x)$ with integer coefficients, such that $f(n)$ is a prime for all natural numbers n greater than 0.
17.* Let A, B, and C be any integers. Then $Ax^2 + Bx + C$ can be factored iff $B^2 - 4AC$ is a square, call it m^2. In that case, let a/b be $2A/(B+m)$ in lowest terms. Then one factor in the unique factorisation is $ax + b$.

1.4 Perfect Numbers

Let n be a natural number with prime factorisation

$$n = {p_1}^{e_1} \dots {p_k}^{e_k}$$

where $p_1 < p_2 < \dots p_k$. Since any factor of n has the form

$${p_1}^{a_1} \dots {p_k}^{a_k}$$

with $0 \le a_j \le e_j$, it follows from combinatorial considerations that the number $t(n)$ of divisors of n is the product

$$(e_1 + 1) \dots (e_k + 1)$$

A typical divisor of n is just a typical term in the sum equal to

$$(1 + p_1 + \dots + {p_1}^{e_1}) \dots (1 + p_k + \dots + {p_k}^{e_k})$$

and so the sum $s(n)$ of the divisors of n is just that product. Note that the j-th factor in the product equals

$$\frac{{p_j}^{e_j+1} - 1}{p_j - 1}$$

As an example,

$$s(12) = (1 + 2 + 2^2)(1 + 3) = 28$$

The sum $s'(n)$ of the *proper* divisors of n is just $s(n) - n$. One of the more venerable games played by number lovers is to compute, for a given natural number n (greater than 1), the sequence

$$n \quad s'(n) \quad s'(s'(n)) \quad s'(s'(s'(n))) \quad \ldots$$

This sequence might be called a 'flight' because the numbers often go up for awhile and then go down to the number 1. For example, with $n = 12$, we have

$$12 \quad 16 \quad 15 \quad 9 \quad 4 \quad 3 \quad 1$$

Other times the sequence comes to a point where it repeats. For example, with $n = 25$, we have

$$25 \quad 6 \quad 6 \quad 6 \quad \ldots$$

If $s'(n) = n$ then we say that n is *perfect.* The first few perfect numbers are 6, 28, 496, and 8128. The sequence might also repeat in blocks of two. For example, we have

$$220 \quad 284 \quad 220 \quad 284 \quad 220 \quad \ldots$$

If n is not perfect and $s'(s'(n)) = n$ then we say that n is *amicable.* Its 'friend' is $s'(n)$ and vice versa. The sequence can repeat in longer blocks too. For example, with $n = 12496$, we have

$$12496 \quad 14288 \quad 15472 \quad 14536 \quad 14264 \quad 12496 \quad \ldots$$

Numbers repeating in blocks of length greater than 2 are *sociable.* Examples of sociable numbers are 14 316 and 1 264 460.

Very little is known about these 'flights'. We do *not* know
1) whether there is an odd perfect number;
2) whether there are infinitely many perfect numbers;
3) whether there are infinitely many amicable numbers;
4) whether there are sociable sequences with arbitrarily long period;
5) whether there are any 'flights' (for example, flight 276) which neither end in 1 nor in a repetition.

Most of what know about these sequences is given by the following two theorems. The first is found in Euclid's *Elements* (300 BC) and the second is due to Leonhard Euler (1707–1783).

Theorem 1.4.1 *If* $2^m - 1$ *is prime then* $2^{m-1}(2^m - 1)$ *is perfect.*

Proof: The factors of $2^{m-1}(2^m - 1)$ are 1, 2, 4, ..., 2^{m-1}, $2^m - 1$, $2(2^m - 1)$, ..., $2^{m-1}(2^m - 1)$. Thanks to unique factorisation, we know there are no other factors. And their sum is $2^m(2^m - 1)$.

Theorem 1.4.2 *Every even perfect numbers is included in Euclid's formula.*

Proof: Suppose n is an even perfect number. We can write n in the form $2^{m-1}q$ with q odd, and m, $q > 1$. Each divisor of n has the form $2^r d$ where $0 \leq r \leq m-1$, and d is a divisor of q. Thus

$$s(n) = (1 + 2 + \cdots + 2^{m-1})s(q) = (2^m - 1)s(q)$$

Since n is perfect,

$$2^m q = s(n) = (2^m - 1)s(q)$$

and hence

$$(2^m - 1)(s(q) - q) = q \tag{1.2}$$

Suppose $s(q) - q > 1$. Then q has distinct factors 1, $s(q) - q$, and q. (If $s(q) - q = q$ then, from Equation 1.2, it follows that $(2^m - 1)q = q$, which is impossible.) Thus

$$s(q) \geq 1 + (s(q) - q) + q = s(q) + 1$$

Contradiction.

Hence $s(q) = q + 1$, so that q is prime. Finally, Equation 1.2 implies that $2^m - 1 = q$.

It is an immediate corollary of this theorem that all even perfect numbers are triangular.

Perfect numbers have always appealed to number mystics. In *De Institutione Arithmetica,* Boethius (475–524) defines a 'superfluous' number as one with $s(n) > 2n$, and a 'diminished' number as one with $s(n) < n$. He writes:

> Between these two kinds of number, as if between two elements unequal and intemperate, is put a number which holds the middle place between the extremes like one who seeks virtue.

In the *City of God,* Augustine (354–430) proclaims:

> Six is a number perfect in itself, and not because God created all things in six days; rather, the converse is true. God created all things in six days because this number is perfect, and it would have been perfect even if the work of the six days did not exist.

Before 1588, only 5 perfect numbers were known. In 1950, only 12 perfect numbers had been discovered. Thanks to the computer, however, we now know of 33 perfect numbers.

Finding even perfect numbers is just a matter of finding primes of the form $2^m - 1$. Primes of this form are called *Mersenne primes* — so named after the priest Marin Mersenne (1588–1648) who correctly stated that the first 8 even perfect numbers are given by $m = 2$, 3, 5, 7, 13, 17, 19, and 31. Mersenne also claimed that $2^{67} - 1$ is prime. Here he was wrong. In 1903, Frank Nelson Cole gave a lecture which consisted of two calculations. First, Cole calculated $2^{67} - 1$. Second, he calculated

$$193,707,721 \times 761,838,257,287$$

He did not say a word as he did this. The two calculations agreed, and Cole received a standing ovation. He had factored the number Mersenne had claimed was prime.

Edouard Lucas (1842–1891), the French artillery officer, found an efficient way of testing whether $2^m - 1$ is prime. His idea was refined by Derrick H. Lehmer (1905–), leading to the following algorithm, which we shall examine in Chapter 4. Let

$$u_1 = 4 \text{ and } u_{n+1} = {u_n}^2 - 2$$

Thus $u_2 = 14$ and $u_3 = 194$. If $m > 2$ then $2^m - 1$ is prime just in case $2^m - 1$ is a factor of u_{m-1}. For example, since $2^5 - 1$ is a factor of

$u_4 = 37,634$, it follows that $2^5 - 1$ is prime, and hence $2^4(2^5 - 1) = 496$ is perfect.

In the following table we give the 33 exponents m which are known to make $2^m - 1$ perfect. There is no even perfect number less than $2^{132048}(2^{132049} - 1)$, other than those given by the table, and there is no odd perfect number less than 10^{300}.

The 33 Exponents Known to Make $2^m - 1$ Prime

2	127	11213
3	521	19937
5	607	21701
7	1279	23209
13	2203	44497
17	2281	86243
19	3217	110503
31	4253	132049
61	4423	216091
89	9689	756839
107	9941	858433

Exercises 1.4

1. What is the smallest natural number with exactly 100 divisors?
2. If $(a, b) = 1$ then $t(ab) = t(a)t(b)$ and $s(ab) = s(a)s(b)$.
3. Show that $s(n)$ is odd iff n is a square or twice a square.
4. Where n is a natural number greater than 1, let $u(n)$ be 2^{k-1} where k is the number of distinct primes dividing n. Prove that the number of ways of factoring n into two relatively prime factors is $u(n)$.

5. If n is not a square then

$$t(n) = 2\sum_{m^2|n} u(n/m^2)$$

6. How many (scale 10) digits are there in the largest known perfect number?
7. Show that $2^m - 1$ is prime only if m is prime.
8. Show that no square is perfect.
9. Prove that every even perfect number ends in 6 or 8.
10. Prove that every even perfect number (except 6) has the form

$$1^3 + 3^3 + 5^3 + \cdots + (2^{n+1} - 1)^3$$

11. The second largest amicable pair was discovered by B. N. I. Paganini in 1866. He was only 16 at the time. Verify Paganini's discovery by showing that 1184 and 1210 are amicable.
12. Show that there are odd amicable numbers by checking 69 615.
13. Thabit Ibn-Qurra (836–901) lived in Baghdad. He discovered the following rule. Let n be a natural number greater than 1. Let $p = 3 \times 2^n - 1$, $q = 3 \times 2^{n-1} - 1$, and $r = 9 \times 2^{2n-1} - 1$. If p, q, and r are primes, then $2^n pq$ and $2^n r$ are amicable. Prove Thabit's rule.
14. What amicable pair does Thabit's rule give with $n = 4$?
15. Prove that if n is a multiple of 3, Thabit's rule will not give an amicable pair.
16. In 1991 Achim Flammenkamp discovered the following chain of sociable numbers:

805984760	2308845400	2525983930
1268997640	3059220620	2301481286
1803863720	3367978564	1611969514

Verify Flammenkamp's discovery.
17. Take flight $3^5 \times 7^2 \times 13 \times 17 \times 19 \times 431$.

1.5 Greatest Integer Function

Where x is any real number, let $[x]$ be the integer n such that $n \leq x < n + 1$. Then $[x]$ is the greatest integer not greater than x. For example, $[-3.1] = -4$ and $[4] = 4$. Note that if m is any integer, then $[x+m] = [x] + m$.

Let m be a positive integer. Let $[x] = qm + r$ where q and r are integers and $0 \leq r < m$. Then

$$[x/m] = \left[\frac{qm + r + x - [x]}{m}\right]$$

$$= q + \left[\frac{r + x - [x]}{m}\right] = q$$

—since $0 \leq x - [x] < 1$. Also $[[x]/m] = q + [r/m] = q$. Hence we have

Theorem 1.5.1 *If m is a positive integer, and x any real number then $[x/m] = [[x]/m]$.*

Another basic property of the greatest integer function is the following.

Theorem 1.5.2 *If m and n are positive integers, $[n/m]$ is the number of integers among $1, 2, \ldots, n$ that are divisible by m.*

Proof: Let jm be the largest multiple of m not exceeding n. Then there are j integers among $1, 2, \ldots, n$ that are divisible by m. Moreover,

$$jm \leq n < (j+1)m$$

so that $j \leq n/m < j+1$, that is, $j = [n/m]$.

It follows from the above that if p is a prime and n a positive integer, the largest integer exponent e such that $p^e | n!$ is

$$\sum_{k=1}^{\infty} \left[\frac{m}{p^k}\right]$$

For there are $[n/p]$ multiples of p among the terms in the product $1 \times 2 \times \ldots \times n$. There are also $[n/p^2]$ multiples of p^2, each of them contributing another factor of p to $n!$ And so on.

As an example, 2 goes into 100! exactly

$$[100/2] + [100/4] + [100/8] + [100/16] + [100/32] + [100/64]$$

$$= 97 \text{ times}.$$

Exercises 1.5

1. There is no integer nearer to x than $[x+\frac{1}{2}]$.
2. Unless x is an integer, $[-x] = -[x] - 1$.
3. If P, Q, and R are positive integers, then

$$\left[\frac{P+\sqrt{R}}{Q}\right] = \left[\frac{P+\left[\sqrt{R}\right]}{Q}\right]$$

4. If y is positive and $x = [x/y]y + r$ then $0 \le r < y$.
5. How many 0's are there at the end of 100! ?
6. If n is a positive integer, let $f(n)$ be the least common multiple of the integers 1, 2, ..., n. For example, $f(6) = 60$. Show that

$$f(n) = \prod_{all\, primes\, p} p^{\left[\frac{\log n}{\log p}\right]}$$

7. If f is defined as in Exercise 6, show that $f(113) < 3^{113}$.
8. Prove that, for all positive integers n,

$$\left[\frac{8n+13}{25}\right] = \left[\frac{n-[(n-17)/25]}{3}\right]$$

1.6 Pythagorean Triangles

Consider the right angled triangle whose two legs are each 1 unit long. As we know from the Theorem of Pythagoras, its hypotenuse x is such that $1^2 + 1^2 = x^2$. That is, $x = \sqrt{2}$.

If this number were rational, we could express it as a fraction a/b, where a and b are relatively prime natural numbers. However, if $x = a/b$ then $2 = a^2/b^2$ and $2b^2 = a^2$. Hence a is even. (Squares of evens are even and squares of odds are odd.) If $a = 2a'$ then $2b^2 = 4a'^2$ or $b^2 = 2a'^2$. But this implies that b is also even — against the assumption that a and b are relatively prime. Contradiction. Hence the length x of the hypotenuse is irrational.

Indeed, in a similar fashion, one can prove that if R is any positive nonsquare integer, then its square root is not a fraction.

The Pythagoreans (500 BC) were a religious group who sought to explain the universe in terms of whole numbers and their ratios. It was a philosophical disaster for them when they discovered the above proof that the length of the hypotenuse of a right angled triangle cannot always be so expressed. In some cases, however, the Pythagoreans were lucky. For example, the hypotenuse of a right angled triangle with legs of lengths 3 and 4 has length 5 — and 5 is a nice rational number.

A right triangle the lengths of whose sides are three natural numbers is a *Pythagorean triangle.* If, moreover, these lengths are relatively prime, it is a *primitive Pythagorean triangle.*

Note that if $a^2+b^2=c^2$ and a prime p divides two of a, b, and c then it divides the third as well. Moreover, its square can be cancelled out of the equation. For example, $9^2+12^2=15^2$ and 3 divides both 9 and 12. Furthermore, 3 divides 15, and we can cancel 3^2 out of the equation to get $3^2+4^2=5^2$. An understanding of primitive Pythagorean triangles thus suffices for an understanding of all Pythagorean triangles.

Note also that there is no Pythagorean triangle both of whose legs a and b are odd. For if $a=2a'+1$ and $b=2b'+1$ then the square on the hypotenuse would be

$$c^2=a^2+b^2=4(a'^2+a'+b'^2+b')+2$$

This number is even and hence c is even. But if $c=2c'$ then $c^2=4c'^2$ is a multiple of 4, whereas the above expression leaves a remainder of 2 if is it divided by 4. Hence a and b cannot both be odd.

In the case of a *primitive* Pythagorean triangle, it cannot be the case that both legs are even. We may take it, then, that exactly one of the legs is even. The next theorem gives a complete characterisation of these triangles.

Theorem 1.6.1 *If a, b, and c are positive integers,*
$a^2+b^2=c^2$ with a even and $(a,b,c)=1$
if and only if
for some positive integers u and v with $u>v$, and u, v not both odd, and $(u,v)=1$,

$$a=2uv,\ b=u^2-v^2 \text{ and } c=u^2+v^2$$

Proof: Let $a = 2a'$. Then, since $a^2 + b^2 = c^2$, we obtain $4a'^2 = (c-b)(c+b)$. Since a is even and b is odd, it follows that c is odd and hence $\frac{1}{2}(c-b)$ and $\frac{1}{2}(c+b)$ are integers. Their product is a'^2. Moreover, they are relatively prime. For if a prime p divided evenly into both of them, it would be a factor of their sum, c, and their difference, b, — against the fact that $(a, b, c) = 1$.

Since $\frac{1}{2}(c-b)$ and $\frac{1}{2}(c+b)$ are relatively prime, and have a product which is a square, it follows that each of them is a square. Let $\frac{1}{2}(c-b) = u^2$ and $\frac{1}{2}(c+b) = v^2$. Then $c = u^2 + v^2$, $b = u^2 - v^2$, and $a = 2a' = 2uv$. Since u^2 and v^2 are relatively prime, so are u and v. Moreover, u and v cannot both be odd, lest $c = u^2 + v^2$ be even, which is impossible.

The converse is straightforward.

Indeed the converse was proved by the ancient Mesopotamians, about 4000 years ago. They used it to compute a table of Pythagorean triangles whose generating numbers u and v have no prime factors other than 2, 3, and 5 (the prime factors of the Mesopotamian scale 60). The first complete, explicit proof of Theorem 1.6.1 was given only in 1738, by C. A. Koerbero.

There are 16 primitive Pythagorean triangles with hypotenuse less than 100. They are listed in the following table.

The Primitive Pythagorean Triangles
with Hypotenuse < 100

3 4 5	20 21 29	11 60 61	13 84 85
5 12 13	12 25 37	16 63 65	36 77 85
8 15 17	9 40 41	33 56 65	39 80 89
7 24 25	28 45 53	48 55 73	65 72 97

The next theorem was first proved by Pierre de Fermat (1601–1665), a lawyer who did Mathematics in his spare time. As we shall see, this theorem is important in the study of 'congruent numbers'.

Theorem 1.6.2 *The area of a Pythagorean triangle is never a square number.*

Proof: Suppose, on the contrary, that there are Pythagorean triangles with square areas. Let w^2 be the smallest area for which such triangles exist. Let x and y be the legs of a Pythagorean triangle with area w^2. Since w is minimal, the triangle is primitive, and, without loss of generality, we may take it that x is odd and y even. By Theorem 1.6.1, there are relatively prime positive integers r and s (not both odd) such that $x = r^2 - s^2$ and $y = 2rs$. Since $w^2 = \frac{1}{2}xy$, it follows that $w^2 = (r-s)(r+s)rs$ and hence $s \leq w^2$. Thus $\frac{1}{2}\sqrt{s} < w$.

Since $r-s$, $r+s$, r, and s are pairwise relatively prime and have a square for a product, it follows that each of them is a square. Thus, for some integers a, b, c, and d, we have

$$r = a^2, \quad s = b^2, \quad a^2 - b^2 = r - s = c^2 \text{ and } a^2 + b^2 = d^2$$

Since r and s are not both odd, and since they are also relatively prime, it follows that c and d are both odd, and relatively prime. Thus $X = \frac{1}{2}(c+d)$ and $Y = \frac{1}{2}(d-c)$ are relatively prime integers, and, moreover, $X^2 + Y^2 = a^2$. Hence there is a Pythagorean triangle with area equal to $\frac{1}{2}XY = (b/2)^2$, which is a square. Hence $b/2 \geq w$. But $b/2 = \frac{1}{2}\sqrt{s} < w$. Contradiction.

In the above proof we assume there is a triangle with a certain property and show that we can always 'descend' to a smaller triangle with the same property (the property of having a square area). This shows that the original triangle cannot exist — since there is a lower limit on triangles with integer sides. This 'method of descent' is one of Fermat's important contributions to Number Theory.

Exercises 1.6

1. Prove that the (real) cube root of 3 is irrational.
2. Let P and P' be any integers. Let Q and Q' be any nonzero integers. Let R be a positive nonsquare integer. Suppose that $\frac{P+\sqrt{R}}{Q} = \frac{P'+\sqrt{R}}{Q'}$ and prove that $Q = Q'$.
3. Let u and v be positive integers with $u > v$, and u and v not both odd, and $(u,v) = 1$. Let u' and v' be positive integers with $u' > v'$, and u' and v' not both odd, and $(u',v') = 1$. Then if $2uv = 2u'v'$ and $u^2 - v^2 = u'^2 - v'^2$, it follows that $u = u'$. Hence primitive Pythagorean triangles are generated from the formulas *without duplication.* (This

was proved first by L. Kronecker, in 1901.)
4. Prove that the area of any Pythagorean triangle is divisible by 6.
5. How many Pythagorean triangles are there with hypotenuse less than 120 ?
6. Find all positive integers a and b such that $a^2 + b^2 = 65^2$.
7. Where a, b, and c are natural numbers,

$$a^2 + 2b^2 = c^2 \text{ and } (a, b, c) = 1$$

iff for some natural numbers u and v with $(u, 2v) = 1$,

$$a = \pm(u^2 - 2v^2), \; b = 2uv \text{ and } c = u^2 + 2v^2$$

8. Show that there are no integers x and y (with $y \neq 0$) such that both $x^2 - y^2$ and $x^2 + y^2$ are squares. (Hint: if $x^2 - y^2 = v^2$ and $x^2 + y^2 = u^2$ then the triangle with sides $u - v$, $u + v$, and $2x$ is a Pythagorean triangle with square area.)
9. Find all Pythagorean triangles with perimeter 1716.
10.* Find a Pythagorean triangle one of whose angles is less than a hundredth of a degree away from 20 degrees.

1.7 Diophantine Equations

In 250 AD, Emperor Decius was executing Christians who refused to sacrifice to pagan gods. In Rome, Plotinus was teaching his version of Platonism. In Alexandria, Diophantus was working on his *Arithmetica,* dedicating it to Dionysius, the Bishop of Alexandria from 247 to 264.

Diophantus studied equations whose variables are rationals, but we none the less give his name to equations whose variables are restricted to being integers. A *Diophantine equation* is an equation whose variables are integers. As an example, if x, y, and z are natural numbers then $x^2 + y^2 = z^2$ is a Diophantine equation.

Some Diophantine equations (such as $x^2 + y^2 = z^2$) have infinitely many solutions. Others, like $2x + 1 = 4y$, have none. And there are some, like $x^2 + y^2 = 8$, which have a nonzero finite number of solutions.

Solving these equations is an art. Indeed, in 1970, Yuri Matijasevich proved that there is no completely general, mechanical method for solving them. No matter how many you can solve already, there is always another one which will require a new, as yet undiscovered approach for its answer.

One technique for solving Diophantine equations is to look at the linear forms of the integers involved. For example, every integer is a multiple of 3, or 1 more than a multiple of 3, or 1 less than a multiple of 3. That is, every integer x has exactly one of the linear forms $3m$, $3m + 1$, and $3m - 1$. As a result, every cube has one of the following forms:

$$x^3 = 9(3m^3) \text{ or } x^3 = 9(3m^3 \pm 3m^2 + m) \pm 1$$

Hence no cube can be 5 more than a multiple of 9. Now if

$$x^3 + 117y^3 = 5$$

then $x^3 = 9(-13y^3) + 5$. Since this is impossible, it follows that the equation $x^3 + 117y^3 = 5$ has no integer solutions. This Diophantine equation was first solved by R. Finkelstein and H. London, in 1971.

Sometimes a larger solution of a Diophantine equation is a linear combination of the next smaller solution. Consider

$$x^2 - 2y^2 = 1$$

Without loss of generality, we can confine our attention to nonnegative integer solutions. Doing so, we note that the values of x and y increase together, and a short computer search reveals that the smallest solutions are (1, 0), (3, 2), (17, 12), (99, 70), and (577, 408). The solution of

$$577 = 99m + 70n$$

$$99 = 17m + 12n$$

is $m = 3$, $n = 4$. Moreover, it is also the case that $17 = 3 \times 3 + 2 \times 4$. This suggests that if (x_n, y_n) is the n-th nonnegative integer solution, then

$$x_{n+1} = 3x_n + 4y_n$$

which can, indeed, be proved to be the case. Similarly, it can be shown that $y_{n+1} = 2x_n + 3y_n$, and, moreover, all the solutions can be obtained from these formulas. Indeed, if (x, y) is a positive integer solution, then $(3x - 4y, 3y - 2x)$ is a nonnegative integer solution. Now

$$x = 3(3x - 4y) + 4(3y - 2x)$$

$$y = 2(3x - 4y) + 3(3y - 2x)$$

so that (x, y) is obtained from a smaller solution, using the linear combinations. This equation was first solved by the Pythagoreans.

Another useful technique for solving Diophantine equations is factoring. To find integers x and y such that $x^3 + y^3 = 2$, we note that this equation is equivalent to

$$(x + y)(x^2 - xy + y^2) = 2$$

Since $x + y$ and $x^2 + xy + y^2$ are integer factors of 2, there are only 4 possibilities:

$$\begin{aligned} x + y &= \pm 1,\ \pm 2 \\ (x + y)^2 - 3xy = x^2 - xy + y^2 &= \pm 2,\ \pm 1 \end{aligned}$$

The only answer is thus $x = 1$ and $y = 1$.

Factorisation can be used to solve the Diophantine equation

$$\frac{1}{x} + \frac{1}{y} = \frac{a}{b}$$

where a and b are given positive integers. This is because the above equation implies that

$$(ax - b)(ay - b) = b^2$$

The reader may wish to check that $1/x + 1/y = 1/8$ has 7 solutions in terms of positive integers.

Factorisation can also be used to solve the simultaneous 'Pell' equations

$$x^2 - Ry^2 = 1$$

$$z^2 - Sy^2 = 1$$

where R and S are given positive nonsquare integers whose product is a square. Suppose $RS = U^2$. If both equations hold, then $Sx^2 - Rz^2 = S - R$ and hence

$$(Sx - Uz)(Sx + Uz) = S^2 - U^2$$

The reader may wish to check that when $R = 2$ and $S = 2312$, the only positive integer solution is with $x = 17$.

Factorisation can be used to solve certain 'conic' Diophantine equations of the form

$$Ax^2 + Bxy + Cy^2 + Dx + Ey = F$$

For this we need the following 'Conic Transformation Theorem'.

Theorem 1.7.1 *Suppose A, B, C, D, E, and F are integers, with* $A \neq 0$.
Let $R = B^2 - 4AC$, $S = BD - 2AE$, *and* $T = 4AF + D^2$. *Suppose* $R \neq 0$.
Then $Ax^2 + Bxy + Cy^2 + Dx + Ey = F$
iff $(Ry + S)^2 - R(2Ax + By + D)^2 = S^2 - RT$.

Proof:

$$Ax^2 + Bxy + Cy^2 + Dx + Ey = F$$

iff

$$4A^2x^2 + 4ABxy + 4ADx + 4ACy^2 + 4AEy = 4AF$$

iff

$$(2Ax + By + D)^2 - (By + D)^2 + 4ACy^2 + 4AEy = 4AF$$

iff

$$(2Ax + By + D)^2 - Ry^2 - 2Sy = T$$

iff

$$R^2y^2 + 2RSy - R(2Ax + By + D)^2 = -RT$$

iff

$$(Ry + S)^2 - R(2Ax + By + D)^2 = S^2 - RT$$

For example, to solve

$$x^2 - xy - 72y^2 + 2x - y = 3$$

we compute $R = 289$, $S = 0$, and $T = 16$. The equation is equivalent to

$$(289y)^2 - 289(2x - y + 2)^2 = -4624$$

Factoring, we obtain

$$(289y - 17(2x - y + 2))(289y + 17(2x - y + 2)) = -4624$$

or

$$(9y - x - 1)(8y + x + 1) = -4$$

Thus, for some factor g of 4,

$$9y - x - 1 = g$$

$$8y + x + 1 = -4/g$$

Hence $17y = g - 4/g$ and it is now easy to show that $y = 0$.

The next three Diophantine equations are important in the solution of the Square Pyramid Problem, which we shall give in Chapter 4.

Theorem 1.7.2 *There are no positive integers x such that $2x^4 + 1$ is a square.*

Proof: To obtain a contradiction, suppose that (x, y) is the least positive integer solution of $2x^4 + 1 = y^2$. Then, for some integer $s > 0$, $y = 2s + 1$ and hence $x^4 = 2s(s + 1)$. If s is odd then s and $2(s + 1)$ are relatively prime, and, for some integers u and v, $s = u^4$ while $2(s + 1) = v^4$. This gives $2(u^4 + 1) = v^4$ with u odd and v even. But then $u^4 + 1$, which has the form $4n + 2$, is divisible by 8. Since this is impossible, s is not odd.

Since s is even, $2s$ and $s + 1$ are relatively prime, and there are integers u and v, both > 1, such that $2s = u^4$ and $s + 1 = v^4$. Since u is even, $w = u/2$ is an integer. Since v^2 is odd, there is a positive integer a such that $v^2 = 2a + 1$. Now

$$u^4/2 + 1 = s + 1 = v^4$$

so that

$$2w^4 = (v^4 - 1)/4 = a(a + 1)$$

As an odd square, v^2 has the form $4n+1$ and hence a is even. Since $2w^4 = a(a+1)$, it follows that there are positive integers b and c such that $a = 2b^4$ and $a+1 = c^4$. Moreover, $2b^4+1 = (c^2)^2$ and hence $y \leq c^2$ — by the minimality of the solution (x, y).

On the other hand, $c^2 \leq a+1 < v^2 \leq s+1 < y$. Contradiction.

Theorem 1.7.3 *There is only one natural number x (namely, 1) such that $2x^2 - 1$ is a fourth power.*

Proof: Suppose that $2x^2 - 1 = y^4$. Squaring, we obtain

$$4y^4 + y^8 - 2y^4 + 1 = 4x^4$$

and hence

$$y^4 + \left(\frac{y^4 - 1}{2}\right)^2 = x^4$$

Since y is odd, y^4 has the form $4n+1$. Thus x is odd. Since x and y are relatively prime, so are $\frac{1}{2}(x^2 - y^2)$ and $\frac{1}{2}(x^2 + y^2)$. Since the product of these two numbers is a square — namely, $\left(\frac{y^4-1}{4}\right)^2$ — it follows that each of them is a square.

Without loss of generality, we may take it that x and y are nonnegative integers and $x \geq y$. If $x = y$, then we have the solution $x = 1$ and $y = 1$.

Suppose $x > y$. Then, since

$$(x-y)^2 + (x+y)^2 = 4\left(\frac{x^2+y^2}{2}\right)$$

is a square, it follows that $x - y$ and $x + y$ are legs of a Pythagorean triangle. Moreover, this triangle has area

$$\frac{1}{2}(x-y)(x+y) = \frac{1}{2}(x^2 - y^2)$$

— another square. However, no Pythagorean triangle has a square area (Theorem 1.6.2). Contradiction.

Theorem 1.7.4 *There is only one positive integer x (namely, 1) such that $8x^4 + 1$ is a square.*

Proof: Suppose $8x^4 + 1 = (2s + 1)^2$. Then $2x^4 = s(s + 1)$.

If s is odd, there are integers u and v such that $s = u^4$ and $s + 1 = 2v^4$, whence $u^4 + 1 = 2v^4$. By Theorem 1.7.3, $2v^4 = 2$ and hence $x = \pm 1$.

If s is even, then we have $s = 2u^4$ and $s+1 = v^4$, whence $2u^4+1 = v^4$ and Theorem 1.7.2 assures us that $u = 0$ and hence $x = 0$.

We conclude this section with a famous result due to Fermat.

Theorem 1.7.5 *No square is the sum of two nonzero fourth powers.*

Proof: Suppose there are positive integers x, y, and z such that $x^4 + y^4 = z^2$. Let us take such a triple with the product xyz minimised. Then $(x, y) = (x, z) = (y, z) = 1$.

Now x and y cannot both be odd (lest x^4 and y^4 both have the form $4n + 1$ and z have the form $4n + 2$). Without loss of generality, let us take it that x is even.

By the Pythagorean Triangle Theorem, there are positive integers u and v, with $u > v$, and u, v not both odd, and $(u, v) = 1$ such that $x^2 = 2uv$ and $y^2 = u^2 - v^2$.

Since $v^2 + y^2 = u^2$ and y is odd, v must be even. By the Pythagorean Triangle Theorem, there are positive integers s and t, with $s > t$ and s, t not both odd, and $(s, t) = 1$ such that $v = 2st$ and $u = s^2 + t^2$.

Hence $x^2 = 2uv = 4st(s^2 + t^2)$, so that, for some positive integers a, b, and c, we have $s = a^2$, $t = b^2$, and $s^2 + t^2 = c^2$. This gives $a^4 + b^4 = c^2$, and hence, by the minimality of xyz, we have $abc \geq xyz$.

However, $(abc)^2 = \frac{1}{4}x^2 < (xyz)^2$. Contradiction.

From the above theorem it follows that there are no positive integers x, y, z, and w such that $x^{4w} + y^{4w} = z^{4w}$.

Exercises 1.7

Solve the following Diophantine Equations.

1. $x^2 + y^2 + z^2 = 8,000,007$.

2. $x^2 - 3y^2 = 1$.
3. $x^2 - y^2 + 4x - 5y = 27$.
4. $3x^2 - 8xy + 7y^2 - 4x + 2y = 109$.
5. $x^4 + 6x^3 + 11x^2 + 6x + 1 = y^2$.
6. $x^2 + 2y^2 = z^2$ simultaneously with $x^2 - 2y^2 = w^2$.
7. $x^4 - 2y^2 = 1$.
8. $4x^4 - 3y^2 = 1$.
9. Prove that no triangular number is a fourth power.
10.* $x^4 - 5y^4 = 1$.
11.* $x^2 + y^4 = 2z^4$. (J. L. Lagrange (1736–1813) gave the first solution to this equation, in 1777.)

1.8 Four Square Theorem *

The numbers 7 and 8 can be written as a sum of four squares:

$$7 = 2^2 + 1^2 + 1^2 + 1^2$$

$$8 = 2^2 + 2^2 + 0^2 + 0^2$$

Note, however, that 7 cannot be written as a sum of *fewer* than 4 integer squares.

What we prove in this section is a result due to Joseph Louis Lagrange (1736–1813): every natural number is a sum of four natural number squares. Lagrange based his work on the following two theorems, which had been proved by Leonhard Euler (1707–1783).

Theorem 1.8.1

$$\begin{aligned} \textit{If}\ p &= ae + bf + cg + dh \\ q &= af - be + ch - dg \\ r &= ag - bh - ce + df \\ s &= ah + bg - cf - de \end{aligned}$$

then

$$(a^2 + b^2 + c^2 + d^2)(e^2 + f^2 + g^2 + h^2) = p^2 + q^2 + r^2 + s^2$$

Hence if every prime is a sum of four squares, then every natural number is a sum of four squares.

Proof: The equation can be verified by straight calculation. The 'hence' follows from the fact that every natural number has a prime factorisation.

Theorem 1.8.2 *For every odd prime p there is an integer m such that $0 < m < p$ and mp is a sum of four squares.*

Proof: The squares

$$0^2,\ 1^2,\ 2^2,\ \ldots,\ \left(\frac{p-1}{2}\right)^2$$

all leave different remainders when divided by p. For suppose $A^2 = ap + r$ and $B^2 = bp + r$, with $A > B$. Then p is a factor of $A^2 - B^2 = (A - B)(A + B)$. However,

$$0 < A - B,\ \ A + B < p$$

so p is a factor neither of $A - B$, nor of $A + B$. Contradiction.

Similarly,

$$-1-0^2,\ -1-1^2,\ -1-2^2,\ \ldots\ -1-\left(\frac{p-1}{2}\right)^2$$

all leave different remainders when divided by p.

Each of the above two lists has $\frac{1}{2}(p+1)$ members.

Altogether, they contain $p + 1$ integers. Since there are only p possible remainders when one divides by p, there is some x^2 from the first list and some $-1 - y^2$ from the second list which leave the same remainder when divided by p. Hence p divides their difference x^2+y^2+1. That is, for some integer m, we have

$$mp = x^2 + y^2 + 1^2 + 0^2$$

Moreover, since $0 \leq x,\ \ y \leq \frac{1}{2}(p-1)$, it follows that $0 < m < p$.

We also have the following.

Theorem 1.8.3 *If p is an odd prime, and m is the least integer such that $0 < m < p$ and $mp = a^2 + b^2 + c^2 + d^2$ for some natural numbers a, b, c, and d, then m is odd.*

Proof: If m is even, then either 0, 2, or 4 of a, b, c, and d are odd. Pairing the odd numbers, we get, say,

$$\left(\frac{a+b}{2}\right)^2 + \left(\frac{a-b}{2}\right)^2 + \left(\frac{c+d}{2}\right)^2 + \left(\frac{c-d}{2}\right)^2 = \frac{1}{2}mp$$

which is an expression of $\frac{1}{2}mp$ as a sum of four natural number squares. Since $\frac{1}{2}m < m$, this is impossible — given m's minimality. So m is odd.

Using the above theorems, Lagrange gave the following, in 1770.

Theorem 1.8.4 *Every natural number theorem is a sum of four natural number squares.*

Proof: Since $2 = 1^2+1^2+0^2+0^2$, Theorem 1.8.1 implies that it suffices to prove that every odd prime p is a sum of four squares.

Let p be any odd prime, and let m be the least integer between 0 and p such that mp is a sum of four squares. (That there is such an m follows from Theorem 1.8.2.) By Theorem 1.8.3, m is odd.

To obtain a contradiction, suppose $m \geq 3$.

Suppose $mp = a^2 + b^2 + c^2 + d^2$, and let x be the integer closest to a/m. Then $|a/m - x| < \frac{1}{2}$ and $x' = a - mx$ is between $-\frac{1}{2}m$ and $\frac{1}{2}m$. Let y, z, and w be the integers closest to b/m, c/m, and d/m respectively. Then $y' = b - my$, $z' = c - mz$, and $w' = d - mw$ are each between $-\frac{1}{2}m$ and $\frac{1}{2}m$.

Let $Z' = x'^2 + y'^2 + z'^2 + w'^2$. Then $Z' < 4(\frac{1}{2}m)^2 = m^2$. Also $Z' \neq 0$, lest m divide each of a, b, c, and d, with the result that m^2 divides $a^2 + b^2 + c^2 + d^2 = mp$. This is impossible because p is prime and $1 < m < p$.

Let $Z = x^2 + y^2 + z^2 + w^2$. Let $T = xx' + yy' + zz' + ww'$. Then the fact that $mp = a^2 + b^2 + c^2 + d^2$ implies that

$$mp = m^2Z + 2mT + Z'$$

(since $a = x' + mx$, etc.)

Let $M = Z'/m = p - mZ - 2T$, an integer. Since $Z' \neq 0$, it follows that $M \neq 0$. Also, since $Z' < m^2$, we have $M < m$. Now

$$\begin{aligned} Mp &= (M/m)mp \\ &= (M/m)(m^2Z + 2mT + Z') \\ &= ZZ' - T^2 + (T + M)^2 \end{aligned}$$

(since $M = Z'/m$).

By Theorem 1.8.1, $ZZ' = T^2 + q^2 + r^2 + s^2$ for some natural numbers q, r, and s. Thus $Mp = q^2 + r^2 + s^2 + (T + M)^2$, a sum of four squares. But $M < m$. Contradiction.

About 1790, Lagrange became subject to fits of depression and loneliness. He no longer wanted to do mathematics. He was rescued from this state by the love of a teenaged girl, Renée Lemonnier, who insisted on marrying him. The marriage took place in 1792, and, for the remaining twenty years of his life, Lagrange was happy.

Exercises 1.8

1. Express 1007 as a sum of four squares.
2. Prove that no natural number of the form $8n + 7$ is a sum of three squares.
3. What is the smallest natural number that can be written as a sum of four positive squares in at least 3 essentially different ways?
4. Let $x = (m - m^3)/6$ where m is an integer. Show that x is an integer. Then show that

$$m = m^3 + (x + 1)^3 + (x - 1)^3 + (-x)^3 + (-x)^3$$

— a sum of 5 cubes.

5. Write 239 as a sum of 9 nonnegative cubes, and show that it cannot be written as a sum of 8 nonnegative cubes.

1.9 Fermat's Last Theorem

Pierre de Fermat (1601–1665) was a councillor for the parliament of Toulouse, and only did mathematics in his spare time. He published only one mathematical article during his lifetime.

In reading Bachet's translation from Greek into Latin of the *Arithmetica* of Diophantus, Fermat came across the equation $x^2 + y^2 = z^2$ (see Book II, Problem 8). In the margin of this translation, Fermat wrote a note to the effect that if $n > 2$ then there are no positive integers x, y, z, and n such that $x^n + y^n = z^n$:

> To divide a cube into two other cubes, a fourth power, or in general any power whatever into two powers of the same denomination above the second is impossible, and I have assuredly found an admirable proof of this, but the margin is too narrow to contain it.

Fermat's assertion is called his 'Last Theorem' because, for a long time, it was the only one of his conjectures which we could neither prove nor disprove. In 1993, the British mathematician Andrew Wiles gave an argument which, it is believed, will soon lead us to a proof that Fermat was, indeed, right.

Fermat himself may have had the proof for the case in which the exponent $n = 3$. As we saw above, he certainly had the proof for the case $n = 4$. In 1823, Legendre disposed of the case with $n = 5$, and, in 1849, Kummer vindicated Fermat's claim for all $n < 100$ — except 37, 59, and 67.

In this section, we prove, among other things, that $x^3 + y^3 = z^3$ has no solution in positive integers. Our proof uses no mathematics not known to Fermat himself. First we need some lemmas which, together with the main result, can give the reader an idea of the sort of Number Theory done in the seventeenth century.

Theorem 1.9.1 *Let A be a given integer. If an integer of the form $a^2 + Ab^2$ is divisible by a prime of the same form then the quotient also has this form.*

Proof: Let the prime be $p^2 + Aq^2$. We have the following two identities.

$$(pb - aq)(pb + aq) = b^2(p^2 + Aq^2) - q^2(a^2 + Ab^2) \quad (1)$$

$$(pa \mp Aqb)^2 + A(pb \pm aq)^2 = (p^2 + Aq^2)(a^2 + Ab^2) \quad (2)$$

If the prime $p^2 + Aq^2$ divides $a^2 + Ab^2$, (1) implies that it divides $pb \pm aq$ for one of the signs, and (2) then implies that it divides $pa \mp Aqb$ for the corresponding sign. Now by (2),

$$\begin{aligned}\frac{a^2 + Ab^2}{p^2 + Aq^2} &= \frac{a^2 + Ab^2}{p^2 + Aq^2}\frac{p^2 + Aq^2}{p^2 + Aq^2} \\ &= \left(\frac{pa \mp Aqb}{p^2 + Aq^2}\right)^2 + A\left(\frac{pb \pm aq}{p^2 + Aq^2}\right)^2\end{aligned}$$

Theorem 1.9.2 *Let x and y be positive integers such that xy has the form $a^2 + 3b^2$ but x does not. If x is odd then y has an odd prime factor not of that form.*

Proof: We have the following identities:

$$\begin{aligned}\frac{a^2 + 3b^2}{4} &= \left(\frac{a - 3b}{4}\right)^2 + 3\left(\frac{a + b}{4}\right)^2 \quad (1) \\ &= \left(\frac{a + 3b}{4}\right)^2 + 3\left(\frac{a - b}{4}\right)^2 \quad (2)\end{aligned}$$

If $xy = a^2 + 3b^2$ is even then a and b have the same parity (i.e. they are both even or both odd), and $a^2 + 3b^2$ has the form $4c$, that is, xy is divisible by 4.

If a and b are even then $xy/4 = (a/2)^2 + 3(b/2)^2$ has the original form.

If a and b are odd then $a = 4m \pm 1$ and $b = 4n \pm 1$. If these take different signs then $\frac{a - 3b}{4}$ and $\frac{a + b}{4}$ are integers, and equation (1) above shows that $xy/4$ has the original form. If a and b take the same sign then $\frac{a + 3b}{4}$ and $\frac{a - b}{4}$ are integers, and equation (2) above shows that $xy/4$ has the original form.

Hence if a and b have the same parity, $xy/4$ has the original form.

From this we see that there is some nonnegative integer k (possibly 0) such that $y/4^k$ is an odd integer, and $xy/4^k$ has the original form.

Now if $y/4^k$ has prime factors only of the riginal form then Theorem 1.9.1 implies that x has this form. Since does not have this form (given), $y/4^k$ has a prime factor, p, not of his form, and p is an odd factor of y.

Theorem 1.9.3 *Let $A = 1,\ 2,$ or 3. Sppose x is an odd positive integer which is a factor of a number of theform a^2+Ab^2 with $(a, b) = 1$. Then x has that same form.*

Proof: Suppose that this theorem is false, and suppose x is the smallest odd positive integer which is a factor of number of the form $a^2 + Ab^2$ with $(a, b) = 1$, without itself having tha form. Then $x > 1$. Dividing a and b by x, we can obtain integers m, n, c, and d such that $a = mx \pm c$, $b = nx \pm d$ with $0 \leq c,\ d < x/2$ (since x is odd). Since x is a factor of $a^2 + Ab^2$,

$$c^2 + Ad^2 = (a - mx)^2 + A(b - nx)^2$$
$$= a^2 + Ab^2 + xz = xy$$

for some integers z and y. Moreover,

$$xy = c^2 + Ad^2 < \frac{1+A}{4}x^2 \leq x^2$$

so that $y < x$. Let $w = (c, d, x)$. Then w is a factor both of a and b. Since $(a, b) = 1$, it follows that $w = 1$. Let $s = (c, d)$. Then $(s, x) = 1$ and

$$x(y/s^2) = (c/s)^2 + A(d/s)^2$$

with c/s, d/s, and y/s^2 all integers, and $(c/s, d/s) = 1$.

If $A = 3$ then Theorem 1.9.2 implies that y/s^2, and hence y, has an odd prime factor p not of the original form.

Suppose $A = 1$ or 2. If all the primes in y/s^2 have the form $a^2 + Ab^2$ then Theorem 1.9.1 implies that x has the same form — against the supposition. Thus y/s^2, and hence y, has a prime factor p not of that form. Since 2 has that form, p is odd.

Thus, whether $A = 1,\ 2,$ or 3, y has an odd prime factor p not of the original form. But since $y < x$, it follows that $p < x$ — against x's minimality. Contradiction.

Theorem 1.9.3 is a useful result. For example, we can employ it to show that there are infinitely many primes of certain kinds.

Theorem 1.9.4 *There are infinitely many primes of the form* $3n + 1$.

Proof: Suppose there are only m such primes, and let their product be Z. Now $3(2Z)^2 + 1$ has an odd prime factor p which is not one of these m primes, and so has the form $3n - 1$. (Of course, p cannot be 3.) By Theorem 1.9.3, p has the form $a^2 + 3b^2$. However, since $a = 3c \pm 1$ for some integer c, a^2 has the form $3d + 1$, with the result that p has the form $3n + 1$. Contradiction.

As another example of the power of Theorem 1.9.3, we prove

Theorem 1.9.5 *The Diophantine equation* $x^2 + 5 = y^3$ *has no solution.*

Proof: Suppose that x and y are integers such that $x^2 + 5 = y^3$. Now $x^2 + 5$ has the form $4n + 1$ or $4n + 2$. Thus y has the form $4n + 1$. This implies that $y^2 + y + 1$ has the form $4n + 3$, and hence it must have a prime factor p of that form.

Since $(y - 1)(y^2 + y + 1) = x^2 + 4$, it follows that p is a factor of $x^2 + 4$. By Theorem 1.9.3, p has the form $a^2 + b^2$. But then it cannot have the form $4n + 3$. Contradiction.

A Diophantine equation of the form $x^2 + k = y^3$ (where k is a given nonzero integer) is called a *Bachet equation* — after Claude-Gaspar Bachet (1581–1638) who wrote poetry and philosophy as well as mathematics. In 1967 there were many integers k for which mathematicians could not solve the equation. However, in the years following 1967, Alan Baker and others developed a method for solving this equation for any given k. (See Ray. P. Steiner's article 'On Mordell's Equation $y^2 - k = x^3$' on pages 703 to 714 in volume 46 of the *Mathematics of Computation* (1986).)

Theorem 1.9.3 also bears fruit in the following result which we shall use in our proof that $x^3 + y^3 = z^3$ has no solution in nonzero integers.

Theorem 1.9.6 *Suppose* $A = 1, 2,$ *or* 3. *Let* a *and* b *be relatively prime integers such that* $a^2 + Ab^2 = s^3$ *for some integer* s. *Then there are integers* u *and* v *such that* $s = u^2 + Av^2$, $a = u^3 - 3Avu^2$ *and* $b = 3u^2v - Av^3$.

Proof: If s and b are both even then a is even — against the fact that $(a,b) = 1$. Hence if s is even, b is odd, and b^2 has the form $8z+1$. Thus if s is even, $a^2 + A$ is divisible by 8. However, a^2 has one of the forms $8w$, $8w+1$, and $8w+4$, so that $a^2 + A$ does not have the form $8w'$. Hence s is odd.

When s has 0 prime factors, $s = 1$, and the result is immediate.

Suppose the theorem is true for all integers s with n prime factors (not necessarily distinct).

Let $s = tp$ where p is an odd prime and t has n prime factors. By Theorem 1.9.3, $p = w^2 + Ax^2$ for some integers w and x. Let

$$c = w^3 - 3Awx^2 \quad \text{and} \quad d = 3w^2x - Ax^3$$

Then $p^3 = c^2 + Ad^2$ and $xc - 3wd = -8w^3x$. Since $p^3 = c^2 + Ad^2$, the only prime which might factor both c and d is p. However, this would imply that p factors w^3x (since p is odd) and hence p is a factor of w or x. But this is impossible since $p = w^2 + Ax^2$. Thus $(c,d) = 1$.

Now

$$\begin{aligned} t^3p^6 = s^3p^3 &= (a^2 + Ab^2)(c^2 + Ad^2) \\ &= (ac \pm Abd)^2 + A(ad \mp bc)^2 \qquad (*) \end{aligned}$$

Since

$$\begin{aligned} (ad - bc)(ad + bc) &= (a^2 + Ab^2)d^2 - b^2(c^2 + Ad^2) \\ &= t^3p^3d^2 - b^2p^3 = p^3(t^3d^2 - b^2) \end{aligned}$$

it follows that p^3 is a factor of $(ad - bc)(ad + bc)$. If p factored both $ad - bc$ and $ad + bc$ then it would factor both ad and bc (taking sum and difference). Since $p^3 = c^2 + Ad^2$ and $(c,d) = 1$, p is not a factor of c and not a factor of d. Hence p would factor both a and b. However, $(a,b) = 1$. Thus only one of $ad - bc$ and $ad + bc$ is divisible by p, and that one is divisible by p^3. Hence, with the appropriate sign, $(*)$ implies that p^3 is a factor of $ac \pm Abd$.

Choose the signs so that

$$e = \frac{ac \pm Abd}{p^3} \quad \text{and} \quad f = \frac{ad \mp bc}{p^3}$$

are both integers. Then $(*)$ becomes

$$t^3p^6 = e^2p^6 + Af^2p^6$$

or $t^3 = e^2 + Af^2$.

Solving for a and b in terms of e and f, we obtain

$$a = ec + Afd \text{ and } \pm b = ed - fc$$

Since $(a, b) = 1$, it follows that $(e, f) = 1$.

By the induction hypothesis, there are integers y and z such that $t = y^2 + Az^2$, $e = y^3 - 3Ayz^2$ and $f = 3y^2z - Az^3$.

Let $u = wy + Axz$ and $v = yx - zw$. Then

$$s = tp = (y^2 + Az^2)(w^2 + Ax^2) = u^2 + Av^2$$

$$\begin{aligned} a &= ec + Afd \\ &= (y^3 - 3Ayz^2)(w^3 - 3Awx^2) + A(3y^2z - Az^3)(3w^2x - Ax^3) \\ &= u^3 - 3Auv^2 \end{aligned}$$

$$\begin{aligned} \pm b &= ed - fc \\ &= (y^3 - 3Ayz^2)(3w^2x - Ax^2) - (3y^2z - Az^3)(w^3 - 3Awx^2) \\ &= 3u^2v - Av^3 \end{aligned}$$

Changing the sign of v if necessary, we obtain the result — using mathematical induction.

Like Theorem 1.9.3, Theorem 1.9.6 allows us to solve certain Diophantine equations.

Theorem 1.9.7 *The Diophantine equation $x^2 + 2 = y^3$ has only 1 solution in natural numbers.*

Proof: If $x^2 + 2 \times 1^2 = y^3$ then, by Theorem 1.9.6, there are integers u and v such that $1 = (3u^2 - 2v^2)v$ and $y = u^2 + 2v^2$. Now $v = \pm 1$ and hence $u = \pm 1$. The only possibility is $x = 5$ and $y = 3$.

Finally, we use Theorem 1.9.6 to prove the main result of this section.

Theorem 1.9.8 *$x^3 + y^3 = z^3$ has no solution in nonzero integers.*

Proof: Suppose it does have a solution in nonzero integers. Among the solutions, pick one which makes $|xyz|$ as small as possible. Then $(x, y) = (x, z) = (y, z) = 1$ since a common divisor could be cancelled out to make $|xyz|$ smaller.

Exactly two of x, y, and z are odd. By rearranging the equation if necessary (and relabelling) we can thus stipulate that x and y are odd, and z is even.

Let $u = \frac{1}{2}(x+y)$ and $w = \frac{1}{2}(x-y)$. Then $x = u+w$ and $y = u-w$. Since $(x, y) = 1$, it follows that $(u, w) = 1$, and u and w are not both odd.

Since $x^3+y^3 = z^3$, it follows that $2u^3+6uw^2 = z^3$, or $2u(u^2+3w^2) = z^3$.

Case 1. u is not divisible by 3.

Since u and w have different parity, u^2+3w^2 is odd. Since $(u, w) = 1$, it follows that $(2u, u^2 + 3w^2) = 1$ and hence there are integers t and s such that $2u = t^3$ and $u^2 + 3w^2 = s^3$.

By Theorem 1.9.6, there are integers a and b such that $u = a^3 - 9ab^2$ and $w = 3a^2b - 3b^3$. Since $(u, w) = 1$, it follows that $(a, 3b) = 1$, and a and $3b$ have different parity. Hence $a + 3b$ and $a - 3b$ are odd. Thus $(a - 3b,\ a + 3b) = 1$.

Now $t^3 = 2u = 2a(a - 3b)(a + 3b)$, so that there are integers c, d, and e such that $2a = c^3$, $a - 3b = d^3$ and $a + 3b = e^3$. Also $c^3 = d^3 + e^3$. Moreover, $cde \neq 0$, lest $\frac{1}{2}(x + y) = u = 0$, and hence $z = 0$. Also

$$|cde|^3 = |2u| = |x + y| < |xyz|^3$$

since z is even. Contradiction.

Case 2. $u = 3v$ for some integer v.

Since $2u(u^2 + 3w^2) = z^3$, it follows that $18v(3v^2 + w^2) = z^3$. Since $u = 3v$ and w have different parity, $3v^2 + w^2$ is odd. Since $(3v, w) = 1$, it follows that $(18v, 3v^2 + w^2) = 1$, and hence there are integers t and s such that $18v = t^3$ and $3v^2 + w^2 = s^3$.

By Theorem 1.9.6, there are integers a and b such that $w = a^3 - 9ab^2$ and $v = 3a^2b - 3b^3$. Since $(w, v) = 1$, it follows that a and b have different parity, so that $a + b$ and $a - b$ are both odd. Also $(a, b) = 1$ and hence $(a + b,\ a - b) = 1$.

Now $t^3 = 18v = 3^3 \times 2b(a - b)(a + b)$ so that, for some integers c, d, and e, we have $2b = c^3$, $a - b = d^3$ and $a + b = e^3$, giving $e^3 = c^3 + d^3$.

Moreover, $cde \neq 0$, lest $\frac{1}{2}(x+y) = u = 3v = 0$. Also

$$|cde|^3 = |2v/3| = |2u/9| = |(x+y)/9| < |xyz|^3$$

Again we have a contradiction.

Exercises 1.9

1. Prove that there are infinitely many primes of the form $3n+2$. (Hint: let Z be the product of all odd primes of the form $3n+2$, if there are only finitely many. Then $3Z+2$ has an odd prime factor of that form.)
2. There are infinitely many primes of the form $4n+1$. (Hint: use Theorem 1.9.3 on $1^2 + (2Z)^2$.)
3. There are infinitely many primes of the form $8n+3$. (Hint: use Theorem 1.9.3 on $Z^2 + 2 \times 1^2$.)
4. Solve the Diophantine equation $x^2 + 1 = y^3$.
5. Solve the Diophantine equation $x^2 + 4 = y^3$.
6. Solve the Diophantine equation $x^2 + 12 = y^3$.
7. Solve the Diophantine equation $x^2 + 81 = y^3$.
8. If x, y, and z are integers such that $x^3 + y^3 = 2z^3$ then $x = \pm y$.
9. Solve the Diophantine equation $x^2 - 1 = y^3$. (Hint: use previous exercise.)
10. Show that no triangular number greater than 1 is a cube.
11. Show that $x^2 + 432 = y^3$ has a unique solution in rational numbers. (Hint: let $x = 36k/n$, $y = 12m/n$, $u = n+k$ and $v = n-k$; then $u^3 + v^3 = (2m)^3$.)

1.10 Congruent Numbers *

A positive integer n is *congruent* if and only if there are integers x and y (with y nonzero) such that both $x^2 + ny^2$ and $x^2 - ny^2$ are squares.

This is the same as saying that there are 3 rational squares in arithmetic progression with common difference n — namely, $(x/y)^2 - n$, $(x/y)^2$ and $(x/y)^2 + n$.

For example, as Leonardo of Pisa (Fibonacci) noted about the year 1220,

$$41^2 + 5 \times 12^2 = 49^2 \text{ and } 41^2 - 5 \times 12^2 = 31^2$$

and hence 5 is congruent. We have $(31/12)^2$, $(41/12)^2$ and $(49/12)^2$ in AP with common difference 5.

From Exercise 1.6 #8, it follows that 1 is not congruent, and from Exercise 1.7 #6, it follows that 2 is not congruent.

If m^2 is the largest square factor of an integer n then n/m^2 is the *square-free part* of n. Note that a positive integer is congruent if and only if its square-free part is congruent. Thus, for example, from the fact that 1 is not congruent, it follows that no square is congruent.

There are exactly 36 square-free integers less than 100 which are congruent. They are listed in the Table.

ALL THE SQUARE-FREE CONGRUENT NUMBERS
< 100

5	21	34	47	69	85
6	22	37	53	70	86
7	23	38	55	71	87
13	29	39	61	77	93
14	30	41	62	78	94
15	31	46	65	79	95

Congruent numbers were discussed as long ago as the tenth century but they are still a lively topic today. For example, they interact with very recent developments in the theory of elliptic curves. If the 'Birch-Swinnerton-Dyer Conjecture' is true then 'Tunnell's Conjecture' is true, and 'Tunnell's Conjecture' gives a necessary and sufficient condition for a number's being congruent. The reader may wish to consult Neal Koblitz's *Introduction to Elliptic Curves and Modular Forms* (New York: Springer-Verlag, 1984).

By Exercise 1.9 #3, there are infinitely many primes of the form $8z+3$. This and the following theorem (first proved by A. Genocchi, in 1882) show that there are infinitely many square-free noncongruent numbers. Later in this section, we shall show that there are infinitely many square-free congruent numbers.

Theorem 1.10.1 *No prime of the form* $8z+3$ *is congruent.*

Proof: Suppose there are such primes and let p be one of them. Let x be the smallest positive integer such that, for some integers y, u, and v, $x^2+py^2=u^2$ and $x^2-py^2=v^2$.

Then $2x^2=u^2+v^2$ and $2py^2=u^2-v^2$. From x's minimality, it follows that $(u,v)=1$ and hence u and v are both odd. Since u and v are both odd, y is even, say, $y=2y'$, and we have

$$2py'^2=\frac{u-v}{2}\frac{u+v}{2}$$

Since

$$\left(\frac{u-v}{2},\frac{u+v}{2}\right)=1$$

there are two possibilities.

Case 1. For some integers s and t, one of $\frac{u-v}{2}$ and $\frac{u+v}{2}$ equals $2s^2$ and the other pt^2.

Then

$$2x^2=(2s^2+pt^2)^2+(2s^2-pt^2)^2$$

so that $x^2=(2s^2)^2+(pt^2)^2$. By the Pythagorean Triangle Theorem, there are relatively prime integers a and b, not both odd, such that $2s^2=2ab$ and $pt^2=a^2-b^2$. But this implies that a and b are squares, say, $a=A^2$ and $b=B^2$, and we have $pt^2=A^4-B^4$. The fourth power of an odd number has the form $8w+1$, whereas the fourth power of an even number has the form $8w$. Since a and b have different parity, A^4-B^4 has the form $8w\pm1$. Since p has the form $8z+3$, and t^2 has one of the forms $8w$, $8w+1$, and $8w+4$, it follows that pt^2 has one of the forms $8w$, $8w+3$, and $8w+4$. Contradiction. Case 1 cannot arise.

Case 2. For some integers s and t, one of $\frac{u-v}{2}$ and $\frac{u+v}{2}$ equals $2ps^2$ and the other t^2.

Then

$$x^2=(2ps^2)^2+(t^2)^2$$

and, by the Pythagorean Triangle Theorem, there are relatively prime integers a and b, not both odd, such that $2ps^2 = 2ab$ and $t^2 = a^2 - b^2$. Hence $a = pA^2$ and $b = B^2$, or $a = A^2$ and $b = pB^2$.

Suppose $a = pA^2$ and $b = B^2$. Then, since $b^2 + t^2 = a^2$ with $(a, b) = 1$, the Pythagorean Triangle Theorem implies that $pA^2 = a = c^2 + d^2$ where c and d have different parity. But pA^2 has the form $4w$ or $4w+3$ whereas $c^2 + d^2$ has the form $4w + 1$. Contradiction.

Suppose $a = A^2$ and $b = pB^2$. Then

$$t^2 = A^4 - p^2B^4 = (A^2 - pB^2)(A^2 + pB^2)$$

Since u is odd and $u = 2ps^2 + t^2$, t is odd, and hence both $A^2 - pB^2$ and $A^2 + pB^2$ are odd. Since $(a, b) = 1$, it follows that

$$(A^2 - pB^2,\ A^2 + pB^2) = 1$$

Hence for some integers e and f, we have $A^2 - pB^2 = e^2$ and $A^2 + pB^2 = f^2$. By x's minimality, $x \leq A$. However,

$$A^2 < 2ab < (2ps^2)^2 + t^4 = x^2$$

Contradiction.

On the other hand,

Theorem 1.10.2 *There are infinitely many square-free congruent numbers.*

Proof: Suppose there are only finitely many square-free congruent numbers, and let $p_1, p_2, \ldots, p_r$ be all the primes which factor at least one of them. Let p be a prime larger than all these.

Now, for any integer n,

$$(4n^2 + 1)^2 \pm (8n^3 - 2n)2^2 = (4n^2 \pm 4n - 1)^2 \quad (*)$$

so that $8n^3 - 2n$ is congruent. In particular, $8p^3 - 2p$ is congruent. Since p^2 is not a factor of $8p^3 - 2p$, it follows that p is a factor of the square-free part of $8p^3 - 2p$. Thus there is a square-free congruent number divisible by p. Contradiction.

Incidentally, (∗) can be used actually to find infinitely many square-free congruent numbers. With $n = 1$, we have 6; with $n = 2$, we have 15 (the square-free part of 60); with $n = 4$, we have 14.

It is not always easy to show that a number is congruent. In his *Recreations in the Theory of Numbers,* Albert Beiler uses the following theorem to show that 23 is congruent.

Theorem 1.10.3 *Let a, b, c and d be any nonzero integers. If there are nonzero integers x, y, z, and w such that $ax^2 + by^2 = cz^2$ and $ax^2 - by^2 = dw^2$ then $|abcd|$ is congruent.*

Proof:

$$\begin{aligned} &(c^2z^4 + d^2w^4)^2 \pm abcd(4xyzw)^2 \\ &= 4(a^2x^4 - b^2y^4)^2 + 16a^2b^2x^4y^4 \pm 16abcdx^2y^2z^2w^2 \\ &= 4(cdz^2w^2 \pm 2abx^2y^2)^2 \end{aligned}$$

For example, with $x = 4$ and $y = 3$, we have a solution to the system $x^2 + y^2 = z^2$ and $x^2 - y^2 = 7w^2$. Hence 7 is congruent. As another example, $x = 5$ and $y = 6$ give a solution to $13x^2 + y^2 = z^2$ and $13x^2 - y^2 = w^2$. Hence 13 is congruent.

To show 23 is congruent, Beiler finds a solution to $x^2 + y^2 = z^2$ and $x^2 - y^2 = 23w^2$, namely, $x = 312$ and $y = 266$.

As the following theorem shows, congruent numbers can also be defined as areas of right triangles with rational sides.

Theorem 1.10.4 *A natural number is congruent iff it is the area of a right triangle whose sides have rational lengths.*

Proof: Suppose n is congruent. Let x and y be integers (with y nonzero) such that $x^2 \pm ny^2$ are both squares. Say $x^2 + ny^2 = u^2$ and $x^2 - ny^2 = v^2$. Then $u/y - v/y$, $u/y + v/y$ and $2x/y$ are rationals which are sides of a right triangle with area n.

Conversely, if a right triangle with rational sides A, B, and C (with C the hypotenuse) has area n, then $C^2 \pm n2^2$ are both squares.

We can go further.

Theorem 1.10.5 *Any congruent number is the area of infinitely many right rational triangles.*

Proof: Let n be congruent. Without loss of generality, we may suppose that n is square-free. Let a and b be positive integers such that $(a, b) = 1$ and a/b is the hypotenuse of a right angled rational sided triangle with area n. Suppose the sides of this triangle are the rationals x and y with $y < x < a/b$. Then $(a/b)^2 = x^2 + y^2$ and $\frac{1}{2}xy = n$.

Let $k = a^4 - 16b^4n^2$. Since $a^2 - 4b^2n = b^2(x - y)^2$ and $a^2 + 4b^2n = b^2(x + y)^2$, it follows that k is a square integer.

Since $b^2(x - y)^2$ is an integer and a square of a rational, it follows that it is a square integer, u^2. Similarly, $b^2(x + y)^2$ is a square integer, v^2.

Suppose a is even, say, $a = 2a'$. Then $a'^2 - b^2n - (u/2)^2$ and $a'^2 + b^2n = (v/2)^2$. Since $(a, b) = 1$, b is odd. Since

$$2nb^2 = \left(\frac{v}{2} - \frac{u}{2}\right)\left(\frac{v}{2} + \frac{u}{2}\right)$$

$u/2$ and $v/2$ have the same parity and hence n is even. Since n is square-free, it has the form $4c + 2$. Since a'^2 has the form $4d$ or $4d + 1$, it follows that $a'^2 + b^2n$ has the form $4e + 2$ or $4e + 3$. But $(v/2)^2$ does not have either of these forms. Contradiction. Hence a is odd.

Let $D = 2ab\sqrt{k}$. Since k is a square integer, D is an integer. Let $A = k/D$, $B = 8a^2b^2n/D$ and $C = (a^4+16b^4n^2)/D$. Then $A^2+B^2 = C^2$ and $\frac{1}{2}AB = n$.

Since a is odd, so is $a^4 + 16b^4n^2$. Thus, since $(a, b) = 1$, we have $(2b, a^4 + 16b^4n^2) = 1$. Thus the numerator of C (when it is expressed as a fraction in lowest terms) is at least

$$\frac{a^4 + 16b^4n^2}{a\sqrt{k}}$$

and this is greater than a, the numerator of the original hypotenuse, as a straightforward calculation reveals.

We can now construct yet another rational right triangle with area n, the numerator of whose hypotenuse is greater still.

It is a corollary to the above that if the system $x^2 + ny^2 = z^2$ and $x^2 - ny^2 = z^2$ has one nontrivial solution, then it has infinitely many.

Exercises 1.10

1. Show that $6(1^2 + 2^2 + \cdots + x^2)$ is congruent. (Hint:

$$(2x^2 + 2x + 1)^2 \pm 4x(x + 1)(2x + 1)$$

are squares.)
2. Find a right angled triangle with rational sides and area 34.
3. Find two right triangles with integer sides and area 210.
4. Show that n is a square-free congruent number iff n is the square-free part of $rs(r^2 - s^2)$ for some positive integers r and s with $r > s$, with $(r, s) = 1$, and with r, s not both odd.
5. A positive integer n is congruent iff the curve $y^2 = x^3 - n^2x$ has infinitely many points with rational coordinates. (Hint: Suppose n is congruent. There are infinitely many rational right triangles with area n. Let their hypotenuses be $C_1, C_2, \ldots$. Then, for any m, $(C_m/2)^2 \pm n$ are squares of rationals. Let $x_m = (C_m/2)^2$. Then $x_m^3 - n^2x_m$ is a square of a rational. Conversely, if $y^2 = x^3 - n^2x$ has a rational point (x, y) with $y \neq 0$ then

$$\left(\frac{x^2 + n^2}{2y}\right)^2 \pm n = \left(\frac{x^2 - n^2 \pm 2xn}{2y}\right)^2 .)$$

1.11 Möbius Function *

In this section we define the Möbius function, and give the Möbius inversion formula. The *Möbius function* is the function μ such that $\mu(1) = 1$, $\mu(n) = 1$ if n is a square-free positive integer with an even number of distinct prime factors, $\mu(n) = -1$ if n is a square-free positive integer with an odd number of distinct prime factors, and $\mu(n) = 0$ if n has a square factor (> 1). For example, $\mu(10) = 1$ and $\mu(100) = 0$.

The Möbius function is so named in honour of August Ferdinand Möbius (1790–1868), the German mathematician who gave us the 'Möbius band'. Möbius published his work on the Möbius function and inversion formula in 1831.

One rather neat property of the Möbius function is the following.

Theorem 1.11.1 *If $n > 1$ then $\sum_{d|n} \mu(d) = 0$.*

Proof: Suppose n has k distinct prime factors. Then the above sum equals

$$1 + \binom{k}{1}(-1) + \binom{k}{2}(-1)^2 + \cdots + \binom{k}{k}(-1)^k = (1-1)^k = 0$$

In order to give a quick proof of the 'inversion formula', we use the following definitions.

If f and g are two functions on the positive integers their *Dirichlet product* is the function

$$(f * g)(n) = \sum_{d|n} f(d)g(n/d) = \sum_{ab=n} f(a)g(b)$$

This product is named after Peter Dirichlet (1805–1859), the great number theorist who was a disciple of Carl Gauss (1777–1855). The brains of these two mathematicians are preserved in the Department of Physiology at Göttingen University.

Now let D be the set of all functions whose domain is the positive integers — excluding functions f such that $f(1) = 0$ — and let I be the element of D such that $I(1) = 1$ and $I(n) = 0$ when $n \neq 1$. Then we have

Theorem 1.11.2 *The set D is an abelian group with respect to the operation $*$. Its identity element is I.*

Proof: To show associativity, note that

$$((f * g) * h)(n) = \sum_{abc=n} f(a)g(b)h(c)$$

Now let $f \in D$. Define $g(1) = 1/f(1)$ and, for $n > 1$,

$$g(n) = -g(1) \sum_{d|n, d \neq n} f(n/d)g(d)$$

For example, $g(2) = (-1/f(1))f(2)(1/f(1))$.

Then $(g*f)(1) = 1 = I(1)$ and

$$(g*f)(2) = \sum_{ab=2} f(a)g(b) = f(1)g(2) + f(2)g(1)$$

$$= -f(2)/f(1) + f(2)g(1) = 0 = I(2)$$

In general, if $n > 0$, $(f*g)(n) = 0$ iff

$$f(1)g(n) = -\sum_{ab=n, a\neq 1} f(a)g(b)$$

which is true. Hence $g = f^{-1}$ with respect to $*$.

For example, if $w(n) = 1$ for all positive integers n, then $\mu * w = I$ (Theorem 1.11.1), so that w is the inverse of μ in the group. The next theorem is the *Möbius Inversion Formula.*

Theorem 1.11.3 *If f, $g \in D$ then*

$$f(n) = \sum_{d|n} g(d) \Longleftrightarrow g(n) = \sum_{d|n} f(d)\mu(n/d)$$

Proof: The left hand side is equivalent to $f = g*w$ or $f*\mu = g*w*\mu$ or $f*\mu = g$, which is the right hand side.

For example, $s(n) = \sum_{d|n} d$ and hence $n = \sum_{d|n} s(d)\mu(n/d)$.

The Möbius function is related to the Prime Number Theorem. This theorem states that if $\pi(x)$ is the number of primes less than or equal to x, then

$$\lim_{x\to\infty} \pi(x)\frac{\ln x}{x} = 1$$

It was proved independently by J. Hadamard and C. J. de la Vallée Poussin, in 1896. As proved in, say, T. Apostol's *Introduction to Analytic Number Theory,* the Prime Number Theorem is equivalent to the fact that the 'average value' of μ is 0. More precisely, it is equivalent to the statement

$$\lim_{x\to\infty} \frac{\sum_{n<x} \mu(n)}{x} = 0$$

In Chapter 7 we shall give a proof of Prime Number Theorem which uses the following two facts.

Theorem 1.11.4

$$\left|\sum_{n\le x}\frac{\mu(n)}{n}\right| \le 1$$

Proof: In the sum

$$\sum_{h\le n}\sum_{d|h}\mu(d)$$

$F(d)$ occurs as many times as d has multiples $\le n$, that is $[n/d]$ times. Hence this sum equals

$$\sum_{d\le n}[n/d]\mu(d)$$

By Theorem 1.11.1, the first sum is just 1, and so the second sum is 1 also. Dropping the square brackets introduces an error of at most $n-1$, so, dividing by n,

$$|\sum_{d\le n}\mu(d)/d| \le 1$$

This will not change if we replace n with x.

Theorem 1.11.5

$$f(x) = \sum_{n\le x} g(x/n) \implies g(x) = \sum_{n\le x}\mu(n)f(x/n)$$

Proof:

$$\sum_{n\le x}\mu(n)f(x/n) = \sum_{n\le x}\mu(n)\sum_{m\le x/n} g(x/mn)$$

$$= \sum_{mn\le x}\mu(n)g(x/mn) = \sum_{r\le x} g(x/r)\sum_{d|r}\mu(d) = g(x)$$

by Theorem 1.11.1.

Exercises 1.11

1. What does the Möbius Inversion Formula give in the case of $t(n) = \sum_{d|n} 1$?
2. Solve $\mu(n) + \mu(n+1) + \mu(n+2) = 3$.
3. Calculate $\frac{1}{50} \sum_{n<50} \mu(n)$.
4. Calculate $\pi(100)$, comparing it with $\frac{x}{\ln x}$.
5. Let x be a real number ≥ 1. Noting that there are $[x/d]$ multiples of d from 1 to $[x]$, prove that

$$\sum_{n \leq x} \sum_{d|n} f(d) = \sum_{d \leq x} f(d)[x/d]$$

6. Using the previous exercise, prove that if $x \geq 1$, then

$$\sum_{n=1}^{\infty} \mu(n)[x/n] = 1$$

Chapter 2

Simple Continued Fractions

Simple continued fractions are a powerful mixture of analysis and algebra which is as important in contemporary Number Theory as it was in the work of Lagrange (1736–1813), who used these fractions to give completely general solutions to the Diophantine equations $Ax+By=C$ and $x^2-Ry^2=C$. In this chapter we give Lagrange's solution to the first equation, and, in Chapter 4, we give a solution very much like that of Lagrange to the second equation.

In his *Recreations in the Theory of Numbers,* Albert Beiler makes eerie comments about simple continued fractions. On page 258 he reports that mathematicians often avoid them, and 'take long circuitous routes around and over rather than through the subterranean depths where the convergent goblins gambol'. Beiler is right. Gauss, for example, never uses them in his *Disquisitiones Arithmeticae,* preferring to give a very artificial 'algebraic' solution to the Diophantine equation $x^2-Ry^2=C$. Beiler says about this equation that, if $C>\sqrt{R}$, 'a graceful retirement before this goblin is indicated. Chrystal's *Algebra* will furnish the dauntless mathematical Siegfried the fragments to forge into a sword to attack this monster'. In this book we shall not only attack but also defeat the monster, and it will be in the 'depths' of this chapter that we begin forging the sword to do so.

Some of the basic theory of simple continued fractions is implicit in in work done by the Pythagoreans. The first writer to take them up explicitly was Daniel Schwenter (1618). Many introductory Number Theory books put the chapter on simple continued fractions after the chapter on Gauss's 'congruence'. In this book we have chosen to reverse that order, first because we want to follow the chronological order in which the concepts were developed, and second because we feel that simple continued fractions are just as fundamental to the subject as congruence (which we treat in Chapter 3).

2.1 Convergents and Convergence

We begin by giving recursive definitions for the numerators and denominators of the 'convergents' which are the essence of simple continued fractions.

Let $f_{-1} = 0$ and $f_0 = 1$. Let $g_{-1} = 1$ and $g_0 = 0$. Let

$$a_1, \ldots, a_n, \ldots$$

be a sequence of real numbers all of which, with the possible exception of a_1, are ≥ 1. Let

$$f_{n+1}(a_1, \ldots, a_{n+1}) = a_{n+1} f_n(a_1, \ldots, a_n) + f_{n-1}(a_1, \ldots, a_{n-1})$$

$$g_{n+1}(a_1, \ldots, a_{n+1}) = a_{n+1} g_n(a_1, \ldots, a_n) + g_{n-1}(a_1, \ldots, a_{n-1})$$

For example,

$$f_3(1, 2, 3) = 3f_2(1, 2) + f_1(1) = 3(2 + 1) + 1 = 10$$

and

$$g_3(1, 2, 3) = 3g_2(1, 2) + g_1(1) = 7$$

If the sequence of a's is understood, we write f_n for $f_n(a_1, \ldots, a_n)$ and g_n for $g_n(a_1, \ldots, a_n)$.

Note that $f_1 = a_1 f_0 + f_{-1} = a_1$, and $g_1 = a_1 g_0 + g_{-1} = 1$, and $g_2 = a_2 g_1 + g_0 = a_2$.

Note also that $1 \leq g_2 < g_3 < \ldots$ and $\lim_{n\to\infty} g_n = \infty$.

If the a's are all 1's, g_n is the n-th term of the Fibonacci sequence:

$$1,\ 1,\ 2,\ 3,\ 5,\ 8,\ 13,\ 21,\ 34,\ 55,\ 89,\ \ldots$$

— where each term is obtained from the preceding two by adding them. This sequence, named after the Italian mathematician Leonardo Fibonacci (1170–1250), is famous for its many interesting properties. Using mathematical induction, one can prove, for example, the following:

for any $k < n$, $g_n = g_{k+1}g_{n-k} + g_k g_{n-k-1}$;
$\gcd(g_{n+1}, g_n) = 1$;
$\gcd(g_n, g_m) = g_{\gcd(n,m)}$;

We begin our forging of Siegfried's sword with a theorem which allows us to 'cancel off' the last a in $f_{n+1}(a_1, \ldots, a_n, a_{n+1})$, obtaining an f with only n arguments. This cancellation, as we shall see, is useful in using mathematical induction to prove a basic property of simple continued fractions.

Theorem 2.1.1 *Where n is a positive integer,*

$$f_n(a_1, \ldots, a_{n-1}, a_n + 1/a_{n+1}) = f_{n+1}(a_1, \ldots, a_n, a_{n+1})/a_{n+1}$$

and

$$g_n(a_1, \ldots, a_{n-1}, a_n + 1/a_{n+1}) = g_{n+1}(a_1, \ldots, a_n, a_{n+1})/a_{n+1}$$

Proof:

$$\begin{aligned}
&f_n(a_1, \ldots, a_{n-1}, a_n + 1/a_{n+1}) \\
&= (a_n + 1/a_{n+1})f_{n-1}(a_1, \ldots, a_{n-1}) + f_{n-2}(a_1, \ldots, a_{n-2}) \\
&= a_n f_{n-1}(a_1, \ldots, a_{n-1}) + (1/a_{n+1})f_{n-1}(a_1, \ldots, a_{n-1}) \\
&\quad + f_n(a_1, \ldots, a_n) - a_n f_{n-1}(a_1, \ldots, a_{n-1}) \\
&= f_{n+1}(a_1, \ldots, a_{n+1})/a_{n+1}
\end{aligned}$$

The proof for g_n is similar.

For example, $5 = g_5(1,1,1,1,1) = g_4(1,1,1,2)$.

Where $a_1, \ldots, a_n, \ldots$ is a sequence of real numbers with $a_n \geq 1$ if $n \geq 2$, we follow Euler in writing $(a_1, \ldots, a_n)$ for the following fraction:

$$a_1 + \cfrac{1}{a_2 + \cfrac{1}{\ddots + a_{n-1} + \cfrac{1}{a_n}}}$$

When the a's are all integers, this is a (finite) *simple continued fraction* (SCF), and the a's are its *partial quotients.* The theorem linking our f's and g's to simple continued fractions is the following. We use the previous theorem in its proof.

Theorem 2.1.2

$$f_n/g_n = (a_1, \ldots, a_n)$$

Proof: If $n = 1$, the theorem reads $f_1/g_1 = a_1/1 = (a_1)$, which is true.
Supposing the theorem true for n,

$$\begin{aligned}(a_1, \ldots, a_{n+1}) &= (a_1, \ldots, a_n + 1/a_{n+1}) \\ &= \frac{f_n(a_1, \ldots, a_n + 1/a_{n+1})}{g_n(a_1, \ldots, a_n + 1/a_{n+1})} \\ &= \frac{f_{n+1}(a_1, \ldots, a_{n+1})}{g_{n+1}(a_1, \ldots, a_{n+1})}\end{aligned}$$

(by Theorem 2.1.1). The result follows by MI.

As an immediate consequence of Theorem 2.1.2, we have

Theorem 2.1.3 *If* $x \geq 1$,

$$(a_1, \ldots, a_n, x) = \frac{x f_n(a_1, \ldots, a_n) + f_{n-1}(a_1, \ldots, a_{n-1})}{x g_n(a_1, \ldots, a_n) + g_{n-1}(a_1, \ldots, a_{n-1})}$$

We shall use Theorem 2.1.3 in Section 6 of this chapter to prove Theorem 2.6.2: if x is an irrational, and p/q is a fraction in lowest terms with $q > 0$ then $|x - p/q| < 1/2q^2$ only if p/q is a convergent of x. This latter theorem is then used in our solution of the Diophantine equation $x^2 - Ry^2 = C$, given at the beginning of Chapter 4.

The fractions $f_1/g_1, f_2/g_2, \ldots$ are *convergents* of f_n/g_n. We shall see that, when there are an infinite number of partial quotients (the a's), these convergents converge to a value which is the value of an infinite simple continued fraction. The next theorems give useful properties of f_n and g_n. First we see that the 'denominators' can be expressed in terms of the 'numerators' and vice versa.

Theorem 2.1.4

$$g_n(a_1, \ldots, a_n) = f_{n-1}(a_2, \ldots, a_n)$$

Proof: First note that $g_0 = 0 = f_{-1}$ and $g_1 = 1 = f_0$.
Supposing the theorem true for all natural numbers $< n$,

$$\begin{aligned} g_n(a_1, \ldots, a_n) &= a_n g_{n-1}(a_1, \ldots, a_{n-1}) + g_{n-2}(a_1, \ldots, a_{n-2}) \\ &= a_n f_{n-2}(a_2, \ldots, a_{n-1}) + f_{n-3}(a_2, \ldots, a_{n-2}) \\ &= f_{n-1}(a_2, \ldots, a_n) \end{aligned}$$

and the result follows by MI.

For example, $g_3(1,2,3) = 7 = f_2(2,3)$. Note that, as a corollary, $f_n(a_1, \ldots, a_n) = g_{n+1}(0, a_1, \ldots, a_n)$.

The next theorem, like Theorem 2.1.1, is a cancellation theorem. We shall use it, in the proof of Theorem 2.1.6, to show that, as far as the 'numerators' are concerned it does not matter if the partial quotients are written forwards or backwards! The proof of Theorem 2.1.5 is by mathematical induction.

Theorem 2.1.5 *Where n is a positive integer, and $a_1 \neq 0$,*

$$f_n(a_1, \ldots, a_n) = a_1 f_{n-1}(a_2 + 1/a_1, a_3, \ldots, a_n)$$

Proof: First note that

$$f_1(a_1) = a_1 = a_1 f_0$$

and

$$f_2(a_1, a_2) = a_1 a_2 + 1 = a_1 f_1(a_2 + 1/a_1)$$

Supposing the theorem to be true for all positive integers $< n$,

$$f_n(a_1, \ldots, a_n) = a_n f_{n-1}(a_1, \ldots, a_{n-1}) + f_{n-2}(a_1, \ldots, a_{n-2})$$

$$= a_n a_1 f_{n-2}(a_2 + 1/a_1, a_3, \ldots, a_{n-1}) + a_1 f_{n-3}(a_2 + 1/a_1, a_3, \ldots, a_{n-2})$$

$$= a_1 f_{n-1}(a_2 + 1/a_1, a_3, \ldots, a_n)$$

and the result follows by MI.

Sometimes a sequence will read the same backwards as forwards. For example, it does not matter if we read the sequence 247742 from the left or from the right. Such sequences are *palindromic.* Most sequences are not palindromic, but the next theorem tells us that, in the case of the f's, it does not matter. Whether the sequence of a's is read from the left or from the right, one obtains the same value for f. Thus, for example, the numerator of

$$5+\frac{1}{3+\frac{1}{1+\frac{1}{7}}}$$

has to equal the numerator of

$$7+\frac{1}{1+\frac{1}{3+\frac{1}{5}}}$$

This curious result lies behind the material on palindromic simple continued fractions, found in Section 10 of this chapter.

Theorem 2.1.6

$$f_n(a_n,\ldots,a_1)=f_n(a_1,\ldots,a_n)$$

Proof: Clearly the theorem is true when $n=1$. Supposing it true for n, and using Theorem 2.1.5 and Theorem 2.1.1,

$$f_{n+1}(a_{n+1},a_n,\ldots,a_1)=a_{n+1}f_n(a_n+1/a_{n+1},a_{n-1},\ldots,a_1)$$

$$=a_{n+1}f_n(a_1,\ldots,a_n+1/a_{n+1})=f_{n+1}(a_1,\ldots,a_{n+1})$$

The result follows by MI.

As another example, $f_3(1,2,3)=10=f_3(3,2,1)$. A similar result holds for g: we have

Theorem 2.1.7

$$g_n(a_1,\ldots,a_n)=g_n(a_{n+1},\ldots,a_2)$$

$$g_n(a_n,\ldots,a_1)=g_n(a_0,\ldots,a_{n-1})$$

Proof: Use Theorem 2.1.4 and Theorem 2.1.6.

The previous theorems lead to further curious results. For example, we have

Theorem 2.1.8

$$\frac{f_n(a_1,\ldots,a_n)}{f_{n-1}(a_1,\ldots,a_{n-1})} = (a_n,\ldots,a_1)$$

$$\frac{g_n(a_1,\ldots,a_n)}{g_{n-1}(a_1,\ldots,a_{n-1})} = (a_n,\ldots,a_2)$$

Proof: Use Theorem 2.1.2, Theorem 2.1.4, and Theorem 2.1.6.

The next theorem will be used in Section 10 of this chapter, in the proof of a theorem characterising repeating, palindromic simple continued fractions. Unless otherwise stated, it is assumed that the f's and g's take their partial quotient arguments in the forwards order. For example, f_n written alone means $f_n(a_1,\ldots,a_n)$.

Theorem 2.1.9 *If* $x \geq 1$,

$$(a_n,\ldots,a_1,x) = \frac{xf_n + g_n}{xf_{n-1} + g_{n-1}}$$

Proof: Using Theorem 2.1.2, and then Theorems 2.1.6 and 2.1.7, we have

$$(a_n,\ldots,a_1,x) = \frac{xf_n(a_n,\ldots,a_1) + f_{n-1}(a_n,\ldots,a_2)}{xg_n(a_n,\ldots,a_1) + g_{n-1}(a_n,\ldots,a_2)}$$

$$= \frac{xf_n(a_1,\ldots,a_n) + f_{n-1}(a_2,\ldots,a_n)}{xg_n(a_0,\ldots,a_{n-1}) + g_{n-1}(a_1,\ldots,a_{n-1})}$$

and the result follows by Theorem 2.1.4.

The following theorem is so important we give it a name. We call it **Plato's Theorem** — in honour of the Greek philosopher Plato (427–347 BC) who did so much to encourage Mathematics. Plato's mathematical contemporaries did, in effect, use simple continued fractions, so it is possible that Plato was aware of this result.

Theorem 2.1.10 (Plato's Theorem)

$$f_n g_{n-1} - f_{n-1} g_n = (-1)^n$$

Proof: First note that the theorem holds for $n = 0$:

$$f_0 g_{-1} - f_{-1} g_0 = 1 = (-1)^0$$

Supposing the theorem true for $n - 1$,

$$\begin{aligned} f_n g_{n-1} - f_{n-1} g_n &= a_n f_{n-1} g_{n-1} + f_{n-2} g_{n-1} - f_{n-1} a_n g_{n-1} - f_{n-1} g_{n-2} \\ &= f_{n-2} g_{n-1} - f_{n-1} g_{n-2} = -(-1)^{n-1} \\ &= (-1)^n \end{aligned}$$

and the result follows by MI.

If the sequence $a_1, \ldots, a_n, \ldots$ consists only of integers then f_n and g_n are integers. Since, by Theorem 2.1.10, any factor common to f_n and g_n would be a factor of ± 1, it follows that f_n and g_n are relatively prime: $(f_n, g_n) = 1$. The convergent f_n/g_n is thus in lowest terms. Similarly,

$$(f_n, f_{n-1}) = (g_n, g_{n-1}) = 1$$

As an example of Plato's Theorem,

$$f_3(1,2,3)g_2(1,2) - f_2(1,2)g_3(1,2,3) = 10 \times 2 - 3 \times 7 = (-1)^3$$

As another example, note that, in the Fibonacci sequence,

$${g_n}^2 - g_{n-1} g_{n+1} = g_n f_{n-1} - g_{n-1} f_n$$

(Theorem 2.1.4). Hence it follows from Plato's Theorem that, in the Fibonacci sequence, ${g_n}^2 - g_{n-1} g_{n+1} = (-1)^{n+1}$. For example, we have $5^2 - 3 \times 8 = 1$.

Note that if we had a way of finding the convergents to a rational a/b then Plato's Theorem would give us a way to solve the Diophantine equation $ax - by = \pm 1$. We shall exploit this idea in Section 5 below.

With the next two theorems, we show that $(a_1, \ldots, a_n) = f_n/g_n$ tends to a limit as n tends to infinity. This will allow us to talk about 'infinite simple continued fractions'.

Theorem 2.1.11
f_1/g_1, f_3/g_3, f_5/g_5, ... is a strictly increasing sequence.
f_2/g_2, f_4/g_4, f_6/g_6, ... is a strictly decreasing sequence.

Proof: Using Plato's Theorem,

$$f_n g_{n-2} - f_{n-2} g_n$$

$$= a_n f_{n-1} g_{n-2} + f_{n-2} g_{n-2} - f_{n-2} a_n g_{n-1} - f_{n-2} g_{n-2} = a_n(-1)^{n-1}$$

Hence $f_n/g_n - f_{n-2}/g_{n-2} = (-1)^{n-1} a_n/(g_n g_{n-2})$. Since $a_n \geq 1$ (for $n > 1$) and $1 \leq g_2 < g_3 < \ldots$, it follows that

$$f_n/g_n > f_{n-2}/g_{n-2}$$

if n is odd, and

$$f_n/g_n < f_{n-2}/g_{n-2}$$

if n is even.

Theorem 2.1.12 *f_n/g_n tends to a limit as n tends to infinity.*

Proof: Let

$$D = f_{2n}/g_{2n} - f_{2n-1}/g_{2n-1} = (-1)^{2n}/g_{2n-1}g_{2n} > 0$$

using Plato's Theorem. Then, by Theorem 2.1.11,

$$a_1 + 1/a_2 = f_2/g_2 > f_{2n}/g_{2n} > f_{2n-1}/g_{2n-1} > f_1/g_1 = a_1$$

Since f_1/g_1, f_3/g_3, f_5/g_5, ... is a strictly increasing sequence bounded above by $a_1 + 1/a_2$, it has a limit. Similarly, f_2/g_2, f_4/g_4, f_6/g_6, ... has a limit. Moreover, since $g_1, g_2, g_3, \ldots$ tends to infinity as n tends to infinity, $\lim_{n\to\infty} D = 0$. Hence the two sequences f_1/g_1, f_3/g_3, f_5/g_5, ... and f_2/g_2, f_4/g_4, f_6/g_6, ... tend to the same limit.

From Theorem 2.1.2 — which states that $f_n/g_n = (a_1, \ldots, a_n)$ — it now follows that $(a_1, \ldots, a_n)$ tends to a limit as n tends to infinity. We denote this limit by

$$(a_1, a_2, \ldots)$$

The sequence (a_1), (a_1, a_2), (a_1, a_2, a_3), ...together with its limit is what we call an *infinite simple continued fraction.* That sequence is the *SCF expansion* of the number which is its limit. As we shall see, every real number has an SCF expansion (i.e. is the limit of such a sequence). However, as we shall prove, in the case of rationals the expansion is finite.

Using the basic properties of limits, we have the following theorem.

Theorem 2.1.13

$$(a_1, \ldots, a_{n-1}, (a_n, a_{n+1}, \ldots)) = (a_1, \ldots, a_n, \ldots)$$

We also need the following four theorems. The first is used in the proof of Theorem 2.9.5, giving us a way of shortening our calculations when we solve the Diophantine equation $x^2 - Ry^2 = 1$.

Theorem 2.1.14 *For the SCF* $(a_1, a_2, \ldots, a_n, 2a_1, a_2, \ldots, a_n)$,

$$f_{2n} = {f_n}^2 + (a_1 f_n + f_{n-1})g_n$$

and

$$g_{2n} = f_n g_n + (a_1 g_n + g_{n-1})g_n$$

Proof: Let A be the above expression for f_{2n}, and B the above expression for g_{2n}. By Theorem 2.1.3,

$$\frac{f_{2n}}{g_{2n}} = \frac{x f_n + f_{n-1}}{x g_n + g_{n-1}}$$

where, here, $x = a_1 + f_n/g_n$. Hence $f_{2n}/g_{2n} = A/B$.

Moreover, by Plato's Theorem,

$$f_n B - g_n A = f_n^2 g_n + a_1 f_n g_n^2 + f_n g_{n-1} g_n - f_n^2 g_n - a_1 f_n g_n^2 - f_{n-1} g_n^2$$

$$= g_n(f_n g_{n-1} - f_{n-1} g_n) = \pm g_n$$

and, similarly, $f_{n-1}B - g_{n-1}A = \pm(f_n + a_1 g_n)$. Since $(f_n, g_n) = 1$, we have $(A, B) = 1$.

Our next two theorems have to do with SCF's which are, not exactly palindromic, but close to it. We shall use these theorems in Section 10 below.

Theorem 2.1.15 *For the SCF* $(a_1, a_2, a_3, \ldots, a_n, a_{n+1}, a_n, \ldots, a_3, a_2)$,

$$f_{2n} = f_{n+1}g_n + f_n g_{n-1}$$

and

$$g_{2n} = g_n(g_{n-1} + g_{n+1})$$

Proof: Let A be the expression for f_{2n}, and B the expression for g_{2n}. By Theorem 2.1.3,

$$\frac{f_{2n}}{g_{2n}} = \frac{xf_{n+1} + f_n}{xg_{n+1} + g_n}$$

where $x = g_n/g_{n-1}$ (Theorem 2.1.8). Hence $f_{2n}/g_{2n} = A/B$.

Also $(A, B) = 1$ since $f_nB - g_nA = \pm g_n$ and

$$f_{n-1}B - g_{n-1}A = \pm(a_{n+1}g_n + g_{n-1}) = \pm g_{n+1}$$

Theorem 2.1.16 *For the SCF* $(a_1, a_2, \ldots, a_n, a_n, \ldots, a_2)$,

$$f_{2n-1} = f_n g_n + f_{n-1}g_{n-1}$$

and

$$g_{2n-1} = {g_n}^2 + {g_{n-1}}^2$$

Proof: Let A be the expression for f_{2n-1} and B the expression for g_{2n-1}. By Theorem 2.1.3,

$$\frac{f_{2n-1}}{g_{2n-1}} = \frac{xf_n + f_{n-1}}{xg_n + g_{n-1}}$$

where $x = g_n/g_{n-1}$ (Theorem 2.1.8). Hence $f_{2n-1}/g_{2n-1} = A/B$.

Moreover, $(A, B) = 1$ since $f_nB - g_nA = \pm g_{n-1}$ and $f_{n-1}B - g_{n-1}A = \pm g_n$.

We close this section with a theorem about palindromic simple continued fractions with an even number of partial quotients. We shall use this theorem in Chapter 3, Section 7, to prove the 'Two Square Theorem'.

Theorem 2.1.17 *For the SCF* $(a_1, a_2, \ldots, a_n, a_n, \ldots, a_2, a_1)$,

$$f_{2n} = {f_n}^2 + {f_{n-1}}^2$$

and

$$g_{2n} = f_n g_n + f_{n-1} g_{n-1}$$

Proof: The proof is left to the reader.

Exercises 2.1

1. Show that $g_6(3,1,3,4,2,5) = 207$.
2. Show that in the Fibonacci sequence, $(g_n, g_m) = g_{(n,m)}$. (Hint:

$$(g_n, g_m) = (g_{n-m+1}g_m + g_{n-m}g_{m-1}, g_m)$$

$$= (g_{n-m}g_{m-1}, g_m) = (g_{n-m}, g_m)$$

and, using an induction hypothesis, $(g_n, g_m) = g_{(n-m,m)} = g_{(m,n)}$.)
3. Prove Theorem 2.1.17.
4. In the Fibonacci sequence, $g_{2n+1} = {g_n}^2 + {g_{n+1}}^2$.
5. Show that $(1, 1, 1, \ldots) = \frac{1}{2}(1 + \sqrt{5})$.
6. Verify that $(9, 2, 3, 5)$ and $(5, 3, 2, 9)$ have the same numerator.

2.2 Uniqueness of SCF Expansions

Let r be any real number. Let $X_1 = r$, and $X_{n+1} = 1/(X_n - [X_n])$, provided X_n is not an integer. Then X_n is called the n-th *complete quotient* of r.

Note that if r is irrational then so is every complete quotient (by mathematical induction) and hence the sequence of complete quotients is infinite.

Note also that $X_n \leq 1$ only if $n = 1$, and that, for all n, $X_n = [X_n] + 1/X_{n+1}$.

From the latter observation it follows that, for any n such that none of the first n complete quotients is an integer, r equals the simple continued fraction $([X_1], \ldots, [X_n], X_{n+1})$. For we have

$$r = X_1 = [X_1] + \frac{1}{X_2} = [X_1] + \frac{1}{[X_2] + \frac{1}{X_3}}$$

and so on.

If the sequence of complete quotients is finite, having as its last term the integer X_n, the SCF expansion of r is just $([X_1], \ldots, [X_{n-1}], [X_n])$. However, what if there are infinitely many complete quotients? This question is answered by the following theorem.

Theorem 2.2.1 *If the sequence of complete quotients of X_1 is infinite,*

$$X_1 = ([X_1], \ldots, [X_n], \ldots)$$

Proof: Since $X_1 = ([X_1]), \ldots, [X_n], X_{n+1})$, Theorem 2.1.3 implies that

$$X_1 = \frac{X_{n+1}f_n + f_{n-1}}{X_{n+1}g_n + g_{n-1}}$$

Hence

$$\begin{aligned} ([X_1], \ldots, [X_n]) - X_1 &= \frac{f_n}{g_n} - \frac{X_{n+1}f_n + f_{n-1}}{X_{n+1}g_n + g_{n-1}} \quad \text{(Theorem 2.1.2)} \\ &= \frac{f_n g_{n-1} - f_{n-1}g_n}{(X_{n+1}g_n + g_{n-1})g_n} \\ &= \frac{\pm 1}{(X_{n+1}g_n + g_{n-1})g_n} \quad \text{(Plato's Theorem)} \end{aligned}$$

Since $X_{n+1} > 1$ (for $n > 0$) and $\lim_{n\to\infty} g_n = \infty$, it follows that

$$\lim_{n\to\infty} ([X_1], \ldots, [X_n]) - X_1 = 0$$

From Theorem 2.2.1, we have the following.

Theorem 2.2.2 *Every real number r can be expressed as a simple continued fraction in the sense that r is either equal to a finite simple continued fraction, or else is equal to the limit of a sequence of finite simple continued fractions in an infinite simple continued fraction.*

For example, if $r = \sqrt{2}$ then

$$X_1 = \sqrt{2}$$
$$X_2 = \frac{1}{\sqrt{2}-1} = 1 + \sqrt{2}$$
$$X_3 = \frac{1}{1+\sqrt{2}-2} = 1 + \sqrt{2}$$

and so on. Hence $\sqrt{2} = (1, 2, 2, 2, \ldots)$.

Thanks to Theorems 2.1.11 and 2.1.12, we can use convergents of $\sqrt{2}$ as approximations to it. Convergents are

$$\frac{1}{1}, \frac{3}{2}, \frac{7}{5}, \frac{17}{12}, \ldots$$

and each of these is a more accurate approximation to $\sqrt{2}$ than its predecessor. In Section 6 below, we shall prove that if f_n/g_n is a convergent of x, then

$$|x - f_n/g_n| \leq 1/g_n^2$$

The fact that convergents of $\sqrt{2}$ give increasingly better approximations to it was, in effect, known to the Pythagoreans, who also noted that if f/g is a convergent of $\sqrt{2}$ then f^2 and $2g^2$ differ by 1. This fact provides the clue for a solution of the Diophantine equation $x^2 - Ry^2 = 1$. See Section 9 below.

Is it possible that $\sqrt{2}$ have some other SCF expansion, with some other sequence of convergents? The answer is no, as the following theorem shows.

Theorem 2.2.3 *With the stipulation that, in the case of a finite simple continued fraction, the final partial quotient not be 1, every real number has exactly one expression as a simple continued fraction.*

Proof: First note that the stipulation is necessary: for $(2) = 2 = (1, 1)$.

Suppose $(a_1, a_2, \ldots) = r = (b_1, b_2, \ldots)$. If a_1 is the only partial quotient in the first SCF expansion, then r is an integer, and either $b_1 = a_1$, or $b_1 = a_1 - 1$ and $b_2 = 1$. Otherwise, we would have $a_1 = b_1 + 1/(b_2, \ldots)$ (using Theorem 2.1.13 for the infinite case), and this is impossible if $(b_2, \ldots) > 1$. Thus, without loss of generality, we may assume that the two SCF expansions each have more than 1 partial quotient.

From this it follows that

$$a_1 + \frac{1}{(a_2, \ldots)} = b_1 + \frac{1}{(b_2, \ldots)}$$

Given our stipulation, $1/(a_2, \ldots)$ and $1/(b_2, \ldots)$ are both less than 1. Since their difference is an integer, it follows that they are equal. Hence $a_1 = b_1$ and $(a_2, \ldots) = (b_2, \ldots)$. Similarly, $a_2 = b_2$ and so on. The result follows by mathematical induction.

Since the SCF expression of a real number r is (in the above sense) unique, we may define the *n-th convergent* of r to be the fraction $f_n(a_1, \ldots, a_n)/g_n(a_1, \ldots, a_n)$ where $r = (a_1, a_2, \ldots)$ is the unique SCF expansion of r (with the final partial quotient not equal to 1 in the finite case — unless otherwise stipulated). The first convergent of r is $[r]$ and, in the finite case, the final convergent is r itself.

Exercises 2.2

1. Find the SCF equal to $\sqrt{3}$.
2. Find the SCF equal to $4/7$.
3. Find the first 5 partial quotients of π.

2.3 SCF Expansions of Rationals

Since an irrational has an infinite number of complete quotients, its SCF expansion is infinite. Are there any rationals whose SCF expansion is infinite? The answer is no:

Theorem 2.3.1 *A real number is rational iff it has a finite SCF expansion.*

Proof: If X_1 is rational, every complete quotient X_n is rational (by mathematical induction). If $n > 1$ then $X_n > 1$ and the equation

$X_{n+1} = 1/(X_n - [X_n])$ has the form $b/d = 1/(a/b - [a/b])$ where a, b, and d are positive integers and $a > b > d$. (We have $b > d$ since b/d is a complete quotient and hence > 1.) Since the positive integers which are the numerators and denominators of the complete quotients thus grow smaller (as n increases), and since this cannot continue indefinitely, there is a complete quotient which is an integer (so that the complete quotient calculations halt). Hence the SCF expansion of a rational is finite.

The converse is immediate.

The equation in the above proof is equivalent to

$$a = [a/b]b + d$$

and d is just the remainder obtained by dividing b into a. For example, suppose $X_1 = 43/30$. Since 30 goes into 43 once with remainder 13, the first partial quotient of 43/30 is $a_1 = 1$, and the second complete quotient is $X_2 = 30/13$. Furthermore, 13 goes into 30 twice with remainder 4. Thus the second partial quotient is $a_2 = 2$, and the third complete quotient is $X_3 = 13/4$. And so on. We have

$$\begin{aligned} X_1 &= 43/30 \\ X_2 &= \frac{1}{\frac{43}{30} - 1} = 30/13 \\ X_3 &= \frac{1}{\frac{30}{13} - 2} = 13/4 \\ X_4 &= \frac{1}{\frac{13}{4} - 3} = 4/1 \end{aligned}$$

and $43/30 = (1, 2, 3, 4)$.

Note also that if $a = [a/b]b + d$ then $(a, b) = (b, d)$. Hence the above procedure can be used to find greatest common divisors. For example, if $X_1 = 86/60$ we get

$$\begin{aligned} X_1 &= 86/60 \\ X_2 &= 60/26 \\ X_3 &= 26/8 \\ X_4 &= 8/2 \end{aligned}$$

and the final integer, namely, 2, is the gcd of 86 and 60.

The above procedure for finding gcd's is called 'Euclid's Algorithm'. It was developed by the Pythagoreans and is found in Euclid's *Elements* (300 BC).

With Theorem 2.2.3, we showed that a real number has only one expression as a simple continued fraction, provided the final partial quotient — if there is one — is not allowed to be 1. The complete quotient calculations (see above) always give the 'right' SCF expansion, since $X_n > 1$ for $n > 1$. (For $r = 1$, the 'right' SCF expansion is (1), not (0, 1).) However, what if we do allow the final partial quotient of the SCF expansion of a rational to be 1? As well as having

$$10/7 = (1, 2, 3)$$

we could have

$$10/7 = (1, 2, 2, 1)$$

Because $(a_1, \ldots, a_n + 1) = (a_1, \ldots, a_n, 1)$, there is, in every finite case, a choice. We can end the simple continued fraction with partial quotient 1 or not. Moreover, we can stipulate that the number of partial quotients in the SCF expansion of a rational be even or odd. For example, consider –8/3. With an odd number of partial quotients, –8/3 = (–3, 2, 1). With an even number of partial quotients, –8/3 = (–3, 3). This choice will prove useful.

Theorem 2.3.1 also has the following, important corollary. Let A and B be any integers with $B > 0$. Let $a = A/(A, B)$ and $b = B/(A, B)$. Suppose $a/b = (a_1, \ldots, a_n)$ (Theorem 2.3.1). By Plato's Theorem, $ag_{n-1} - bf_{n-1} = \pm 1$. Hence we have

Theorem 2.3.2 *Where A and B are any integers, not both 0, there are integers s and t such that $As + Bt = (A, B)$.*

In the exercises at the end of this section, we indicate a proof of this theorem that does not involve simple continued fractions.

In *A Mathematician's Apology,* G. H. Hardy claims that Number Theory's 'very remoteness from human activities should keep it gentle and clean'. To discover that Hardy was wrong, the reader need only look at Neal Koblitz's *A Course in Number Theory and Cryptography* (New York: Springer-Verlag, 1987). As the military establishment well realises, Number Theory is useful for enciphering and deciphering messages. We shall not dwell on the many possibilities, but we shall give one example, an original one, based on the SCF expansion of rationals.

SECRET CIPHER

If the letter A occurs at the end of a word, associate it with 27. Otherwise associate with each letter the number of the place it occupies in the alphabet. Each word is then associated with a unique finite sequence a_1, ..., a_n of positive integers, the last one not being 1. We cipher the word into the rational number $(a_1, \ldots, a_n)$. To decipher it, we use Euclid's Algorithm. For example, 'reason' is ciphered as (18, 5, 1, 19, 15, 14) = 457,708/25,193. In the first exercise at the end of this section, we give the reader an opportunity to decipher a message that has been coded in this fashion.

The next theorem is useful for calculating SCF expansions of rationals, but it applies to all real numbers.

Theorem 2.3.3 *Let* $r = (a_1, a_2, a_3, a_4, \ldots)$.

$$If\ \ a_2 \neq 1, \quad -r = (-a_1 - 1,\ 1,\ a_2 - 1,\ a_3,\ a_4,\ \ldots)$$

$$If\ \ a_2 = 1, \quad -r = (-a_1 - 1,\ a_3 + 1,\ a_4,\ \ldots)$$

Proof: Unless r is an integer, $[-r] = -[r] - 1$, so that

$$\begin{aligned} -r &= [-r] - r + [r] + 1 \\ &= [-r] + \cfrac{1}{1 + \cfrac{r - [r]}{1 - r + [r]}} \\ &= [-r] + \cfrac{1}{1 + \cfrac{1}{\cfrac{1}{r - [r]} - 1}} \\ &= -a_1 - 1 + \cfrac{1}{1 + \frac{1}{X_2 - 1}} \end{aligned}$$

If $a_2 > 1$, the result is immediate. If $a_2 = 1$, then

$$X_3 = \frac{1}{X_2 - [X_2]} = \frac{1}{X_2 - 1}$$

so that $-r = (-a_1 - 1, 1 + X_3)$, and the result follows.

For example, $\sqrt{2} = (1, 2, 2, 2, \ldots)$ and $-\sqrt{2} = (-2, 1, 1, 2, 2, 2, \ldots)$. As a consequence of Theorem 2.3.3, we have

Theorem 2.3.4 *Let a and b be positive integers such that $\frac{1}{2} < a/b < 1$. If $1 - a/b = (0, a_2, a_3, \ldots)$ then*

$$a/b = (0, 1, a_2 - 1, a_3, \ldots)$$

Proof: If $1 - a/b = (0, a_2, a_3, \ldots)$ then $-a/b = (-1, a_2, a_3, \ldots)$. Since $\frac{1}{2} < a/b < 1$, it follows that $1 - a/b < \frac{1}{2}$, so that $a_2 \neq 1$. Thus, by Theorem 2.3.3,

$$a/b = (-(-1) - 1,\ 1,\ a_2 - 1,\ a_3,\ \ldots)$$

For example, suppose we have calculated $3/20 = (0,\ 6,\ 1,\ 2)$. Then we can immediately conclude that $17/20 = (0,\ 1,\ 5,\ 1,\ 2)$.

Exercises 2.3

1. Decipher the following secret message:

$$\frac{75880}{3771}\ \frac{172}{19}\ \frac{63886}{4239}\ \frac{1070}{71}\ \frac{23137}{17424}\ \frac{455557196}{50223473}\ \frac{61}{57}\ \frac{41}{5}\ \frac{172}{19}\ \frac{431}{61}$$

2. Where r is any real number, let $r = (a_1, a_2, a_3, \ldots)$. Then $a_2 = 1$ iff $r - [r] > \frac{1}{2}$.
3. Show that if the second partial quotient, a_2, of a real number r is not 1 then, when $n > 1$, the n-th convergent of $-r$ is the negative of the $(n-1)$-th convergent of r. What happens if $a_2 = 1$?
4. Let A and B be given integers (not both 0). Let S be the set of

all positive integers of the form $As + Bt$ (with s and t any integers). Let d be the least element of S. Show that $d = (A, B)$, thus giving a proof of Theorem 2.3.2. (Hint: if $A = qd + r$ and $d = As' + Bt'$ then $r = A(1 - qs') + B(-qt')$. By d's minimality, $r = 0$. Similarly, d is also a factor of B.)

2.4 Farey Series *

We should like to list the SCF expansions of all positive proper fractions with denominator not greater than 10. What are these fractions? Is there an easy way to compute them in increasing order?

The answer is provided by the theory of the Farey series. Although it was first investigated by C. Haros, it is named after John Farey, who published a note on it in 1816 in the *London, Edinburgh and Dublin Philosophical Magazine*.

The *Farey series* F_n of order n is the ascending sequence of positive proper fractions in lowest terms whose denominators are not greater than n. For example, F_5 is the sequence

$$1/5 \quad 1/4 \quad 1/3 \quad 2/5 \quad 1/2 \quad 3/5 \quad 2/3 \quad 3/4 \quad 4/5$$

Where x/y, x'/y' and x''/y'' are three successive terms of F_5, notice that $x'/y' = (x + x'')/(y + y'')$. Is this true for F_{10}? For any Farey series? The various questions about Farey series can be answered with what we know about simple continued fractions. This is because, as we shall prove in Section 6 below, SCF expansions provide sharp approximations, giving us a precision instrument for filling in any terms missing from a given Farey series. The key theorem is the following.

Theorem 2.4.1 *Let $x/y = (a_1, \ldots, a_{2m+1})$ be a member of F_n. Let*

$$r = \left[\frac{n - g_{2m}}{y}\right] \quad \textit{and } s = \left[\frac{n + g_{2m}}{y}\right]$$

Then $\gcd(sx - f_{2m}, sy - g_{2m}) = 1$ *and the term just less than x/y in F_n is*

$$\frac{sx - f_{2m}}{sy - g_{2m}}$$

Also $\gcd(rx + f_{2m}, ry + g_{2m}) = 1$ *and the term just greater than* x/y *in* F_n *is*

$$\frac{rx + f_{2m}}{ry + g_{2m}}$$

(If x/y *is the first term in* F_n *then* $(sx - f_{2m})/(sy - g_{2m}) = 0$*. If* x/y *is the last term in* F_n *then* $(rx + f_{2m})/(ry + g_{2m}) = 1$*.)*

Proof: By Plato's Theorem,

$$x(sy - g_{2m}) - (sx - f_{2m})y = 1$$

so that $(sx - f_{2m},\ sy - g_{2m}) = 1$ and

$$\frac{sx - f_{2m}}{sy - g_{2m}} = \frac{x}{y} - \frac{1}{(sy - g_{2m})y} \quad (*)$$

Since $n \geq y$, it follows that $s \geq 1$ and hence

$$sx - f_{2m} \geq x - f_{2m} \geq 0$$

and $sy - g_{2m} > 0$. Hence, by (*), $(sx - f_{2m})/(sy - g_{2m}) < x/y$.

Let $e = ((n + g_{2m})/y) - s$. Then $0 \leq e < 1$. Also

$$sy - g_{2m} = \left(\frac{n + g_{2m}}{y} - e\right) y - g_{2m} = n - ey \leq n$$

Thus $(sx - f_{2m})/(sy - g_{2m})$ is a nonnegative fraction in lowest terms whose denominator is not greater than n.

Let $x'' = sx - f_{2m}$ and $y'' = sy - g_{2m}$. As we have just noted, x''/y'' is a member of F_n (or 0). Also, as we showed above, $xy'' - x''y = 1$.

Suppose x'/y' is a member of F_n between x/y and x''/y''. Since $x/y > x'/y'$, it follows that $x/y - x'/y' \geq 1/yy'$. Similarly, $x'/y' - x''/y'' \geq 1/y'y''$. Adding the last two inequalities, we obtain

$$\frac{x}{y} - \frac{x''}{y''} \geq \frac{y + y''}{yy'y''}$$

or $xy'' - x''y \geq (y + y'')/y'$, and hence — since $xy'' - x''y = 1$ — we have $1 \geq (y + y'')/y'$. Thus

$$y' \geq y + y'' = y + n - ey > n$$

Hence x'/y' is not really a member of F_n. Contradiction.

The proof of the second assertion is similar.

For example, $2/5 = (0,2,2)$ is a member of F_5. Since $g_2 = 2$, $s = [7/5] = 1$. The term just less than $2/5$ in F_5 is $(2-1)/(5-2) = 1/3$.

The following theorem is an immediate consequence of Theorem 2.4.1.

Theorem 2.4.2 *If x/y and x'/y' are two successive terms in a Farey series, then $x'y - xy' = 1$.*

The next theorem, also a consequence of Theorem 2.4.1, gives a recursive formula for the terms of F_n. As we shall see, it also allows us to answer the question we raised at the beginning of this section.

Theorem 2.4.3 *If x/y, x'/y' and x''/y'' are three successive terms in F_n then*

$$x'' = \left[\frac{n+y}{y'}\right] x' - x$$

and

$$y'' = \left[\frac{n+y}{y'}\right] y' - y$$

Proof: Applying Theorem 2.4.1 to x'/y',

$$\begin{aligned} r + s &= \left[\frac{n - g_{2m}}{y'}\right] + s \\ &= \left[\frac{n + sy' - g_{2m}}{y'}\right] \\ &= \left[\frac{n+y}{y'}\right] \end{aligned}$$

Hence

$$x'' = rx' + f_{2m} = \left[\frac{n+y}{y'}\right] x' - (sx' - f_{2m}) = \left[\frac{n+y}{y'}\right] x' - x$$

The proof for y'' is similar.

At the beginning of this section we noted that if x/y, x'/y' and x''/y'' are three successive terms of F_5, then $x'/y' = (x+x'')/(y+y'')$. Then we asked if this was true for any Farey series. Thanks to Theorem 2.4.3, we can now answer that question in the affirmative:

Theorem 2.4.4 *If x/y, x'/y' and x''/y'' are three successive terms in F_n then $x'/y' = (x+x'')/(y+y'')$.*

Starting with 0/1 and 1/10 we use Theorem 2.4.3 to compute F_{10} as follows.

$$x_2 = \left[\frac{10+1}{10}\right] \times 1 - 0 = 1$$
$$y_2 = \left[\frac{10+1}{10}\right] \times 10 - 1 = 9$$
$$x_3 = \left[\frac{10+10}{9}\right] \times 1 - 1 = 1$$
$$y_3 = \left[\frac{10+10}{9}\right] \times 9 - 10 = 8$$

and so on, obtaining the following table.

F_{10} AND THE SCF EXPANSIONS OF ITS MEMBERS

1/10	(0, 10)	5/9	(0, 1, 1, 4)
1/9	(0, 9)	4/7	(0, 1, 1, 3)
1/8	(0, 8)	3/5	(0, 1, 1, 2)
1/7	(0, 7)	5/8	(0, 1, 1, 1, 2)
1/6	(0, 6)	2/3	(0, 1, 2)
1/5	(0, 5)	7/10	(0, 1, 2, 3)
2/9	(0, 4, 2)	5/7	(0, 1, 2, 2)
1/4	(0, 4)	3/4	(0, 1, 3)
2/7	(0, 3, 2)	7/9	(0, 1, 3, 2)
3/10	(0, 3, 3)	4/5	(0, 1, 4)
1/3	(0, 3)	5/6	(0, 1, 5)
3/8	(0, 2, 1, 2)	6/7	(0, 1, 6)
2/5	(0, 2, 2)	7/8	(0, 1, 7)
3/7	(0, 2, 3)	8/9	(0, 1, 8)
4/9	(0, 2, 4)	9/10	(0, 1, 9)
1/2	(0, 2)		

Exercises 2.4

1. Show that the term following 5/7 inf F_{1000} is 713/998.
2. Prove the second part of Theorem 2.4.1.
3. How many members does F_{15} have?
4. Let x/y and x'/y' be two consecutive numbers in a Farey series. Let c be the circle with radius $1/2y^2$ which touches the number line at x/y, and let c' be the circle with radius $1/2y'^2$ which touches the number line (on the same side of it) at x'/y'. Prove that c and c' are tangent to each other.

2.5 $Ax + By = C$

In this section we give a simple continued fraction solution to the Diophantine equation $Ax+By = C$. We begin with an puzzle, due to Sam Loyd, that leads to such an equation.

A cow was standing on a railroad bridge almost 100 cow-lengths long. Suddenly she saw a train just five times the length of the bridge away from its end. If she had run away from the train, she would have failed to escape by 1 cow-length, but she made a dash towards the train, and saved herself by 10 cow-lengths. If all the distances are in whole numbers of cow-lengths, how far did Betsy run?

Let x be the length of the bridge. Let y be the distance to safety, in the direction of the train. Let t be the time it would have taken the train to hit Betsy had she run away from it. Let t' be the time it took her to get off the tracks. Then, where a is Betsy's speed, and b the train's,

$$t = \frac{x-y-1}{a} = \frac{5x+x-1}{b}$$

and

$$t' = \frac{y}{a} = \frac{5x-10}{b}$$

Adding, we obtain $(x-1)/a = (11x-11)/b$, so that $b/a = 11$. From the equation with t', we have $11y = 5x - 10$ or

$$5x - 11y = 10$$

To solve Diophantine equations of the form $Ax+By = C$, we can use SCF expansions of rationals. If the greatest common divisor, (A, B), of A and B is not a divisor of C then the equation has no integer solutions. However, if (A, B) is a divisor of C, we can divide it out to get an equation in which $(A, B) = 1$. Without loss of generality, then, let us take it that $(A, B) = 1$ — and $A > 0$.

To solve $Ax+By = C$, we find $B/A = (a_1, \ldots, a_{2n+1})$ — with an odd number of partial quotients. Where K is any integer, let $x = f_{2n}C+BK$ and $y = -g_{2n}C - AK$. Then, by Plato's Theorem,

$$A(f_{2n}C + BK) + B(-g_{2n}C - AK) = -C(-1)^{2n+1} = C$$

or $Ax + By = C$.

Moreover, there are no other solutions than those given above. For if

$$Ax + By = C = Af_{2n}C - Bg_{2n}C$$

then

$$x = f_{2n}C - B(g_{2n}C + y)/A$$

Since $(A, B) = 1$, A is a factor of $g_{2n}C + y$. Hence for

$$K = -(g_{2n}C + y)/A$$

we have $x = f_{2n}C + BK$.

As an example, let us discover how far Betsy ran. Here $A = 5$ and $B = -11$. We have $B/A = (-3, 1, 4)$, and $f_2/g_2 = -2/1$, and $x = -20 - 11K$. To get x to be almost 100, we must take $K = -10$. From this it follows that Betsy ran 40 cow-lengths to safety.

Suppose A, B and C are positive integers, with $(A, B) = 1$, and suppose we want only *positive* integer solutions to $Ax + By = C$. Then we must restrict K so that $f_{2n}C + BK$ and $-g_{2n}C - AK$ are both positive. This is equivalent to having $f_{2n}C/B > -K > g_{2n}C/A$.

The length of the interval in which $-K$ must fall is

$$f_{2n}C/B - g_{2n}C/A = C/AB$$

— by Plato's Theorem. Thus if $C \leq AB$, there is at most one positive integer solution.

Let us find the positive integer solutions of

$$17x + 19y = 320$$

Since $320 < 323 = 17 \times 19$, there is at most one positive integer solution. $19/17 = (1, 8, 2)$ with $f_2 = 9$ and $g_2 = 8$. The general solution is $x = 2880 + 19K$ and $y = -2560 - 17K$. For a positive solution, $-K$ must fall between 150.59 and 151.58. Taking $K = -151$, we obtain the unique positive integer solution: $x = 11$ and $y = 7$.

Exercises 2.5

1. A grocer bought an equal number of fat puppies and rats, paying twice as much for the puppies as for the rats. Although he marked then all up the same ten per cent, the rats sold faster. If he received back the amount of his initial outlay when he had disposed of all but seven animals, how many did he buy at the start?
2. Queen Saranya used to divide her maids into two companies, one which would follow her five abreast, and the other which would follow her seven abreast — both companies in rectangular formation. These companies, moreover, would consist of different numbers of maids on each of nine different days. What is the smallest number of maids Saranya could have had?
3. What exact postages can you *not* pay if you have only 4 and 7 cent stamps?
4. What numbers leave remainder 2 if you divide by 13, at the same time as leaving remainder 3 if you divide by 53 ?

2.6 SCF Approximations

The convergents of a number make good approximations to it. Conversely, any good rational approximation to an irrational is a convergent of it. We make these thoughts precise in the following theorems.

Theorem 2.6.1 *If n is a positive integer, and f_n/g_n is the n-th convergent of the real number x, then*

$$\frac{1}{g_n g_{n+2}} < \left| x - \frac{f_n}{g_n} \right| \le \frac{1}{g_n g_{n+1}}$$

Proof: Note that in writing 'g_{n+2}' we are assuming that f_n/g_n is not the penultimate or ultimate convergent of x. By Theorem 2.1.3,

$$\begin{aligned} x - \frac{f_n}{g_n} &= \frac{X_{n+1}f_n + f_{n-1}}{X_{n+1}g_n + g_{n-1}} - \frac{f_n}{g_n} \\ &= \frac{-(-1)^n}{(X_{n+1}g_n + g_{n-1})g_n} \end{aligned}$$

Since $a_{n+1} = [X_{n+1}]$, it follows that $X_{n+1} < a_{n+1} + 1$ and

$$X_{n+1}g_n + g_{n-1} < (a_{n+1}+1)g_n + g_{n-1} = g_{n+1} + g_n \le g_{n+2}$$

Hence $1/g_n g_{n+2} < |x - f_n/g_n|$. Also

$$g_{n+1} = a_{n+1}g_n + g_{n-1} \le X_{n+1}g_n + g_{n-1}$$

and hence $|x - f_n/g_n| \le 1/g_n g_{n+1}$.

It follows at once from Theorem 2.6.1 that the convergents of x are successively closer to x. For example, in the SCF expansion of $\sqrt{2}$, $f_4/g_4 = 17/12$, $f_5/g_5 = 41/29$, and $f_6/g_6 = 99/70$. Moreover,

$$1/840 < |17/12 - \sqrt{2}| \le 1/348$$

The next theorem is crucial in our treatment of the Diophantine equation $x^2 - Ry^2 = C$.

Theorem 2.6.2 *Let x be any irrational. Let p/q be a fraction in lowest terms with $q > 0$. If $|x - p/q| < 1/2q^2$ then p/q is a convergent of x.*

Proof: Let $p/q = (a_1, \dots, a_n)$ where n is even iff $x < p/q$. Let $p' = f_{n-1}(a_1, \dots, a_{n-1})$ and $q' = g_{n-1}(a_1, \dots, a_{n-1})$.

Let $w = (xq' - p')/(-xq + p)$ so that $x = (wp + p')/(wq + q')$.

By Plato's Theorem,

$$\begin{aligned}\frac{p}{q} - x &= \frac{p}{q} - \frac{wp + p'}{wq + q'} \\ &= \frac{pq' - qp'}{q(wq + q')} \\ &= \frac{(-1)^n}{q(wq + q')}\end{aligned}$$

Given the choice for n, we have $wq + q' > 0$. Also if $|x - p/q| < 1/2q^2$ then $2 < (wq + q')/q$, or $w > 2 - q'/q \geq 1$ (since $q' = g_{n-1} \leq g_n = q$). By Theorem 2.1.3,

$$(a_1, \ldots, a_n, w) = \frac{wp + p'}{wq + q'} = x$$

Since the SCF expansion of x is unique, $p/q = (a_1, \ldots, a_n)$ is a convergent of x.

Since, for example, $355/113 - \pi = 0.000000266 < 1/(2 \times 113^2)$, it follows that $355/113$ is a convergent of π.

Theorem 2.6.3 *Suppose x, y, A and B are positive integers, and C any nonzero integer such that AB is nonsquare and $C^2 < AB$.*
If $Ax^2 - By^2 = C$ then x/y is a convergent of $\sqrt{B/A}$.

Proof: First suppose C is a positive integer. Then $Ax^2 > By^2$ and $\sqrt{A}x/y > \sqrt{B}$, and hence

$$\frac{x\sqrt{A} + y\sqrt{B}}{2y} > \sqrt{B}$$

Since $C^2 < AB$, it follows that $C < \sqrt{AB}$, and

$$|Ax^2 - By^2| < \frac{\sqrt{A}(x\sqrt{A} + y\sqrt{B})}{2y}$$

Dividing by $x\sqrt{A} + y\sqrt{B}$, we eventually get

$$\left|\frac{x}{y} - \frac{\sqrt{B}}{\sqrt{A}}\right| < \frac{1}{2y^2}$$

and, by Theorem 2.6.2, it follows that x/y is a convergent of $\sqrt{B/A}$.

Now suppose C is a negative integer. $Ax^2-By^2 = C$ iff $By^2-Ax^2 = -C$. By the previous result, y/x is a convergent of $\sqrt{A/B}$. Hence x/y is a convergent of $\sqrt{B/A}$.

For example, all the solutions of $x^2 - 2y^2 = 1$ can be found by looking at the convergents of $\sqrt{2}$. We showed above that

$$\sqrt{2} = (1, 2, 2, 2, \ldots)$$

The first few convergents are 1/1, 3/2, and 7/5. The smallest solution of the equation is with $x = 3$ and $y = 2$.

Exercises 2.6

1. Find the smallest positive integer solution of $x^2 - 61y^2 = 1$. (This was first done by Bhaskara (1114–1185), the author of the poetical mathematics book *Lilavati*.)
2. Let p/q be a fraction in lowest terms with $q > 0$. Let f_n/g_n (with $n > 1$) be a convergent of x. Show that if p/q is closer to x than f_n/g_n is, then $q > g_n$.

2.7 SCF Expansions of Quadratic Surds

We now turn to the SCF expansions of numbers of the form $(P+\sqrt{R})/Q$ where R is a positive nonsquare integer, and P and Q are integers such that Q is a factor of $P^2 - R$. (The latter condition entails no loss of generality since it can be achieved by multiplying the numerator and denominator of $(P + \sqrt{R})/Q$ by $\pm Q$.) These numbers are interesting because they have infinite SCF expansions which are repeating. They are important because their SCF expansions can be used to solve any Diophantine equation of the form $x^2 - Ry^2 = C$.

If $X_1 = (P + \sqrt{R})/Q$, and $a = [X_1]$ then

$$X_2 = \frac{1}{(P + \sqrt{R})/Q - a} = \frac{aQ - P + \sqrt{R}}{(R - (aQ - P)^2)/Q}$$

(rationalising). Hence if $P_2 = aQ - P$ and $Q_2 = (R - P_2{}^2)/Q$ then $(P_2 + \sqrt{R})/Q_2$ is the second complete quotient of X_1.

By mathematical induction, it follows that the n-th complete quotient X_n of X_1 is $(P_n + \sqrt{R})/Q_n$ where P_n and Q_n are given by the recursive formulas $P_1 = P$, $Q_1 = Q$, and

$$P_{n+1} = [(P_n + \sqrt{R})/Q_n]Q_n - P_n$$
$$Q_{n+1} = (R - P_{n+1}{}^2)/Q_n$$

Note that if P_n and Q_n are integers such that Q_n is a factor of $P_n{}^2 - R$ then P_{n+1} and Q_{n+1} are integers such that Q_{n+1} is a factor of $P_{n+1}{}^2 - R$ (since $(R - P_{n+1}{}^2)/Q_{n+1} = Q_n$, an integer). Thus it follows by mathematical induction that all the members of the *PQ sequence* (P_1, Q_1), (P_2, Q_2), ... are ordered pairs of integers, and Q_n is a divisor of $P_n{}^2 - R$ for all n.

Note also that we can calculate Q_{n+1} without using the division operation. Since $Q_nQ_{n-1} = R - P_n{}^2$ and $Q_nQ_{n+1} = R - P_{n+1}{}^2$, it follows that

$$Q_n(Q_{n+1} - Q_{n-1}) = (P_n - P_{n+1})(P_n + P_{n+1}) = (P_n - P_{n+1})a_nQ_n$$

where $a_n = [(P_n + \sqrt{R})/Q_n]$. Hence

$$Q_{n+1} = Q_{n-1} + (P_n - P_{n+1})a_n$$

Since a_n has to be computed anyway (to find P_{n+1}) this formula allows for faster calculation of the PQ sequence when the numbers are large.

As an example, let $R = 13$, $P = 100$ and $Q = 3$. Note that 3 is a divisor of $100^2 - 13$. The PQ sequence is

$$(100, 3)\ (2, 3)\ (1, 4)\ (3, 1)\ (3, 4)\ (1, 3)\ (2, 3)\ (1, 4)\ \ldots$$

Note that it repeats.

Henceforth we shall write PQ sequences in table form, adding as a bottom row the sequence of partial quotients $a_n = [(P_n + \sqrt{R})/Q_n]$. For example, where $R = 13$, $P = 100$ and $Q = 3$, we have the following table.

P	100	2	1	3	3	1	2	1	...
Q	3	3	4	1	4	3	3	4	...
	34	1	1	6	1	1	1	1	...

As we shall prove, every PQ sequence eventually repeats a term and is thereafter periodic. In any period there is a smallest Q (possibly repeated with different P's), and, for that Q, there is a unique pair (P, Q) such that P is minimised. A period of a PQ sequence which begins with this minimum (P, Q) is called an *SCF ending* for R.

In the above example, the SCF ending, written with the partial quotients, is

P	3	3	1	2	1
Q	1	4	3	3	4
	6	1	1	1	1

If there is only one SCF ending for R (as P and Q vary, subject only to the restriction that Q is a factor of $P^2 - R$), then R is *single hearted.*

We shall not make much use of the concept of single-heartedness in this book, but it is important for more advanced Number Theory because, when R has the form $4n+2$ or $4n+3$, then the 'real quadratic field' $\mathbf{Q}(\sqrt{\mathrm{R}})$ has 'class number' 1 just in case R is single hearted. We do not know if there are infinitely many single hearted numbers. The reader may wish to consult Anglin's McGill University MSc Thesis 'Simple Continued Fractions and the Class Number' (1985).

Now let

$$\underline{a_1}, \ldots, \underline{a_n}$$

denote the infinite periodic sequence

$$a_1, \ldots, a_n, a_1, \ldots, a_n, a_1, \ldots$$

If all the a's are equal, we denote this sequence by $\underline{a_1}$. For example,

$$(100 + \sqrt{13})/3 = (34, \underline{1}, 1, 6, 1, \underline{1})$$

In solving $x^2 - Ry^2 = C$, it is important to know when in the PQ sequence we have $Q = 1$ (if ever). Related to this is the following theorem.

Theorem 2.7.1 *If in the PQ sequence of $(P + \sqrt{R})/Q$ there is some $Q_n = 1$ then the PQ sequence is thereafter identical to that of $\sqrt{R}$.*

Proof: If $Q_n = 1$ then

$$P_{n+1} = [(P_n + \sqrt{R})/1] \times 1 - P_n = [\sqrt{R}]$$

and

$$Q_{n+1} = (R - {P_{n+1}}^2)/1 = R - [\sqrt{R}]^2$$

Moreover, the PQ sequence of $\sqrt{R}$ begins $(0,1)$, $([\sqrt{R}], R - [\sqrt{R}]^2)$, ... and all the succeeding terms are uniquely determined by these first terms.

We conclude this section by giving the SCF expansions of several classes of numbers of the form $(P + \sqrt{R})/Q$ where $Q = 1$. Let a be a positive integer. Since $a^2 < a^2 + 1 < a^2 + 2a + 1$, it follows that

$$a < \sqrt{a^2+1} < a + 1$$

and hence $[\sqrt{a^2+1}] = a$. Using this fact, we obtain the following PQ sequence for $a + \sqrt{a^2+1}$.

P	a	a
Q	1	1
	$2a$	$2a$

In other words, $a+\sqrt{a^2+1} = (\underline{2a})$. For example, $1+\sqrt{2} = (2,2,2,\ldots)$.

Similarly, we can derive the following PQ sequence for $a + \sqrt{a^2+2}$.

P	a	a	a
Q	1	2	1
	$2a$	a	$2a$

For example, $1 + \sqrt{3} = (2,1,2,1,\ldots)$.

For $a - 1 + \sqrt{a^2-1}$ (with $a > 1$) we have

P	$a-1$	$a-1$	$a-1$
Q	1	$2a-1$	1
	$2a-2$	1	$2a-2$

Finally, for $a-1+\sqrt{a^2-2}$ (with $a>2$) we have the following.

$$\begin{array}{cccccc} P & a-1 & a-1 & a-2 & a-2 & a-1 \\ Q & 1 & 2a-3 & 2 & 2a-3 & 1 \\ & 2a-2 & 1 & a-2 & 1 & 2a-2 \end{array}$$

For example, $2+\sqrt{7}=(4,1,1,1,4,1,1,1,4,1,\ldots)$.

Exercises 2.7

1. Derive the SCF expansion for $a-1+\sqrt{a^2-2}$.
2. Show that 13 is not single hearted.
3. In a PQ sequence, Q_n is even and $2Q_n$ is a factor of P_n^2-R iff Q_{n+1} is even and $2Q_{n+1}$ is a factor of P_{n+1}^2-R.
4. All the Q_n's in the PQ sequence for $(P+\sqrt{R})/Q$ are even iff Q is even and $2Q$ is a factor of P^2-R.
5. Express $\sqrt{13}$ as a simple continued fraction, using the $\underline{a}$ notation.
6. Where a is a positive integer, find the PQ sequences for

$$3a+1+\sqrt{(3a+1)^2+2a+1}$$

and

$$3a+5+\sqrt{(3a+6)^2-6}$$

2.8 Periodic SCF Expansions

All the immediately preceding SCF expansions are periodic. When does this occur? When are the SCF expansions periodic right from the beginning? When do they become periodic later on? We answer these questions in this section.

If u and v are rationals, and R is a positive nonsquare integer, the *conjugate* of $w=u+v\sqrt{R}$ is $w'=u-v\sqrt{R}$. By rationalising the denominators, one can show that the conjugate of

$$\frac{u_1+v_1\sqrt{R}}{u_2+v_2\sqrt{R}} \quad \text{is} \quad \frac{u_1-v_1\sqrt{R}}{u_2-v_2\sqrt{R}}$$

We use this fact to prove the following theorem.

Theorem 2.8.1 *The SCF expansion of $(P+\sqrt{R})/Q$ is periodic after a certain point.*

Proof: By Theorem 2.1.3,

$$X_1 = (P+\sqrt{R})/Q = \frac{X_n f_{n-1} + f_{n-2}}{X_n g_{n-1} + g_{n-2}}$$

Taking conjugates, we obtain

$$X_1' = \frac{X_n' f_{n-1} + f_{n-2}}{X_n' g_{n-1} + g_{n-2}}$$

and hence

$$X_n' = -\frac{g_{n-2}}{g_{n-1}} \frac{X_1' - \frac{f_{n-2}}{g_{n-2}}}{X_1' - \frac{f_{n-1}}{g_{n-1}}}$$

Since $\lim_{n\to\infty} f_n/g_n = X_1$, and since the g_n's are positive, it follows that X_n' is negative for all sufficiently large n. Since $X_n > 1$ (for $n > 1$), we have $X_n - X_n' > 1$ or $2\sqrt{R}/Q_n > 1$ and hence $2\sqrt{R} > Q_n > 0$ — for all sufficiently large n. Since $Q_{n+1} = (R - P_{n+1}^2)/Q_n$, it follows that, for all sufficiently large n, $\sqrt{R} > |P_n|$. As there is only a finite number of possible values for the ordered pairs of integers (P_n, Q_n) — for n sufficiently large — there is a repetition of a complete quotient $(P_n + \sqrt{R})/Q_n$ for some n.

Corollary: For all sufficiently large n in a PQ sequence, $Q_n > 0$, $\sqrt{R} + P_n > Q_n$ (since $X_n > 1$) and $\sqrt{R} > P_n$.

Theorem 2.8.1 is the deepest theorem in this section. It was first proved by Joseph Lagrange, in 1769, about two years after he married Vittoria Conti. The next theorem gives a necessary condition for a real number's having a purely periodic SCF expansion.

Theorem 2.8.2 *If $y = (\underline{a_1}, \ldots, \underline{a_s})$ then*
(1) y is a root of $h(y) = g_s y^2 + (g_{s-1} - f_s)y - f_{s-1}$
and (2) $-1 < y' < 0$.

Proof: By Theorem 2.1.3,

$$y = (a_1, \ldots, a_s, y) = \frac{yf_s + f_{s-1}}{yg_s + g_{s-1}}$$

so that $g_s y^2 + (g_{s-1} - f_s)y - f_{s-1} = 0$. The other root of $h(y)$ is thus the conjugate, y'. Now $h(0) = -f_{s-1} < 0$ (since $y > 1$) and

$$h(-1) = (g_s - g_{s-1}) + (f_s - f_{s-1}) > 0$$

so that $h(y)$ has a root between -1 and 0. This root cannot be y, which is > 1.

For example, if $y = (2, 2, 2, \ldots)$ then for any positive integer s, y is a root of $g_s y^2 + (g_{s-1} - f_s)y - f_{s-1}$. If $s = 1$, this polynomial is $y^2 - 2y - 1$ which has root $1 + \sqrt{2}$.

PQ sequences eventually become periodic, and the P's and Q's then obey what might be called the 'Galois condition for pure periodicity' (see Theorem 2.8.5 below). More precisely, we have:

Theorem 2.8.3 *For all sufficiently large n in a PQ sequence,*

$$\sqrt{R} + P_n > Q_n > \sqrt{R} - P_n > 0$$

Proof: When n is sufficiently large, $Q_n > 0$ and X_n has a purely periodic SCF expansion (Theorem 2.8.1). Thus, by Theorem 2.8.2, $X_n' > -1$ (for all sufficiently large n) and hence $P_n - \sqrt{R} > -Q_n$ or $Q_n > \sqrt{R} - P_n$.

The result now follows using the corollary to Theorem 2.8.1.

Corollary: Where n is sufficiently large,

$$\sqrt{R} > P_n > 0, \quad 2\sqrt{R} > Q_n > 0 \quad \text{and} \quad 2\sqrt{R} > X_n > 1$$

For example, in the SCF expansion of $\sqrt{2}$, we have, for large n, $P_n = Q_n = 1$, and these obey the above inequalities.

The following theorem shows that the numbers whose SCF's eventually repeat are precisely our 'quadratic surds'.

Theorem 2.8.4 *The SCF expansion of an irrational r is periodic after a certain point iff* $r = (P + \sqrt{R})/Q$ — *where P and Q are integers, and R is a positive nonsquare integer.*

Proof: Suppose $r = (a_1, \ldots, a_m, \underline{a_{m+1}}, \ldots, \underline{a_{m+n}})$.

Let $y = (\underline{a_{m+1}}, \ldots, \underline{a_{m+n}})$. From Theorem 2.1.3, we have

$$r = \frac{yf_m + f_{m-1}}{yg_m + g_{m-1}}$$

By Theorem 2.8.2, $y = (A + \sqrt{D})/B$ for some integers A, B, and D (with D nonsquare). Rationalising the expression for r in terms of A, B, and $\sqrt{D}$, we find that r does have the form $(P + \sqrt{R})/Q$ for some integers P, Q, and R (with R nonsquare).

The converse follows by Theorem 2.8.1.

The next theorem, due to Evariste Galois (1811–1832), gives a necessary and sufficient condition for pure periodicity.

Galois, incidentally, died in a duel with Pesheux d'Herbinville. Galois's father had committed suicide, Galois's mathematical article had been rejected, and Galois's lover, Stéphanie Dumotel, had jilted him.

Theorem 2.8.5

$$(P+\sqrt{R})/Q = (\underline{a_1, \ldots, a_k}, (P+\sqrt{R})/Q) \text{ iff } \sqrt{R}+P > Q > \sqrt{R}-P > 0$$

Proof: Note that we are assuming that R is nonsquare. The left to right implication follows from Theorem 2.8.3.

Suppose $\sqrt{R} + P > Q > \sqrt{R} - P > 0$. Suppose

$$(P + \sqrt{R})/Q = (a_1, \ldots, a_r, \underline{a_{r+1}}, \ldots, \underline{a_{r+s}})$$

— where the period begins at a_{r+1} (Theorem 2.8.1).

Let $Y_n = Q_n/(-P_n + \sqrt{R})$. Since $Q > \sqrt{R} - P > 0$ it follows that $Y_1 = Q/(-P + \sqrt{R}) > 1$. Now

$$Y_{n+1} = \frac{Q_{n+1}}{-P_{n+1} + \sqrt{R}}$$

$$\begin{aligned}
&= \frac{R - P_{n+1}^2}{Q_n(-P_{n+1} + \sqrt{R})} \\
&= \frac{\sqrt{R} + P_{n+1}}{Q_n} \\
&= \frac{\sqrt{R} + a_n Q_n - P_n}{Q_n} \\
&= a_n + \frac{1}{Y_n}
\end{aligned}$$

Since $\sqrt{R} + P > Q > 0$, it follows that $a_1 \geq 1$ — as is always the case for the other partial quotients. Thus, by mathematical induction, $Y_n > 1$ for all $n \geq 1$, and hence $[Y_{n+1}] = a_n$.

Suppose $r > 0$ (so that the expansion is not purely periodic). Then

$$a_{r+s} = [Y_{r+s+1}] = [Y_{r+1}] = a_r$$

so that the period begins not at a_{r+1}, as indicated, but at a_r (or earlier). Contradiction.

Corollary: The SCF expansion of $[\sqrt{R}] + \sqrt{R}$ is purely periodic.

Exercises 2.8

1. Let P, Q, and R be integers. Suppose that R is nonnegative and Q is nonzero. Then $(P + \sqrt{R})/Q$ is rational iff R is a square.
2. Find the period of the SCF expansion of $(27 + \sqrt{28})/29$.
3. Show that there are exactly 2 SCF endings for 13.
4. The complete quotient immediately preceding $(P_{n+1} + \sqrt{R})/Q_{n+1}$ in a purely periodic SCF expansion is given by

$$\begin{aligned}
Q_n &= \frac{R - P_{n+1}^2}{Q_{n+1}} \\
a_n &= \left[\frac{P_{n+1} + \sqrt{R}}{Q_n}\right] \\
P_n &= a_n Q_n - P_{n+1}
\end{aligned}$$

5. Let $(P_n+\sqrt{R})/Q_n$ be a complete quotient in a purely periodic SCF ending. Then $P_{n+1}=P_n$ iff $Q_n|2P_n$.
6. Show that 7 is the only single hearted number of the form $9n^2-2$.
7. For any purely repeating SCF with period length s,

$$f_{sk}=(f_s-g_{s-1})(g_{sk}/g_s)+g_{sk-1}$$

2.9 Pell Equation

After a long journey in the theory of simple continued fractions, we have at last reached the green valley of second degree Diophantine equations. In this section (and the next) we show how to solve the *Pell equation* $x^2-Ry^2=1$ (where R is a given nonsquare positive integer).

The Diophantine equation $x^2-Ry^2=1$ goes back to the Pythagoreans, who solved it for the case $R=2$. With $R=410,286,423,278,424$, we have the key equation in the *Cattle problem* of Archimedes (250 BC). This Cattle Problem was solved for the first time only in 1965. The first published solution was due to Harry L. Nelson. See 'A Solution to Archimedes' Cattle Problem' in the *Journal of Recreational Mathematics,* 13 (1980-81), 164-76. The Diophantine equation $x^2-Ry^2=1$ was also of interest to Bhaskara (1114–1185) who solved it for the case $R=61$. The first fully general and complete solution was given by Joseph Lagrange in 1766. The reason it is called the 'Pell equation' is that Leonhard Euler (1707–1783) mistakenly thought that John Pell (1611–1685) had had something to do with it.

Throughout this section, R is a positive nonsquare integer, and P_1 and Q_1 are integers such that Q_1 is a factor of P_1^2-R. (If R were a square, the Pell equation could be solved by factoring. The only solution in that case is with $y=0$.)

We need a preliminary result:

Theorem 2.9.1 *Let $X_1=(P_1+\sqrt{R})/Q_1$. Let f_n/g_n be the n-th convergent of X_1, and let $X_n=(P_n+\sqrt{R})/Q_n$ be the n-th complete quotient of X_1. Then*

$$(-1)^{n-1}P_n=P_1(f_{n-1}g_{n-2}+f_{n-2}g_{n-1})-Q_1f_{n-1}f_{n-2}+g_{n-1}g_{n-2}\frac{R-P_1^2}{Q_1}$$

and

$$(-1)^{n-1}Q_nQ_1 = (Q_1f_{n-1} - P_1g_{n-1})^2 - Rg_{n-1}^2$$

Proof: By Theorem 2.1.3,

$$(P_1 + \sqrt{R})/Q_1 = \frac{X_nf_{n-1} + f_{n-2}}{X_ng_{n-1} + g_{n-2}}$$

We solve this equation for X_n, and then rationalise the denominator. The result follows by Plato's Theorem, and by equating the rational and irrational parts of the resulting equation. The details are left as an exercise.

As corollaries of Theorem 2.9.1, we have the following.

Theorem 2.9.2

$$g_{n-1}P_n + g_{n-2}Q_n = Q_1f_{n-1} - P_1g_{n-1}$$

Proof:

$$\begin{aligned} &g_{n-1}(-1)^{n-1}P_n + g_{n-2}(-1)^{n-1}Q_n \\ &= P_1g_{n-1}f_{n-1}g_{n-2} + P_1g_{n-1}^2f_{n-2} - Q_1g_{n-1}f_{n-1}f_{n-2} + g_{n-2}Q_1f_{n-1}^2 \\ &\quad -2g_{n-2}f_{n-1}P_1g_{n-1} \\ &= Q_1f_{n-1}(-g_{n-1}f_{n-2} + g_{n-2}f_{n-1}) + P_1g_{n-1}(-f_{n-1}g_{n-2} + g_{n-1}f_{n-2}) \\ &= Q_1f_{n-1}(-1)^{n-1} - P_1g_{n-1}(-1)^{n-1} \end{aligned}$$

Theorem 2.9.3

$$f_{n-1}P_n + f_{n-2}Q_n = P_1f_{n-1} + g_{n-1}(R - P_1^2)/Q_1$$

Proof: The proof is similar to that of the previous theorem.

The next theorem is the simple continued fraction solution of the Pell equation. As an algorithm for generating solutions, it was known to the ancient Greeks, but the theory behind the algorithm was not understood before Lagrange made a study of it.

Theorem 2.9.4 *Let s be the length of the period of the SCF expansion of $\sqrt{R}$.*

If s is even then (x, y) is a positive integer solution of $x^2 - Ry^2 = 1$ iff for some positive integer k, $x = f_{ks}$ and $y = g_{ks}$ (where f_{ks}/g_{ks} is the ks-th convergent of $\sqrt{R}$).

If s is odd then (x, y) is a positive integer solution of $x^2 - Ry^2 = 1$ iff for some positive integer k, $x = f_{2ks}$ and $y = g_{2ks}$.

Hence $x^2 - Ry^2 = 1$ has infinitely many solutions.

Proof: By Theorem 2.9.1, with $P_1 = 0$ and $Q_1 = 1$,

$$(-1)^{n-1}Q_n = f_{n-1}^2 - Rg_{n-1}^2$$

Except for the first complete quotient, the SCF expansion of $\sqrt{R}$ is exactly like the purely periodic SCF expansion of $[\sqrt{R}]+\sqrt{R}$ (Theorems 2.7.1 and 2.8.5). Thus $Q_n = 1$ iff for some nonnegative integer k, $n = ks + 1$.

Thus, if s is even, $f_{ks}^2 - Rg_{ks}^2 = 1$.

Conversely, by Theorem 2.6.3, if (x, y) is a positive integer solution of the equation, then x/y is a convergent of $\sqrt{R}$. Hence, by Theorem 2.9.1, $x = f_{n-1}$ and $y = g_{n-1}$ where $(-1)^{n-1}Q_n = 1$. Hence $n = ks + 1$.

The result follows similarly when s is odd.

Corollary: If s is even, the least positive integer solution of $x^2 - Ry^2 = 1$ is $x = f_s$ and $y = g_s$.
If s is odd, the least positive integer solution of $x^2 - Ry^2 = 1$ is $x = f_{2s}$ and $y = g_{2s}$.

In what follows, (a, b) shall denote the least positive integer solution of $x^2 - Ry^2 = 1$ (for a given R).

The next theorem, due to B. Carrara (1890), gives a way of shortening the calculation of a and b when s is odd.

Theorem 2.9.5 *If s is the length of the period of the SCF expansion of $\sqrt{R}$, $f_{2n} = f_s^2 + g_s^2 R$ and $g_{2s} = 2f_s g_s$.*

Proof: By Theorem 2.7.1 and Theorem 2.8.5,

$$\sqrt{R} = (a_1, a_2, \ldots, a_s, 2a_1, a_2, \ldots)$$

where $a_1 = [\sqrt{R}] = P_{s+1}$. By Theorem 2.1.14,

$$f_{2s} = f_s^2 + (a_1 f_s + f_{s-1})g_s$$
$$g_{2s} = f_s g_s + (a_1 g_s + g_{s-1})g_s$$

and the result follows by Theorems 2.9.3 and 2.9.2.

Since, by Theorem 2.9.2, $f_s = [\sqrt{R}]g_s + g_{s-1}$ (in the SCF expansion of $\sqrt{R}$), it is not actually necessary to calculate the numerators of the convergents, only the denominators. Furthermore, when s is odd, $f_s^2 + g_s^2 R = 2f_s^2 + 1$ (Theorem 2.9.1), and we have the following.

Theorem 2.9.6 *If the length s of the period of the SCF expansion of $\sqrt{R}$ is odd, then the least positive solution of $x^2 - Ry^2 = 1$ is*

$$a = 2([\sqrt{R}]g_s + g_{s-1})^2 + 1, \quad b = 2([\sqrt{R}]g_s + g_{s-1})g_s$$

For example, suppose we want to solve $x^2 - 13y^2 = 1$. We compute the PQ sequence for $\sqrt{13}$ — adding a row for the g's — continuing until we get some $Q = 1$:

P	0	3	1	2	1	3
Q	1	4	3	3	4	1
a	3	1	1	1	1	6
g	1	1	2	3	5	

From the above table we see that $s = 5$, an odd number. From Theorem 2.9.6, it follows that the least positive solution of $x^2 - 13y^2 = 1$ is

$$x = 2(3g_5 + g_4)^2 + 1 = 2 \times 18^2 + 1 = 649$$

$$y = 2(3g_5 + g_4)g_5 = 2 \times 18 \times 5 = 180$$

Exercises 2.9

1. Find a and b when $R = 89$.
2. Solve $x^2 - Ry^2 = -1$.
3. Solve $x^2 - Ry^2 = 2$.

4. Let p be an odd prime. Then $x^2 - py^2 = -1$ has a solution iff p has the form $4n + 1$.
5. Let R be a positive nonsquare integer. Let f_n/g_n be the n-th convergent of $\sqrt{R}$. Then, for all positive integers n, there is a positive integer x such that $x < 2\sqrt{R}$ and R is a factor of $f_n^2 - x$.
6. Find the 4 smallest triangles with consecutive integer sides and an integer area.
7. Find the 4 smallest Pythagorean triangles whose two legs are consecutive integers.
8. Sheik Noshack prefers to arrange his gold coins in a perfect equilateral triangle, but, occasionally, he separates them into 23 equal squares. How many coins does he have?
9. If $a_1 = [\sqrt{R}]$ and $a_{n+1} = (a_n + R/a_n)/2$ then $|a_n - \sqrt{R}| < 1/(g_{2^n})^2$. This fact lies behind the ancient Babylonian method for approximating square roots.

2.10 Prefaced Palindromes *

A *palindrome* is an expression which reads the same backwards as forwards. An infamous example is: MADAM, I'M ADAM. In this section we examine certain kinds of palindromic SCF's, partly for fun, and partly to obtain some shortcuts in solving $x^2 - Ry^2 = 1$.

An SCF expansion of the form

$$(\underline{a_1}, a_2, a_3, \ldots, a_3, \underline{a_2})$$

is called a *prefaced palindrome* — 'prefaced' because the part that reads the same backwards as forwards is prefaced by a_1. For example, $(\underline{249}, 1, \underline{1})$ is a prefaced palindrome. Also, for any positive integers m and n, $(\underline{m})$, and $(\underline{m}, \underline{n})$ are prefaced palindromes. In order to give a necessary and sufficient condition for a number to have an SCF expansion which is a prefaced palindrome, we use the following theorem.

Theorem 2.10.1

$$(\underline{a_1}, \ldots, \underline{a_s}) = (P + \sqrt{R})/Q \quad \text{iff} \quad (\underline{a_s}, \ldots, \underline{a_1}) = \frac{Q}{-P + \sqrt{R}}$$

Proof: The notation is understood to mean that the period of the second SCF expansion is the reverse of the period of the first.

Let y be the first SCF and x the second. By Theorem 2.8.2, y is a root of

$$h(z) = g_s z^2 + (g_{s-1} - f_s)z - f_{s-1}$$

(the convergents being convergents of y). From Theorem 2.1.9, it follows that $x = (xf_s + g_s)/(xf_{s-1} + g_{s-1})$ and hence $-1/x$ is also a root of $h(z)$. Hence y and $-1/x$ are conjugates.

Theorem 2.10.2 *If the PQ sequence for* $(P_1 + \sqrt{R})/Q_1$ *is*

$$\begin{array}{ccccccc} P_1 & P_2 & P_3 & \cdots & P_s & P_1 & P_2 & \cdots \\ Q_1 & Q_2 & Q_3 & \cdots & Q_s & Q_1 & Q_2 & \cdots \\ a_1 & a_2 & a_3 & \cdots & a_s & a_1 & a_2 & \cdots \end{array}$$

(with period length s) then the PQ sequence for $(P_2 + \sqrt{R})/Q_1$ *is*

$$\begin{array}{cccccccc} P_2 & P_1 & P_s & \cdots & P_3 & P_2 & P_1 & \cdots \\ Q_1 & Q_s & Q_{s-1} & \cdots & Q_2 & Q_1 & Q_s & \cdots \\ a_1 & a_s & a_{s-1} & \cdots & a_2 & a_1 & a_s & \cdots \end{array}$$

Proof: By Theorem 2.9.7,

$$(\underline{a_k}, a_{k-1}, \ldots, a_1, a_s, a_{s-1}, \ldots, \underline{a_{k+1}}) = \frac{Q_{k+1}}{-P_{k+1} + \sqrt{R}} = \frac{P_{k+1} + \sqrt{R}}{Q_k}$$

We call the second PQ sequence in the statement of Theorem 2.10.2 the *reflection* of the first, and we also say that $(P_2 + \sqrt{R})/Q_1$ itself is the reflection of $(P_1 + \sqrt{R})/Q_1$. From Theorem 2.10.2, it follows that the reflection of a reflection is the original PQ sequence. A PQ sequence is sometimes its own reflection. We have, for example, for $(249 + \sqrt{62501})/2$:

P	249	249	1	249
Q	2	250	250	2
a	249	1	1	249

Since $P_2 = 249 = P_1$, the above number is its own reflection. We name such numbers *self-reflections.*

Theorem 2.10.3 *Where $(P+\sqrt{R})/Q$ is purely periodic, the following 3 conditions are equivalent:*
(1) $(P+\sqrt{R})/Q$ is a prefaced palindrome
(2) $Q|2P$
(3) $(P+\sqrt{R})/Q$ is a self-reflection .

Proof:

$$\begin{aligned}\text{If } (P+\sqrt{R})/Q &= (\underline{a_1}, a_2, a_3, \ldots, a_3, \underline{a_2}) \\ &= a_1 + 1/(\underline{a_2}, a_3, \ldots, a_3, a_2, \underline{a_1}) \\ &= a_1 + (-P+\sqrt{R})/Q\end{aligned}$$

(using Theorem 2.10.1), then $a_1 = 2P/Q$ so that Q is a divisor of $2P$. Hence (1) implies (2).

Suppose Q is a divisor of $2P$. Since $Q > \sqrt{R} - P > 0$ (Theorem 2.8.5), it follows that

$$(P+Q)/Q > \sqrt{R}/Q > P/Q$$

so that

$$1 + 2P/Q > (P+\sqrt{R})/Q > 2P/Q$$

Hence $a_1 = [(P+\sqrt{R})/Q] = 2P/Q$ and

$$P_2 = a_1Q_1 - P_1 = a_1Q - P = 2P - P = P = P_1$$

Hence (2) implies (3).

That (3) implies (1) follows from Theorem 2.10.2.

We can now characterise the PQ sequence for $[\sqrt{R}]+\sqrt{R}$ and hence the PQ sequence for $\sqrt{R}$. This will allow us some shortcuts in solving $x^2 - Ry^2 = 1$.

Theorem 2.10.4 *$[\sqrt{R}]+\sqrt{R}$ is a prefaced palindrome.*
If s is the length of its period, and n is an integer such that $1 < n < s+1$ then
(1) $Q_n \neq 1$
(2) $a_n < \sqrt{R}$
(3) $P_n = P_{n+1}$ iff s is even and $n = \frac{1}{2}s + 1$
(4) $Q_n = Q_{n+1}$ iff s is odd and $n = \frac{1}{2}s + \frac{1}{2}$.

Proof: If $Q_n = 1$ then, by Theorem 2.8.5, $P_n = [\sqrt{R}]$ and the period has started over. But the period only starts over at the $(s+1)$-th complete quotient.

Since $Q_n \neq 1$ and $0 < P_n < \sqrt{R}$ (Theorem 2.8.5), a_n cannot be greater than $[(\sqrt{R}+\sqrt{R})/2]$.

If $P_n = P_{n+1}$ then $(P_n + \sqrt{R})/Q_n$ is a self-reflection and hence $Q_{n+m} = Q_{n-m}$ for all integers m such that $0 < m < n$ (Theorem 2.10.2). With $m = n-1$, we obtain $Q_{2n-1} = Q_1 = 1$. By (1) above it follows that $2n - 1 = s + 1$ and hence $n = \frac{1}{2}s + 1$.

If s is even, the SCF has the form

$$(\underline{a_1}, a_2, a_3, \ldots, a_{\frac{1}{2}s}, a_{\frac{1}{2}s+1}, a_{\frac{1}{2}s}, \ldots, a_3, \underline{a_2}),$$

and

$$(\underline{a_{\frac{1}{2}s+1}}, a_{\frac{1}{2}s}, \ldots, a_3, a_2, a_1, a_2, a_3, \ldots, \underline{a_{\frac{1}{2}s}})$$

is a prefaced palindrome, and hence a self-reflection (Theorem 2.10.3). Thus $P_{\frac{1}{2}s+1} = P_{\frac{1}{2}s+1+1}$.

Suppose $Q_n = Q_{n+1}$. By Theorem 2.10.1,

$$(\underline{a_n}, a_{n-1}, \ldots, a_2, a_1, a_s, \ldots, a_{n+2}, \underline{a_{n+1}}) = Q_{n+1}/(-P_{n+1} + \sqrt{R})$$

$$= (P_{n+1} + \sqrt{R})/Q_{n+1}$$

(since $Q_{n+1} = Q_n$). Thus the immediately preceding SCF expansion is palindromic. By (2), a_1 is the one and only partial quotient greater than $\sqrt{R}$, and so it must be dead in the middle of the period of the SCF. Thus s is odd, and $n = \frac{1}{2}s + \frac{1}{2}$.

If s is odd then $s - 1$ is even, and, by Theorem 2.10.2, $Q_{\frac{1}{2}s+\frac{1}{2}} = Q_{\frac{1}{2}s+\frac{1}{2}+1}$.

The shortcuts for solving $x^2 - Ry^2 = 1$ come out of the following corollary to the above theorem.

Theorem 2.10.5 *Let s be the length of the period of the SCF expansion of $\sqrt{R}$, and let f_n/g_n be its n-th convergent.*
If s is even then $g_s = g_{\frac{1}{2}s}(g_{\frac{1}{2}s-1} + g_{\frac{1}{2}s+1})$.
If s is odd then $g_s = {g_{\frac{1}{2}s-\frac{1}{2}}}^2 + {g_{\frac{1}{2}s+\frac{1}{2}}}^2$.

Proof: If s is even then the result follows by Theorem 2.1.15 (taking $n = \frac{1}{2}$) (Theorem 2.10.4). If s is odd the result follows from Theorem 2.1.16 (taking $n = \frac{1}{2}s + \frac{1}{2}$) (Theorem 2.10.4).

For example, to solve $x^2 - 91y^2 = 1$, we do not have to calculate all the entries in the following table:

P	0	9	1	8	7	7	8	1	9
Q	1	10	9	3	14	3	9	10	1
a	9	1	1	5	1	5	1	1	
g	1	1	2	11	13	76	89	165	

It is enough to calculate until we get the repeating 7's in the top row (Theorem 2.10.4). These occur in columns $n = 5$ and $n + 1 = 6$, so $s = 2(n - 1) = 8$ and

$$y = g_s = g_4(g_3 + g_5) = 11 \times 15 = 165$$

In other words, to find a solution of the Pell equation $x^2 - 91y^2 = 1$, it is enough to calculate the following portion of the above table:

P	0	9	1	8	7	7
Q	1	10	9	3	14	
a	9	1	1	5	1	
g	1	1	2	11	13	

We close this section by finding, for all nonsquare positive integers

R from 2 to 99, the smallest positive integer y making Ry^2+1 a square.

R	y	R	y
2	2	53	9100
3	1	54	66
5	4	55	12
6	2	56	2
7	3	57	20
8	1	58	2574
10	6	59	69
11	3	60	4
12	2	61	226153980
13	180	62	8
14	4	63	1
15	1	65	16
17	8	66	8
18	4	67	5967
19	39	68	4
20	2	69	936
21	12	70	30
22	42	71	413
23	5	72	2
24	1	73	267000
26	10	74	430
27	5	75	3
28	24	76	6630
29	1820	77	40
30	2	78	6
31	273	79	9
32	3	80	1
33	4	82	18
34	6	83	9
35	1	84	6
37	12	85	30996
38	6	86	1122
39	4	87	3
40	3	88	21
41	320	89	53000
42	2	90	2
43	531	91	165
44	30	92	120
45	24	93	1260
46	3588	94	221064
47	7	95	4
48	1	96	5
50	14	97	6377352
51	7	98	10
52	90	99	1

Exercises 2.10

1. Show that $(P+\sqrt{R})/Q = (\underline{a_1}, \ldots, \underline{a_s})$ is purely palindromic iff $Q_s = Q$.
2. Show that the smallest nontrivial square of the form $1621y^2+1$ is $A^2+1621B^2$, where $A = 10G^2+78GG'-10G'^2$, and $B = G^2+G'^2$ — with $G = g_{40}$ and $G' = g_{39}$.
3. Verify one of the entries in the above table.

Chapter 3

Congruence

Carl Friedrich Gauss begins the *Disquisitiones Arithmeticae* (1801):

> If a number a divides the difference of the numbers b and c, b and c are said to be *congruent relative to a;* if not, b and c are *noncongruent.* The number a is called the *modulus.* If the numbers b and c are congruent, each of them is called a *residue* of the other.

Thus if a is a factor of $b - c$, one writes $b \equiv c \pmod{a}$, which is read 'b is congruent to c mod a'. For example, $23 \equiv 17 \pmod{3}$. Note that 23 and 17 both leave remainder 2 when divided by 3, and 2 is a residue of them both. In general, two integers leave the same positive integer remainder when divided by another integer if and only if it is a factor of their difference.

In order to solve the Diophantine equation $x^2 - Ry^2 = C$, we shall need to know how to solve the congruence equation $z^2 \equiv R \pmod{C}$.

3.1 Basic Properties

It is easy to show that $\equiv$ is an equivalence relation. Moreover, if $a \equiv b \pmod{n}$ and $c \equiv d \pmod{n}$ then $a + c \equiv b + d \pmod{n}$, $a - c \equiv b - d \pmod{n}$, and $ac \equiv bc \equiv bd \pmod{n}$. Hence if $a \equiv b \pmod{n}$ then, for any positive integer m, $a^m \equiv b^m \pmod{n}$, and, more generally, if $f(x)$ is any polynomial with integer coefficients, and $a \equiv b \pmod{n}$ then $f(a) \equiv f(b) \pmod{n}$.

The theorem for division involves the greatest common divisor (a, n) of the integers a and n.

Theorem 3.1.1 $ab \equiv ac \pmod{n}$ *iff* $b \equiv c \pmod{n/(a,n)}$.

Proof: By Theorem 2.3.2 there are integers s and t such that $as - nt = (a, n)$. Since

$$(a,n)(b-c) = a(b-c)s - n(b-c)t$$

it follows that n is a factor of $ab - ac$ only if n is a factor of $(a,n)(b-c)$. Thus if $ab \equiv ac \pmod{n}$ then $b \equiv c \pmod{n/(a,n)}$.

The converse follows from the fact that (a, n) is a divisor of a.

We can now solve the linear congruence equation $ax \equiv b \pmod{n}$. Let $c = (a, n)$. If the equation has a solution s then, for some integer q, $as - b = qn$ and, using the distributive law, c is a factor of b. Thus if c is not a factor of b, the equation has no solution. If, however, c is a factor of b, we can divide it out of the equation using Theorem 3.1.1. This will leave an equation in which the coefficient of x is relatively prime to the modulus. Thus there is no loss of generality if, from the beginning, we insist that a and n be relatively prime.

Theorem 3.1.2 *Let a be a positive integer and n a nonzero integer. Suppose $(a, n) = 1$. Let s and t be integers such that $as - nt = 1$. (For example, let $n/a = (a_1, \ldots, a_{2m+1})$ and let $s = f_{2m}$ and $t = g_{2m}$.) Then $ax \equiv b \pmod{n}$ iff $x \equiv sb \pmod{n}$.*

Proof: Using Theorem 3.1.1, we see that the following are equivalent, mod n:

$$\begin{aligned} ax &\equiv b \\ ax &\equiv b(as - nt) \\ ax &\equiv asb \\ x &\equiv sb \end{aligned}$$

The equation $ax \equiv 1 \pmod{n}$ has a solution just in case $(a, n) = 1$. This solution is unique modulo n, and is the *inverse* a^{-1} of a with respect to the modulus n. For example, 5 is the inverse of 3 with respect to the modulus 14.

Another basic result, due to Lagrange, is the following. We shall use it, in Section 3, in our treatment of 'primitive roots'.

Theorem 3.1.3 *Let $f(x) = c_0x^n + c_1x^{n-1} + \cdots + c_{n-1}x + c_n$ be a polynomial with integer coefficients. Let p be a prime which is not a factor of c_0. Then $f(x) \equiv 0 \pmod p$ has at most n solutions which are distinct modulo p.*

Proof: By Theorem 3.1.2, the result is true for degree 1 polynomials.

Suppose it true for polynomials of degree $n-1$. If $f(a) \equiv 0 \pmod p$ then, mod p,

$$\begin{aligned} f(x) &\equiv f(x) - f(a) \\ &= c_0(x^n - a^n) + c_1(x^{n-1} - a^{n-1}) + \cdots + c_{n-1}(x-a) \\ &= (x-a)g(x) \end{aligned}$$

where $g(x)$ is a polynomial of degree $n-1$. Since p is prime, any root of $f(x)$ is thus either a root of $x-a$ or a root of $g(x)$. By the induction hypothesis, $g(x)$ has at most $n-1$ roots. Hence $f(x)$ has at most n roots.

Corollary: Where p is prime, $x^d \equiv 1 \pmod p$ has at most d solutions.

One of the great theorems of classical Number Theory is Legendre's Theorem. This theorem is named after its discoverer, Adrien-Marie Legendre (1752–1833), and it gives a simple necessary and sufficient condition for the Diophantine equation $ax^2 + by^2 + cz^2 = 0$ to have nontrivial solutions. Happily, there is an elementary proof of this result, and we include it in this book (as Theorem 3.7.4). The following theorem is one of the lemmas we shall use in our proof of Legendre's Theorem.

Theorem 3.1.4 *Let r, s, t be positive reals whose product, n, is a natural number. Let a, b, c be any integers. Then $ax + by + cz \equiv 0 \pmod n$ has a nontrivial solution with $|x| \leq r$, $|y| \leq s$, and $|z| \leq t$.*

Proof: If $0 \leq x \leq [r]$, and $0 \leq y \leq [s]$, and $0 \leq z \leq [t]$, then there are more than n possibilities for the triple (x, y, z). Hence at least two of them, say (x_1, y_1, z_1) and (x_2, y_2, z_2), are such that $ax_1 + by_1 + cz_1 \equiv ax_2 + by_2 + cz_2 \pmod{n}$. But then

$$a(x_1 - x_2) + b(y_1 - y_2) + c(z_1 - z_2) \equiv 0 \pmod{n}$$

with $|x_1 - x_2| \leq r$, $|y_1 - y_2| \leq s$, and $|z_1 - z_2| \leq t$.

Exercises 3.1

1. Use the theory of congruence to explain the fact that a natural number is divisible by 9 just in case the sum of its digits is.
2. Show that every even perfect number ends in the digit 6 or 8.
3. Solve $172x \equiv 20 \pmod{52}$.
4. Find a solution of $11x + 12y + 13z \equiv 0 \pmod{60}$ such that $|x| \leq 3$, $|y| \leq 4$, and $|z| \leq 5$.

3.2 Euler's ϕ-Function

Let $\phi(n)$ be the number of positive integers not greater than n and relatively prime to it. This function is Euler's 'fee-function'. For example, $\phi(1) = 1$, $\phi(6) = 2$, and, if p is any prime, $\phi(p) = p - 1$.

The ϕ-function has many uses in mathematics. We shall show, for example, that a regular n-gon can be constructed using only ruler and compass iff $\phi(n)$ is a power of 2. We begin our treatment of the ϕ-function by showing that it is 'multiplicative'.

Theorem 3.2.1 *If* $(m, n) = 1$ *then* $\phi(mn) = \phi(m)\phi(n)$.

Proof:

$$\begin{array}{cccc} 1 & 2 & \cdots & m \\ m+1 & m+2 & \cdots & m+m \\ 2m+1 & 2m+2 & \cdots & 2m+m \\ \vdots & \vdots & & \vdots \\ (n-1)m+1 & (n-1)m+2 & \cdots & (n-1)m+m \end{array}$$

In the above array, either a column contains only elements relatively prime to m or only elements not relatively prime to m. The number of columns containing only elements relatively prime to m is $\phi(m)$.

Since $(m, n) = 1$, it follows by Theorem 3.1.1 that the members of a column are all distinct modulo n. Hence each column contains $\phi(n)$ numbers relatively prime to n. (For $(x, n) = 1$ iff $(x + an, n) = 1$.)

Thus the number of entries in the array relatively prime to both m and n is $\phi(m)\phi(n)$.

From this we obtain

Theorem 3.2.2 $\phi(n) = n\prod(1 - 1/p)$ *where the product is taken over all primes dividing* n.

Proof: Where p is a prime, $\phi(p^a) = p^a - p^{a-1} = p^a(1 - 1/p)$ and the result follows by Theorem 3.2.1.

The next theorem was first proved by Leonhard Euler (1707–1783), the Swiss mathematician who also proved that every even perfect number is of the type given in Euclid.

Theorem 3.2.3 *The* $\phi(n)$ *residues which are relatively prime to* n *form a multiplicative group. Hence if* $(a, n) = 1$ *then* $a^{\phi(n)} \equiv 1 \pmod{n}$.

Proof: Use Theorem 3.1.2 with $b = 1$.

As a special case of Euler's Theorem, we have Fermat's Theorem:

Theorem 3.2.4 (Fermat's Little Theorem) *If* p *is prime then, for all integers* a *such that* $\gcd(a, p) = 1$, *we have* $a^{p-1} \equiv 1 \pmod{p}$.

Fermat's Little Theorem has many uses. For example, we can use it to show that if p is a prime of the form $6n + 5$ then $x^3 \equiv a \pmod{p}$ has exactly 1 solution (modulo p), namely, $(a^{-1})^{2n+1}$. For

$$\begin{aligned} ((a^{-1})^{2n+1})^3 &\equiv (a^{-1})^{p-2}a^{-1}a \\ &\equiv (a^{-1})^{p-1}a \\ &\equiv a \pmod{p} \end{aligned}$$

Moreover, if any of the $p-1$ equations $x^3 \equiv k \pmod{p}$ — with k an integer from 1 to $p-1$ inclusive — if any of these equations had more than one solution, there would not be enough solutions to go around. Hence each such equation has exactly 1 solution.

The converse of Fermat's Little Theorem is not true. This is thanks to the existence of a set of integers discovered by Robert Daniel Carmichael (1879–1967). A positive integer m is a *Carmichael number* iff m is composite, and $a^{m-1} \equiv 1 \pmod{m}$ for any integer a such that $\gcd(a, m) = 1$. The smallest Carmichael number is 561. It is now known that there are infinitely many Carmichael numbers.

The Euler ϕ-function can be used for sending secret messages. Let p and q be two large primes, so large that there is no practical way of factoring their product pq (if you do not already know the factors). Let m be a positive integer ($< pq$) which you want to send in secret. (It might, for example, be your bid on a project.) You calculate $n = m^{23} \pmod{pq}$ — the 23 is arbitrary — and send the equation

$$x^{23} \equiv n \pmod{pq}$$

If someone intercepts this message, they will not be able to find out what the x is (unless they already have the factorisation of pq).

The way the equation is solved is this. First, knowing the factorisation of pq, the person you sent the message to calculates $\phi(pq) = (p-1)(q-1)$ She then solves $23y \equiv 1 \pmod{(p-1)(q-1)}$. Next (with the help of a computer), she raises both sides of the message congruence to the power y. This gives

$$x^1 \equiv m^{23y} \equiv m \pmod{pq}$$

recovering the original number m.

For example, if the large primes are 3 and 5 — let us imagine they are large — and you want to send the message 7, you calculate $13 = 7^3 \pmod{15}$, and send the equation

$$x^3 \equiv 13 \pmod{15}$$

Since 15 is such a large number, we are supposing, someone who intercepts this message will not be able to use, say, trial and error to find out what x is. The intended receiver, however, knows that $15 = 3 \times 5$, and calculates $\phi(15) = (3-1)(5-1) = 8$. She then solves

$$3y \equiv 1 \pmod 8$$

obtaining $y = 3$. Next, she calculates $13^3 \pmod{15}$, recovering the original number 7.

Exercises 3.2

1. $\sum \phi(d) = n$ where the sum is taken over all divisors d of n.
2. Given a positive integer N, find an upper bound for the set of natural numbers x such that $\phi(x) \leq N$.
3. Find the largest solution of $\phi(x) = 480$.
4. What is the smallest positive integer which quadruples when its last digit is moved back to become its first digit?
5. Prove the theorem of John Wilson (1741–1793): if $n > 1$ then n is prime iff $(n-1)! \equiv -1 \pmod n$. (This was first proved by Lagrange, in 1773.)
6. Solve $x^{19} \equiv 4282 \pmod{9991}$.
7. Prove that 561 is a Carmichael number.
8.* Let p be a prime > 3. Let

$$g(x) = (x-1)(x-2)\ldots(x-p+1) - x^{p-1} + 1$$

Then $g(x)$ has degree at most $p-2$. By Fermat's Theorem, it has $p-1$ roots, mod p. Hence all its coefficients are 0, mod p. Explain the 'hence'.
9.* Conclude from the previous exercise that if

$$g(x) = c_{p-2}x^{p-2} + \cdots + c_2x^2 + c_1x + c_0$$

then $c_0 = (p-1)! + 1$ and c_1 is divisible by p^2. Hint: let $x = p$ to get $-p^{p-1} = c_{p-2}p^{p-2} + \cdots + c_2p^2 + c_1p$ with p^3 dividing c_2p^2.
10.* Conclude from the previous exercises that $g(2p) \equiv c_0 \pmod{p^3}$ and hence

$$(2p-1)(2p-2)\ldots(2p-p+1) \equiv (p-1)! \pmod{p^3}$$

Thus $\binom{2p-1}{p-1} \equiv 1 \pmod{p^3}$.

11.* Show that if p is a prime > 3 then $\binom{2p}{p} \equiv 2 \pmod{p^3}$.

12.* Show that there are $\sum_{n=2}^{n} \phi(n)$ terms in the Farey series F_n.

13.* Show that $\sum_{d|n} \mu(n/d)d = \phi(n)$.

3.3 Primitive Roots

By Theorem 3.2.3 (Euler's Theorem), the $\phi(n)$ residues which are relatively prime to n form a multiplicative group modulo n. If this group is cyclic, its generators are called *primitive roots* of n. With the next few theorems we determine which integers n have primitive roots, i.e. which integers n are such that the group of Theorem 3.2.3 is cyclic.

As an example, every nonzero residue modulo 13 can be expressed as a power of 2. Hence 2 is a primitive root modulo 13. Indeed, we have $2^1 \equiv 2$, $2^2 \equiv 4$, $2^3 \equiv 8$, $2^4 \equiv 3$, $2^5 \equiv 6$, $2^6 \equiv 12$, $2^7 \equiv 11$, $2^8 \equiv 9$, $2^9 \equiv 5$, $2^{10} \equiv 10$, $2^{11} \equiv 7$ and $2^{12} \equiv 1$.

Our first primitive root theorem states that every prime has a primitive root. The following proof is a counting argument, based on the fact that

$$\sum_{d|(p-1)} \phi(d) = p - 1$$

(See Exercises 3.2 # 1.) The result was first proved by Gauss.

Theorem 3.3.1 *Every prime has a primitive root.*

Proof: Where p is a prime, and d is a factor of $p-1$, let $h(d)$ be the number of positive integers less than p with order d.

(A positive integer a has *order* d if $a^d \equiv 1 \pmod{p}$ and there is no positive integer e less than d such that $a^e \equiv 1 \pmod{p}$. It follows from Group Theory that the order is a factor of $p-1$.)

If $h(d) > 0$ there is some integer a with order d. The residues a, a^2, ..., a^{d-1}, a^d are all solutions of $x^d \equiv 1 \pmod{p}$ and, by Theorem 3.1.3, these are the only solutions of that equation. Since any residue b of order d solves that equation, it will be found among the powers of a.

Now, among the powers of a, a^i has order d just in case $(i,d) = 1$. For let $g = (i,d)$. Then $(a^i)^{d/g} \equiv (a^{i/g})^d \equiv 1 \pmod{p}$. So if $g \neq 1$ then a^i does not have order d. Conversely, if a^i does not have order d then $(a^i)^s \equiv 1 \pmod{p}$ for some integer s such that d is not a factor of s. Since d is a factor of is — because a has order d — it follows that $g \neq 1$.

Hence if $h(d) > 0$ then $h(d) = \phi(d)$. Since

$$\sum_{d|(p-1)} h(d) = p - 1 = \sum_{d|(p-1)} \phi(d)$$

it follows that $h(d)$ is never 0. In particular, $h(p-1) \neq 0$, that is, p has a primitive root.

We now extend the above result to powers of odd primes. Our proof uses the Binomial Theorem.

Theorem 3.3.2 *Every power of an odd prime has a primitive root.*

Proof: Let a be a primitive root of an odd prime p. Let

$$k = (a^{p-1} - 1)/p$$

By Fermat's Theorem, k is an integer. Let $b = a$ if p is not a factor of k, but let $b = a + p$ if p is a factor of k.

If $p|k$ then $a^{p-1} \equiv 1 \pmod{p^2}$ and

$$\begin{aligned} b^{p-1} &\equiv (a+p)^{p-1} \\ &\equiv a^{p-1} + (p-1)a^{p-2}p \\ &\equiv 1 + (p-1)a^{p-2}p \pmod{p^2} \end{aligned}$$

Thus, whether p is a factor of k or not, $b^{p-1} = 1 + pn_1$ where p is not a factor of n_1.

Suppose that $b^{(p-1)p^{j-1}} = 1 + p^j n_j$ where p is not a factor of n_j. Raising both sides of this equation to the power p, we obtain

$$b^{(p-1)p^j} \equiv 1 + pp^j n_j \pmod{p^{j+2}}$$

— since $p > 2$ — and hence

$$\begin{aligned} b^{(p-1)p^j} &= 1 + p^{j+1}(n_j + mp) \\ &= 1 + p^{j+1} n_{j+1} \end{aligned}$$

where p is not a factor of n_{j+1}. Thus, by mathematical induction, it follows that, for all positive integers j,

$$b^{(p-1)p^{j-1}} = 1 + p^j n_j$$

where p is not a factor of n_j.

As a result, the order of b modulo p^e is an integer $p^s d$ where $0 \leq s \leq e - 1$, and d is a factor of $p - 1$. For our proof, it suffices to show that $s = e - 1$ and $d = p - 1$, so that b has order $\phi(p^e)$. Since

$$1 + p^{s+1} n_{s+1} = b^{(p-1)p^s} \equiv 1 \pmod{p^e}$$

it follows that p^e is a factor of p^{s+1} and hence $e \leq s + 1$, so that (since $s \leq e - 1$), $s = e - 1$. Since $(b^{p^s})^d \equiv 1 \pmod{p}$, it follows by Fermat's Theorem that $b^d \equiv 1 \pmod{p}$ and hence $a^d \equiv 1 \pmod{p}$ — so that $d = p - 1$ (since a is a primitive root of p). Hence b is a primitive root of p^e.

Corollary: The double of a power of an odd prime has a primitive root.

Proof of Corollary: Let $c = b$ if b is odd, and let $c = b + p^e$ if b is even. Then c is a primitive root of $2p^e$.

Theorem 3.3.3 *2^n has no primitive root if $n \geq 3$.*

Proof: Using mathematical induction, it is not hard to show that if a is an odd postive integer then $a^{2^{n-2}} \equiv 1 \pmod{2^n}$ if $n \geq 3$. But $\phi(2^n) = 2^{n-1}$.

Theorem 3.3.4 *If $(m, n) = 1$ and m, $n > 2$ then mn has no primitive root.*

Proof: Suppose $(a, mn) = 1$. Since m, $n > 2$, it follows that $\phi(m)$ and $\phi(n)$ are both even. Moreover, by Theorem 3.2.3, $a^{\phi(m)} \equiv 1 \pmod{m}$ and hence $a^{\phi(m)\phi(n)/2} \equiv 1 \pmod{m}$. Also $a^{\phi(n)} \equiv 1 \pmod{n}$ and hence $a^{\phi(m)\phi(n)/2} \equiv 1 \pmod{n}$. Since m and n are relatively prime, this implies that

$$a^{\phi(m)\phi(n)/2} \equiv 1 \pmod{mn}$$

or $a^{\phi(mn)/2} \equiv 1 \pmod{mn}$ (by Theorem 3.2.1). Thus a is not a primitive root of mn.

From the previous theorems we conclude that the only integers with primitive roots are 1, 2, 4, p^e, and $2p^e$, with p an odd prime and e any positive integer.

We shall use primitive roots in the next section, to study decimal expansions.

Exercises 3.3

1. The number of primitive roots of n is either 0 or $\phi(\phi(n))$.
2. Find the smallest primitive root of 71.
3. Let p be a prime greater than 3, and let q be the product of its primitive roots. Then p is a factor of $q - 1$.
4. Let n be a positive integer and let a be an odd positive integer. Then there is a positive integer x such that $5^x \equiv \pm a \pmod{2^n}$.
5. Find all the positive integers less than 50 with primitive root 2.

3.4 Decimal Expansions *

Let m and n be integers such that $n > m$ and $(m, n) = 1$. To find the decimal expansion of m/n we divide n into m, using long division. Since the remainders we obtain must always be less than n, and since we are 'bringing down' only zeros, the calculations eventually repeat. We thus have a repeating period, and, since there are only $n-1$ nonzero remainders possible, the length of the period never exceeds $n-1$. As in the case of $1/7 = .\overline{142857}$, we do sometimes get the maximum period length possible. According to the following theorem, this occurs when, and only when, n is prime and 10 is a primitive root of n. It is not known whether there are infinitely many primes which have 10 as a primitive root. M. Ram Murty came close to proving this, but the question is still open. The reader may wish to consult M. Ram Murty, 'Artin's Conjecture for Primitive Roots,' *The Mathematical Intelligencer,* 10 (1988), 59-67. There are exactly 9 primes less than 100 for which 10 is a primitive root. They are 7, 17, 19, 23, 29, 47, 59, 61, and 97.

Theorem 3.4.1 *Let m and n be integers such that $n > m > 0$ and* $(m, n) = 1$.
The decimal expansion of m/n is purely periodic iff $(n, 10) = 1$.
And in that case, the length of the period equals the order of 10 modulo n. Thus the decimal expansion of m/n is purely periodic with period length $n-1$ iff n is prime and 10 is a primitive root of n.

Proof: Suppose $m/n = .\overline{a}$, with a representing the block of, say, k digits in the repeating period. Then $10^k m/n = a.\overline{a}$ and hence

$$(10^k - 1)m/n = a$$

or $m/n = a/(10^k - 1)$. Hence 2 and 5 are not factors of n and $(n, 10) = 1$.

Also $(10^k - 1)m \equiv 0 \pmod{n}$ so that, since $(m, n) = 1$, $10^k \equiv 1 \pmod{n}$ and thus the order of 10 modulo n does not exceed the length of the period.

Conversely, suppose $(n, 10) = 1$ and let v be the order of 10 modulo n. Then $10^v = 1 + ns$ so that $10^v m/n = m/n + ms$. If $m/n = .b_1 b_2 \ldots b_v b_{v+1} \ldots$, we obtain

$$b_1 b_2 \ldots b_v . b_{v+1} \ldots = .b_1 b_2 \ldots b_v b_{v+1} \ldots + ms$$

Equating the fractional parts, $.b_{v+1} \ldots = .b_1 b_2 \ldots b_v b_{v+1} \ldots$. Thus m/n has a purely periodic decimal expansion.

Also the length of the period does not exceed the order of 10 modulo n.

Exercises 3.4

1. Express 1/17 as a decimal.
2. Note that $1/7 = .\overline{142857}$ and $142 + 857 = 999$. Show that this observation can be generalised to any prime having 10 as a primitive root.
3. To express $.23\overline{545}$ as a fraction, we write

$$\frac{23545 - 23}{99900}$$

with a 9 for each repeating digit, and a 0 for each nonrepeating digit. Show that this method always works.
4. Let a/b be a proper reduced fraction (with a, and b positive integers). Let $e_1 = 60a/b$ and let $e_{n+1} = 60(e_n - [e_n])$ (with n any positive integer). Then the Babylonian sexagesimal expansion for a/b is $.[e_1][e_2][e_3]\ldots$ Prove this.
5. If each letter stands for a different scale 10 digit, solve

$$EVE/DID = .TALKTALKTALKTALK\ldots$$

3.5 $x^2 \equiv R \pmod{C}$

The theorems in this section help us understand the congruence $x^2 \equiv R \pmod{C}$, and they also prepare the way for the results of Section 7, where we answer the question, 'how many ways can a number be written as the sum of two squares?' and also the question, 'when does the Diophantine equation $ax^2 + by^2 + cz^2 = 0$ have a nontrivial solution?' A key theorem in this current section is the Chinese Remainder Theorem.

We begin with a theorem conjectured by John Wilson (1741–1793), and first proved by J. Lagrange. Let p be a prime. By Theorem 3.1.2, each of the numbers $1, 2, \ldots, p-1$ has an inverse modulo p. A number x is its own inverse just in case p factors $x^2 - 1 = (x-1)(x+1)$, that is, just in case $x \equiv \pm 1 \pmod{p}$. Thus each of the numbers $2, 3, \ldots, p-2$ has an inverse which is not itself. Hence their product, modulo p, is just 1. As a result, $(p-1)! \equiv -1 \pmod{p}$.

Moreover, suppose n is composite with prime factor p. If $(n-1)! \equiv -1 \pmod{n}$ then $(n-1)! \equiv -1 \pmod{p}$. But

$$(n-1)! = (n-1)(n-2)\ldots(p+1)p(p-1)\ldots 2 \times 1 \equiv 0 \pmod{p}$$

This gives us

Theorem 3.5.1 (Wilson's Theorem) *Let n be a natural number > 1. Then $(n-1)! \equiv -1 \pmod{n}$ iff n is prime.*

Now suppose p is a prime of the form $4m+1$, and consider the $2m$ congruences

$$\begin{aligned} 4m &\equiv -1 \pmod{p} \\ 4m-1 &\equiv -2 \pmod{p} \\ &\vdots \\ 2m+1 &\equiv -2m \pmod{p} \end{aligned}$$

Multiplying these congruences together, we find that

$$(4m)! \equiv (2m)!^2 \pmod{p}$$

and hence $(2m)!^2 \equiv -1 \pmod{p}$. Thus if $p \equiv 1 \pmod{4}$ then $x^2 \equiv -1 \pmod{p}$ has a solution.

Suppose that, for some positive integer n, $x^2 \equiv -1 \pmod{p^n}$. Let y be the inverse of $2x$ modulo p, so that $2xy = kp+1$ for some integer k. Then

$$\begin{aligned} (x-(1+x^2)y)^2 &= x^2 - 2(1+x^2)xy + (1+x^2)^2y^2 \\ &\equiv x^2 - (1+x^2)(kp+1) \pmod{p^{n+1}} \\ &\equiv x^2 - (1+x^2) \pmod{p^{n+1}} \\ &\equiv -1 \pmod{p^{n+1}} \end{aligned}$$

Hence, by mathematical induction, if $p \equiv 1 \pmod{4}$ then, for all positive integers n, $x^2 \equiv -1 \pmod{p^n}$ has a solution.

The converse is also true. If p is an odd prime and $x^2 \equiv -1 \pmod{p^n}$ has a solution, then if $p = 4m+3$, we have

$$1 \equiv x^{p-1} \equiv (x^2)^{(p-1)/2} \equiv (-1)^{2m+1} \equiv -1 \pmod{p}$$

— which is impossible. Thus we have the following theorem.

Theorem 3.5.2 *Let p be an odd prime and let n be any positive integer. Then $x^2 \equiv -1 \pmod{p^n}$ has a solution iff $p \equiv 1 \pmod 4$.*

How many solutions does $x^2 \equiv -1 \pmod{p^n}$ have, if it has any? To answer this question we have

Theorem 3.5.3 *Let R be an integer and let p be an odd prime which is not a factor of R. If $x^2 \equiv R \pmod{p^n}$ has a solution s, then it has exactly two solutions, namely s and $-s$.*

Proof: Since p is not a factor of R, it follows that $s \not\equiv -s \pmod{p^n}$.

If t is another solution, $t^2 \equiv s^2 \pmod{p^n}$ and hence

$$(t-s)(t+s) \equiv 0 \pmod{p^n}$$

Now p cannot divide both factors lest it divide $2s$ and hence R. Thus $t \equiv \pm s \pmod{p^n}$.

From Theorem 3.5.2 and Theorem 3.5.3, it follows that if $p \equiv 1 \pmod 4$ then $x^2 \equiv -1 \pmod{p^n}$ has exactly 2 solutions modulo p^n.

For the case in which the modulus is a power of the even prime 2, we have

Theorem 3.5.4 *Let R be an odd integer, and let n be an integer ≥ 3. If $x^2 \equiv R \pmod{2^n}$ has a solution s, then it has exactly 4 solutions: s, $-s$, $s + 2^{n-1}$, and $-s + 2^{n-1}$.*

Proof: Clearly these are all solutions, and it follows from the fact that s is odd that they are distinct.

Suppose t is another solution. It too would be odd. Since $t^2 \equiv s^2 \pmod{2^n}$, it follows that

$$\frac{1}{2}(t-s)\frac{1}{2}(t+s) \equiv 0 \pmod{2^{n-2}}$$

The two factors, $\frac{1}{2}(t-s)$ and $\frac{1}{2}(t+s)$, cannot both be even, lest 2 factor their sum t. Hence either $t \equiv s \pmod{2^{n-1}}$ or $t \equiv -s \pmod{2^{n-1}}$. Thus t is congruent to one of $\pm s$ and $\pm s + 2^{n-1}$, modulo 2^n.

What if the modulus is not a power of a prime? For that case we use

Theorem 3.5.5 (Chinese Remainder Theorem) *Let m and n be two relatively prime positive integers. Let s and t be integers such that $ms - nt = 1$.*
Then $x \equiv a \pmod{m}$ and $x \equiv b \pmod{n}$
iff $x \equiv a + ms(b-a) \pmod{mn}$.
Hence the simultaneous congruences $x \equiv a_i \pmod{m_i}$ — with $i = 1, \ldots, k$ — have a solution if the moduli m_i are pairwise relatively prime.

Proof:

$$\begin{aligned} x &\equiv b \\ \text{iff } x &\equiv a + b - a + nt(b-a) \\ \text{iff } x &\equiv a + (1+nt)(b-a) \\ \text{iff } x &\equiv a + ms(b-a) \quad (\text{mod n}) \end{aligned}$$

Also $x \equiv a \pmod{m}$ iff $x \equiv a + ms(b-a) \pmod{m}$. Since $(m,n) = 1$, it follows that $x \equiv b \pmod{n}$ and $x \equiv a \pmod{m}$ iff

$$x \equiv a + ms(b-a) \pmod{mn}$$

For example, to solve $x^2 \equiv -1 \pmod{65}$ we first factor 65 and solve $x^2 \equiv -1 \pmod{5}$ and $x^2 \equiv -1 \pmod{13}$. Pairing the solutions in all possible ways, we obtain 4 systems:

$$x \equiv 2 \pmod{5} \qquad \text{with } x \equiv 5 \pmod{13}$$

$$x \equiv 2 \pmod{5} \qquad \text{with } x \equiv -5 \pmod{13}$$

$$x \equiv -2 \pmod{5} \qquad \text{with } x \equiv 5 \pmod{13}$$

$$x \equiv -2 \pmod{5} \qquad \text{with } x \equiv -5 \pmod{13}$$

Using Theorem 3.5.5 in each case, we get the 4 solutions to the original equation: ± 8 and $\pm 18 \pmod{65}$.

One of the first mathematicians to solve Chinese Remainder Problems was Sun Tsu (400 AD). In particular, he solved the following:

divide by 3, the remainder is 2;
divide by 5, the remainder is 3;
divide by 7, the remainder is 2;
what will be the number?

Sun Tsu also gave a formula for determining the sex of a foetus. If x is the age of the pregnant woman, and y is number of the month in which she will give birth, and

$$z = 49 + y - x - (1 + 2 + 3 + 4 + 5 + 6 + 7 + 8 + 9)$$

then the child will be a son if and only if z is odd. Like Pythagoras, Sun Tsu associated the odd with the masculine, and the even with the feminine. Note that z will usually be a negative number, something mysterious and impressive in the days of Sun Tsu.

Basing ourselves on the Chinese Remainder Theorem, we get the following general result:

Theorem 3.5.6 *Let m and n be relatively prime integers > 1. Let $f(x)$ be a polynomial with integer coefficients. If $f(x) \equiv 0 \pmod{m}$ has solutions $a_1, \ldots, a_p$ (mod m), and $f(x) \equiv 0 \pmod{n}$ has solutions $b_1, \ldots, b_q$ (mod n), then $f(x) \equiv 0 \pmod{mn}$ has exactly pq solutions, namely, those obtainable by applying the Chinese Remainder Theorem to all possible pairs $x \equiv a_i \pmod{m}$ and $x \equiv b_j \pmod{n}$.*

Proof: To show that the pq solutions are distinct modulo mn we argue as follows. If

$$a_1 + ms(b_2 - a_1) \equiv a_3 + ms(b_4 - a_3) \pmod{mn}$$

then $a_1 \equiv a_3 \pmod{m}$ and

$$a_1 + (1 + nt)(b_2 - a_1) \equiv a_3 + (1 + nt)(b_4 - a_3) \pmod{n}$$

so that $b_2 \equiv b_4 \pmod{n}$.

From Theorem 3.5.3 and 3.5.6 we have

Theorem 3.5.7 *Where n is odd and $(a, n) = 1$, and where k is the number of distinct primes in the factorisation of n, $x^2 \equiv a \pmod{n}$ has either no solutions or 2^k solutions.*

From Theorem 3.5.7 and 3.5.2, we obtain

Theorem 3.5.8 *Suppose there are k distinct prime factors of n and all of them are of the form $4m+1$. Then $x^2 \equiv -1 \pmod{n}$ has exactly 2^k solutions.*

The Chinese Remainder Theorem also gives us a lemma we shall need in our proof of Legendre's Theorem:

Theorem 3.5.9 *Suppose that a, b, and c are pairwise relatively prime integers and there are integers g, h, and i such that $g^2 \equiv -bc \pmod{a}$, $h^2 \equiv -ca \pmod{b}$, and $i^2 \equiv -ab \pmod{c}$. Then there are integers a_1, b_1, c_1, a_2, b_2, and c_2 such that*

$$ax^2 + by^2 + cz^2 \equiv (a_1x + b_1y + c_1z)(a_2x + b_2y + c_2z) \pmod{abc}$$

Proof: If b^{-1} is the inverse of b modulo a, we have, mod a,

$$\begin{aligned} ax^2 + by^2 + cz^2 &\equiv by^2 + cz^2 \\ &\equiv b^{-1}(b^2y^2 - g^2z^2) \\ &\equiv (y + b^{-1}gz)(by - gz) \end{aligned}$$

Similarly,

$$ax^2 + by^2 + cz^2 \equiv (c^{-1}hx + z)(-hx + cz) \pmod{b}$$

$$ax^2 + by^2 + cz^2 \equiv (x + a^{-1}iy)(ax - iy) \pmod{c}$$

By the Chinese Remainder Theorem, there is some a_1 such that

$$a_1 \equiv 0 \pmod{a}$$

$$a_1 \equiv c^{-1}h \pmod{b}$$

$$a_1 \equiv 1 \pmod{c}$$

and so on (finding values for b_1, c_1, ... to satisfy the theorem).

Theorem 3.5.9 is used to prove the following theorem, which is another lemma for Legendre's Theorem, given in Section 3.7.

Theorem 3.5.10 *Suppose a is a positive integer, and b and c are negative integers. Suppose that a, b, and c are square-free and pairwise relatively prime. Suppose that b and c are not both -1. Furthermore, suppose the equations*

$$x^2 \equiv -bc \pmod{a}$$

$$x^2 \equiv -ca \pmod{b}$$

$$x^2 \equiv -ab \pmod{c}$$

all have solutions.
Then $ax^2 + by^2 + cz^2 = 0$ has a nontrivial integer solution with $|x| \le -2b\sqrt{-ac}$, $|y| \le 2a\sqrt{bc}$, and $|z| \le -ab$.

Proof: By Theorem 3.5.9, there are integers a_1, b_1, c_1, a_2, b_2, and c_2 such that

$$ax^2 + by^2 + cz^2 \equiv (a_1x + b_1y + c_1z)(a_2x + b_2y + c_2z) \pmod{abc}$$

By Theorem 3.1.4, there are integers x, y, and z, not all zero, such that $|x| \le \sqrt{bc}$, $|y| \le \sqrt{-ac}$, and $|z| \le \sqrt{-ab}$, and $a_1x + b_1y + c_1z \equiv 0 \pmod{abc}$. Thus, for some integers x, y, z, with $x^2 \le bc$, $y^2 \le -ac$, and $z^2 \le -ab$, we have $ax^2 + by^2 + cz^2 \equiv 0 \pmod{abc}$, with x, y, z not all zero.

Given the above inequalities, $ax^2 + by^2 + cz^2$ is either $-2abc$, $-abc$, 0 (as desired), or abc.

If it is abc then $x^2 = bc$, while $y = z = 0$. Since b and c are relatively prime and square-free, this implies that $x^2 = 1$ and $b = c = -1$, against the given.

If $ax^2 + by^2 + cz^2 = -2abc$, then $x = 0$ and $y^2 = -ac$ and $z^2 = -ab$. Since a and c are relatively prime and square-free, this implies that $y^2 = 1$ and $a = 1$ and $c = -1$. Similarly, $b = -1$, violating the given.

If $ax^2 + by^2 + cz^2 = -abc$ then let

$$\begin{aligned} x' &= -by + xz \\ y' &= ax + yz \\ z' &= z^2 + ab \end{aligned}$$

If each of these is 0, then $-ab = z^2$, and hence $a = 1$ and $b = -1$. In that case, $ax^2 + by^2 + cz^2 = 0$ has nontrivial solution $x = 1$, $y = 1$ and $z = 0$. Furthermore, if x', y' and z' are not all 0, then they themselves give a nontrivial solution to the original equation:

$$\begin{aligned}
& ax'^2 + by'^2 + cz'^2 \\
= \;& a(b^2y^2 - 2bxyz + x^2z^2) + b(a^2x^2 + 2axyz + y^2z^2) \\
& +c(z^4 + 2abz^2 + a^2b^2) \\
= \;& ab(ax^2 + by^2 + cz^2) + xyz(-2ab + 2ab) \\
& +z^2(ax^2 + by^2 + cz^2) + cab(z^2 + ab) \\
= \;& -ab(abc) - z^2(abc) + z^2(abc) + (abc)ab \\
= \;& 0
\end{aligned}$$

The bounds on $|x|$, $|y|$, and $|z|$ now follow.

Exercises 3.5

1. Solve $x^2 \equiv -1 \pmod{97}$.
2. Find the smallest natural number which leaves remainder 1 when divided by 3, remainder 2 when divided by 5, and remainder 3 when divided by 7.
3. Pursued by a lion, Diana and her guide are dashing up the steps of a pyramid. Diana takes 5 steps at a time, the guide 6, and the lion 7. Towards the end of this tale, Diana is 1 step from the top, the guide 9, and the lion 19. How many steps are there in the pyramid?
4. How many solutions has $x^2 \equiv 9 \pmod{2^{17}5^{32}7^3}$?
5. Prove that every prime of the form $4m + 1$ is a sum of two squares. (Hint: use Theorem 1.9.3.)
6. Find a nontrivial integer solution of $3x^2 - 5y^2 - 7z^2 = 0$.
7. If R has a prime factor of the form $4m + 3$ then the period length s of the SCF expansion of $\sqrt{R}$ is even.

3.6 Palindromic SCF's *

Since there was no one in the world who would have introduced him to the young lady, our first father simply went up to her and said, 'madam, I'm Adam'. This phrase is palindromic: it reads the same if one reverses the order of the letters. This first palindrome had wondrous consequences and so do palindromic simple continued fractions. They will help answer the question, 'in how many ways can a number be written as the sum of two relatively prime squares?'

A finite simple continued fraction is *palindromic* if its sequence of partial quotients reads the same forwards as backwards. $(1,2,1)$ and $(3,1,1,3)$ and (9) are palindromic, but $(1,2,3)$ and $(1,1,11)$ and $(-2,2)$ are not. In the next theorem we give a necessary and sufficient condition for an SCF with an even number of partial quotients to be palindromic.

Theorem 3.6.1 *Suppose x and y are relatively prime integers with $x > y > 0$. Then*

$$x/y = (a_1, a_2, \ldots, a_k, a_k, \ldots, a_2, a_1)$$

iff $y^2 \equiv -1 \pmod{x}$.

Proof: By Theorem 2.1.8,

$$\frac{f_n(a_1,\ldots,a_n)}{f_{n-1}(a_1,\ldots,a_{n-1})} = (a_n,\ldots,a_1)$$

Thus the left hand side of the equivalence implies that $x/f_{2k-1} = x/y$ and hence, by Plato's Theorem, $xg_{2k-1} - y^2 = 1$, so that $y^2 \equiv -1 \pmod{x}$.

Conversely, suppose $y^2 \equiv -1 \pmod{x}$. Let $x/y = (a_1,\ldots,a_n)$ where n is even. Since $xg_{n-1} - f_{n-1}y = (-1)^n$, we have $-f_{n-1}y \equiv 1 \pmod{x}$ and it follows that $y^2 \equiv -1 \equiv f_{n-1}y \pmod{x}$ and hence $y \equiv f_{n-1} \pmod{x}$ (Theorem 3.1.1). Since $x > f_{n-1} > 0$, and $x > y > 0$, we have $y = f_{n-1}$. Thus

$$(a_1,\ldots,a_n) = x/y = x/f_{n-1} = (a_n,\ldots,a_1)$$

by Theorem 2.1.8. Hence $(a_1, \ldots, a_n)$ is palindromic.

For example, $y^2 \equiv -1 \pmod{4225}$ has solutions 268, 1282, 2943, and 3957. Moreover,

$$\begin{aligned} 4225/268 &= (15,1,3,3,1,15) \\ 4225/1282 &= (3,3,2,1,1,1,1,2,3,3) \\ 4225/2943 &= (1,2,3,2,1,1,1,1,2,3,2,1) \\ 4225/3957 &= (1,14,1,3,3,1,14,1) \end{aligned}$$

In the same way, we have a necessary and sufficient condition for an SCF with an odd number of partial quotients to be palindromic.

Theorem 3.6.2 *If x and y are relatively prime integers with $x > y > 0$ then*

$$x/y = (a_1, a_2, \ldots, a_k, a_{k+1}, a_k, \ldots, a_2, a_1)$$

iff $y^2 \equiv 1 \pmod{x}$.

Exercises 3.6

1. Find the two integers less than 100 which can be expressed as a sum of two relatively prime squares in exactly two ways.
2. If $x/y = (1,2,3,4,4,3,2,1)$, show that x factors $y^2 + 1$.

3.7 Sums of Two Squares *

There are many puzzles involving sums of two squares.

> The Marshall of Noland can march his soldiers in two square formations in exactly 12 ways. What is the smallest possible number of soldiers in his army?

To solve such puzzles, we need the following theorems.

Theorem 3.7.1 *Let x be a positive integer. The number of decompositions of x as the sum of two relatively prime squares equals the number of solutions of $y^2 \equiv -1 \pmod{x}$ with $0 \le y \le x/2$.*

Proof: The theorem is true for $x = 1$ or 2. Let $x > 2$.

Every decomposition of x as the sum of two relatively prime squares leads to a solution of the congruence equation: for let $x = r^2 + s^2$ with $r > s$ be such a decomposition. Let $r/s = (a_k, \ldots, a_1)$ with $a_1 > 1$. Then

$$r = f_k(a_k, \ldots, a_1) = f_k(a_1, \ldots, a_k)$$

(by Theorem 2.1.6) and

$$s = g_k(a_k, \ldots, a_1) = f_{k-1}(a_{k-1}, \ldots, a_1) = f_{k-1}(a_1, \ldots, a_{k-1})$$

(Theorem 2.1.4 and 2.1.6). Hence, by Theorem 2.1.17,

$$x = f_{2k}(a_1, a_2, \ldots, a_k, a_k, \ldots, a_2, a_1)$$

Let

$$y = g_{2k}(a_1, a_2, \ldots, a_k, a_k, \ldots, a_2, a_1)$$

By Theorem 3.6.1, $y^2 \equiv -1 \pmod{x}$, and, since $a_1 \geq 2$, it follows that $x/y \geq 2$ and hence $0 \leq y \leq x/2$.

Moreover, two different decompositions of x into a sum of two relatively prime squares cannot lead in this way to the same solution y of the congruence equation. For suppose

$$g_{2k}(a_1, a_2, \ldots, a_k, a_k, \ldots, a_2, a_1) = y = g_{2m}(b_1, b_2, \ldots, b_m, b_m, \ldots, b_2, b_1)$$

Since $f_{2k} = x = f_{2m}$, it follows that $f_{2k}/g_{2k} = f_{2m}/g_{2m}$ and the two SCF expansions, neither ending in 1, are identical.

From the above two paragraphs, we may conclude that the number of solutions to the congruence equation, in the given range, is not less than the number of decompositions.

Now every solution of the congruence equation leads to a decomposition of x as a sum of two relatively prime squares: for let $y^2 \equiv -1 \pmod{x}$ with $0 \leq y \leq x/2$. Then $(x, y) = 1$ and, by Theorem 3.6.1,

$$x/y = (a_1, a_2, \ldots, a_k, a_k, \ldots, a_2, a_1)$$

with $a_1 > 1$. Hence, by Theorem 2.1.17, $x = f_k^2 + f_{k-1}^2$ with $(f_k, f_{k-1}) = 1$.

Moreover, two different solutions of $y^2 \equiv -1 \pmod{x}$ with $0 \leq y \leq x/2$ can never lead in this way to the same decomposition of x as a sum of two relatively prime squares: for suppose $f_k = f_m$ and $f_{k-1} = f_{m-1}$. Then

$$(a_k, \ldots, a_1) = f_k/f_{k-1} = f_m/f_{m-1} = (b_m, \ldots, b_1)$$

(Theorem 2.1.8), and, since $a_1 > 1$ and $b_1 > 1$, the two SCF expansions are identical. Hence $g_{2k} = g_{2m}$.

From the above two paragraphs, we may conclude that the number of decompositions is not less than the number of solutions of $y^2 \equiv -1 \pmod{x}$ with $0 \leq y \leq x/2$.

For example, $x^2 \equiv -1 \pmod{997}$ has exactly one solution between 0 and 997/2, namely, 161. (In Section 11 below we give a fast way of finding such solutions.) Furthermore,

$$997/161 = (6, 5, 5, 6)$$

and $f_1 = 6$, $f_2 = 31$. Finally, $997 = 6^2 + 31^2$. This is the only decomposition of 997 as a sum of two squares.

Theorem 3.7.2 *Let n be a positive integer greater than 1, with exactly k distinct prime factors, all of the form $4m + 1$. Then the number of ways n, or $2n$, can be expressed as a sum of two relatively prime squares is 2^{k-1}.*

Proof: From Theorem 3.5.8 and Theorem 3.6.1 and the fact that $x^2 \equiv -1 \pmod{n}$ iff $(-x)^2 \equiv -1 \pmod{n}$, it follows that n is a sum of relatively prime squares in exactly 2^{k-1} ways.

Moreover, by Theorems 3.5.6 and 3.5.8, $x^2 \equiv -1 \pmod{2n}$ has 2^k solutions and hence 2^{k-1} solutions with $0 \leq x \leq n$. By Theorem 3.6.1, $2n$ has 2^{k-1} decompositions as a sum of two relatively prime squares.

Corollary: Every prime of the form $4m + 1$ has exactly one decomposition as a sum of two squares.

The reader is now in a position to answer the question, 'in how many ways can a number be written as the sum of two relatively prime squares?' (Hint: you could use Theorem 1.9.3.)

The next theorem addresses the question of writing a number as a sum of two squares which are not necessarily relatively prime. Recall from Section 1.4 that if n is a positive integer, $t(n)$ is the number of positive integer divisors of n. Recall also that if n is a positive integer greater than 1, $u(n)$ was defined as 2^{k-1}, where k is the number of distinct primes dividing n. For convenience, let us say that $u(1) = 1$. We also define

$$C(n) = \sum_{g^2|n} u(n/g^2)$$

If n is not a square then $t(n) = 2C(n)$, and if n is a square then $t(n) = 2C(n) - 1$. (See Exercises 1.4, # 5 — there is an answer at the back.) Let n be a positive integer with exactly k distinct prime factors, all of the form $4m + 1$. By Theorem 3.7.2, n, or $2n$, can be expressed as a sum of two relatively prime squares in exactly $u(n)$ ways. This is also true when $n = 1$. If $n = e^2 + f^2$ and $g = \gcd(e, f)$, then

$$\frac{n}{g^2} = \left(\frac{e}{g}\right)^2 + \left(\frac{f}{g}\right)^2$$

with e/g and f/g relatively prime integers. Hence the number of ways n can be expressed as a sum of two squares, not necessarily relatively prime, is $C(n)$ — since Theorem 3.7.2 applies to n/g^2. Since $g^2|2n$ iff $g^2|n$, the same is true of $2n$. Thus the number of ways n, or $2n$, can be written as a sum of two squares is $\frac{1}{2}t(n)$ if n is not a square, and $\frac{1}{2}t(n) + \frac{1}{2}$ if n is a square — assuming that the prime factors of n all have the form $4m + 1$. This can be generalised as follows.

Theorem 3.7.3 *Let $N = 2^a RS$ where a, R, and S are nonnegative integers, and all the prime factors of R have the form $4m + 3$, and all the prime factors of S have the form $4m + 1$. Then N can be written as a sum of two squares iff R is a square. In that case, the number of expressions of N as a sum of two squares is $\frac{1}{2}t(S)$ if S is not a square, and $\frac{1}{2}t(S) + \frac{1}{2}$ if S is a square.*

Proof: Suppose $N = x^2 + y^2$. If p is a prime factor of R then $x^2 + y^2 \equiv 0 \pmod{p}$. If p is not a factor of y then y has an inverse y^{-1} modulo p, and $(xy^{-1})^2 \equiv -1 \pmod{p}$ — against Theorem 3.5.2. Hence p is a factor of y, and thus also of x. From this it follows that p^2 is a factor of N, and we obtain

$$\frac{N}{p^2} = \left(\frac{x}{p}\right)^2 + \left(\frac{y}{p}\right)^2$$

If there is any other prime factor of R (possibly p again), its square can also be factored out in the above fashion. Hence R itself is a square, say $R = r^2$, and $r|x$ and $r|y$.

Suppose $R = r^2$. Now $2 = 1^2 + 1^2$, and if p is a prime of the form $4m + 1$, it can be written as a sum of two squares (Theorem 3.7.2). Moreover, it follows from the identity first given by Abu Ja'far al-Khazin (950 AD), that if two numbers can be written as a sum of two squares, so can their product:

$$(a^2 + b^2)(c^2 + d^2) = (ac + bd)^2 + (ad - bc)^2$$

Hence N can be written as a sum of two squares.

Suppose $N = x^2 + y^2$. We have seen that $R = r^2$ and $r|x$ and $r|y$. Thus the number of ways N can be written as a sum of two squares is just the number of ways $2^a S$ can be so written.

Moreover, if $a \geq 2$, and $2^a S = x^2 + y^2$ then x and y are both even. We then have

$$2^{a-2} S = \left(\frac{x}{2}\right)^2 + \left(\frac{y}{2}\right)^2$$

Thus the number of ways $2^a S$ can be written as a sum of two squares is just the number of ways $2^b S$ can be so written — where $b = 0$ if a is even, and $b = 1$ is a is odd.

The result now follows from the remarks preceding the theorem.

Corollary: If R is a square, $N = 2^a RS$ can be expressed as a sum of 2 unequal nonzero squares in exactly $\left[\frac{t(S)}{2}\right]$ ways.

For example, $25 = 2^0 \times 1 \times 5^2 = 0^2 + 5^2 = 3^2 + 4^2$ can be expressed as a sum of two squares in exactly $\frac{1}{2}t(5^2) + \frac{1}{2} = 2$ ways.

The next theorem was discovered and proved by Adrien Marie Legendre (1752–1833).

Theorem 3.7.4 (Legendre's Theorem) *Let a, b, and c be squarefree nonzero integers which are pairwise relatively prime. Then*

$$ax^2 + by^2 + cz^2 = 0$$

has a nontrivial integer solution
iff (1) a, b, and c do not all have the same sign, and (2) the following equations all have solutions:

$$x^2 \equiv -bc \pmod{a}$$

$$x^2 \equiv -ca \pmod{b}$$

$$x^2 \equiv -ab \pmod{c}$$

Proof: First suppose there is a nontrivial integer solution. Then (1) obviously holds. Furthermore, the fact that there is a nontrivial integer solution implies that there is a nontrivial integer solution with x, y, and z pairwise relatively prime (since a, b, and c are squarefree). Now let p be a prime factor of the squarefree integer a. Then

$$c^2z^2 \equiv -bcy^2 \pmod{p}$$

for some relatively prime integers z and y. Also p is not a factor of y, for then it would be a factor of c or z: it cannot be a factor of c, since it is already a factor of a and $\gcd(a, c) = 1$; nor can it be a factor of z if it is already a factor of y. Hence y has an inverse modulo p, and hence $x^2 \equiv -bc \pmod{p}$ has a solution. Hence, by the Chinese Remainder Theorem, $x^2 \equiv -bc \pmod{a}$ has a solution. Thus (2) follows.

Now suppose (1) and (2) hold. Without loss of generality, we may take it that $a > 0$ and $b, c < 0$.
Case 1. $b = c = -1$.
In this case $x^2 \equiv -1 \pmod{a}$ has a solution. Hence, by Theorem 3.7.1, a is a sum of two relatively prime squares: $a = y^2 + z^2$. Hence we have $a \times 1^2 + (-1)y^2 + (-1)z^2 = 0$.
Case 2. b and c are not both -1.
In this case the result follows from Theorem 3.5.10.

Exercises 3.7

1. Show that if $2x = y^2 + z^2$ then

$$x = \left(\frac{y+z}{2}\right)^2 + \left(\frac{y-z}{2}\right)^2$$

2. In how many ways can a number be written as the sum of two relatively prime squares?
3. Find the smallest length which is the hypotenuse of exactly 8 primitive Pythagorean triangles.
4. Let $h = p_1^{e_1} \dots p_k^{e_k}$ where the p's are distinct primes all of the form $4m+1$. Then h is the hypotenuse of exactly 2^{k-1} primitive Pythagorean triangles.
5. 'My second raise of $120 per lecture,' exclaimed the professor, 'and for the third time in a row my fee is a square number of dollars!' What did the overpaid braggart now earn?
6. What is the smallest possible number of soldiers in the Marshall of Noland's army?
7. Check Theorems 3.7.1 and 3.7.2 in the case of $x = 4225$.
8. Consider the following equations:

$$\begin{aligned} 3x^2 - 5y^2 + 7z^2 &= 0 \\ x^2 + 2y^2 + 3z^2 &= 0 \\ -x^2 + y^2 - 3z^2 &= 0 \end{aligned}$$

Which ones do, and which ones do not have a nontrivial solution? Solve those which have nontrivial solutions.

3.8 Quadratic Residues

In order to solve congruences of the form $x^2 \equiv a \pmod{p}$ where p is a large prime, it is helpful to use the theory of 'quadratic residues'.

An integer a is a *quadratic residue* modulo n iff $(a, n) = 1$ and $x^2 \equiv a \pmod{n}$ has a solution. An integer a is a *quadratic nonresidue* modulo n iff $(a, n) = 1$ and $x^2 \equiv a \pmod{n}$ has no solution.

For example, 1 and 4 are quadratic residues modulo 5, while 2 and 3 are quadratic nonresidues modulo 5. However, 5 is neither a quadratic residue nor a quadratic nonresidue modulo 5.

If p is an odd prime and $(a, p) = 1$, the *Legendre symbol*

$$\left(\frac{a}{p}\right)$$

is defined as 1 if a is a quadratic residue mod p, and -1 if a is a quadratic nonresidue mod p. For example, by Theorem 3.5.2,

$$\left(\frac{-1}{p}\right) = 1$$

iff p has the form $4m + 1$.

The first theorem in this section leads to a formula for $\left(\frac{a}{p}\right)$.

Theorem 3.8.1 *If p is an odd prime, the solutions of*

$$x^{\frac{p-1}{2}} \equiv 1 \pmod{p}$$

are just the quadratic residues modulo p.

Proof: The numbers $1^2, 2^2, \ldots, \left(\frac{p-1}{2}\right)^2$ are all distinct mod p — for if $a^2 \equiv b^2 \pmod{p}$ then p factors $a \pm b$, and, since $-(p-1) < a \pm b < p-1$, we have $a \pm b = 0$.

By Fermat's Theorem, the numbers $1^2, 2^2, \ldots, \left(\frac{p-1}{2}\right)^2$ all solve the congruence equation $x^{(p-1)/2} \equiv 1 \pmod{p}$. By Theorem 3.1.3, that equation has at most $(p-1)/2$ solutions.

If a is an integer not divisible by the odd prime p, then, by Fermat's Theorem, $a^{\frac{p-1}{2}} \equiv \pm 1 \pmod{p}$. By Theorem 3.8.1, the quadratic residues give the $+1$, while the quadratic nonresidues give the -1. We thus have the following formula for $\left(\frac{a}{p}\right)$:

Theorem 3.8.2 *If p is an odd prime which does not factor a then*

$$\left(\frac{a}{p}\right) \equiv a^{\frac{p-1}{2}} \pmod{p}$$

From this we obtain

Theorem 3.8.3 *If p is an odd prime which factors neither a nor b then*

$$\left(\frac{a}{p}\right)\left(\frac{b}{p}\right) = \left(\frac{ab}{p}\right)$$

Also if $a \equiv b \pmod{p}$ then $\left(\frac{a}{p}\right) = \left(\frac{b}{p}\right)$.

Another result following from Theorem 3.8.2 is

Theorem 3.8.4 *If p is an odd prime, $\left(\frac{2}{p}\right) = 1$ iff $p \equiv \pm 1 \pmod{8}$.*

Proof: First suppose that p is an odd prime of the form $4m+3$. Then

$$2 \times 4 \times 6 \times \ldots \times (p-1)$$

$$\equiv 2 \times 4 \times 6 \times \ldots \times (2m)(-(2m+1))(-(2m-1)) \times \ldots \times (-3)(-1)$$
$$\equiv (-1)^{m+1}(2m+1)! \pmod{p}$$

so that

$$2^{\frac{p-1}{2}}(2m+1)! \equiv (-1)^{m+1}(2m+1)! \pmod{p}$$

and, by Theorem 3.8.2,

$$\left(\frac{2}{p}\right) = (-1)^{\frac{p+1}{4}}$$

Similarly, if p is an odd prime of the form $4m+1$ then

$$\left(\frac{2}{p}\right) = (-1)^{\frac{p-1}{4}}$$

The result now follows.

Theorem 3.8.4 can be used to solve the Diophantine equation $x^2 + 6 = y^3$.

Theorem 3.8.5 *The Diophantine equation $x^2 + 6 = y^3$ has no solutions.*

Proof: Suppose there is a solution (x, y). Then integer x is odd (lest y be even and hence 4 be a factor of 6). Considering the equation modulo 8, we obtain $y \equiv -1 \pmod 8$. Now

$$x^2 - 2 = (y-2)(y^2 + 2y + 4)$$

and $y^2 + 2y + 4 \equiv 3 \pmod 8$. Hence $y^2 + 2y + 4$ has a prime factor p congruent to ± 3 mod 8. (If all the odd primes factoring $y^2 + 2y + 4$ had the form $8m \pm 1$ then $y^2 + 2y + 4$ would have the same form.) Now for this prime p, $x^2 \equiv 2 \pmod p$ has a solution. But that contradicts Theorem 3.8.4.

Exercises 3.8

1. For what primes p does $x^2 \equiv -2 \pmod p$ have a solution?
2. Every prime of the form $8m+1$ can be expressed in the form $a^2 + 2b^2$ in exactly one way.
3.* Prove that if $2^m + 1$ is prime then every quadratic nonresidue of $2^m + 1$ is a primitive root.

3.9 Theorema Aureum

The 'Golden Theorem' is the Law of Quadratic Reciprocity, which we shall prove in this section. It was discovered by Euler, and first proved by Gauss (in 1796).

In simple terms, what it says is that if you have two odd primes, p and q, and at least one of them has the form $4m+1$, then $x^2 \equiv p \pmod q$ has a solution just in case $x^2 \equiv q \pmod p$ does. Moreover, if both primes have the form $4m + 3$, then $x^2 \equiv p \pmod q$ has a solution just in case $x^2 \equiv q \pmod p$ does *not*. For example, since $x^2 \equiv 71 \pmod 5$ has a solution (namely, 1), and since 5 has the form $4m + 1$, it follows from the Law of Quadratic Reciprocity that $x^2 \equiv 5 \pmod{71}$ has a solution. That is, there is a square of the form $71m + 5$.

To prove the Law of Quadratic Reciprocity, we use a couple of 'counting' lemmas, discovered by Gauss.

Theorem 3.9.1 *Let p be an odd prime which does not factor the positive integer a. Consider the integers*

$$a, \ 2a, \ 3a, \ \ldots, \ \frac{1}{2}(p-1)a$$

and their least positive residues modulo p. Let

$$r_1, \ r_2, \ \ldots, \ r_n$$

be those residues which exceed $\frac{1}{2}p$, and let

$$s_1, \ s_2, \ \ldots, \ s_k$$

be the others. Then the integers

$$p-r_1, \ p-r_2, \ \ldots, \ p-r_n, \ s_1, \ s_2, \ ,\ldots, \ s_k$$

are just the $\frac{1}{2}(p-1)$ integers from 1 to $\frac{1}{2}(p-1)$, and $\left(\frac{a}{p}\right) = (-1)^n$.

Proof: To obtain a contradiction, suppose $p - r_i = s_j$. Suppose $r_i \equiv ba \pmod{p}$ and $s_j \equiv ca \pmod{p}$ with $1 \leq b, \quad c \leq \frac{1}{2}(p-1)$. Then $-ba \equiv ca \pmod{p}$, and p factors $b+c$. Contradiction. This establishes the first assertion of the theorem.

From this it follows that

$$(p-r_1)\ldots(p-r_n)s_1\ldots s_k = \left(\frac{p-1}{2}\right)!$$

so that

$$(-1)^n a^{\frac{p-1}{2}} \left(\frac{p-1}{2}\right)! \equiv \left(\frac{p-1}{2}\right)! \pmod{p}$$

and thus

$$a^{\frac{p-1}{2}} \equiv (-1)^n \pmod{p}$$

The second result now follows from Theorem 3.8.2.

Theorem 3.9.2 *Let p be an odd prime not dividing the positive odd integer a. Let*

$$t = \left[\frac{a}{p}\right] + \left[\frac{2a}{p}\right] + \cdots + \left[\frac{\frac{1}{2}(p-1)a}{p}\right]$$

Then $\left(\frac{a}{p}\right) = (-1)^t$.

Proof: We use the notation of the preceding theorem. Let j be a positive integer not exceeding $\frac{1}{2}(p-1)$. The least positive residue modulo p of ja is $ja - \lfloor ja/p \rfloor p$. Thus

$$\begin{aligned} a + 2a + \cdots + \frac{1}{2}(p-1)a &= \left[\frac{a}{p}\right]p + \left[\frac{2a}{p}\right]p + \cdots + \left[\frac{\frac{1}{2}(p-1)a}{p}\right]p \\ &\quad + r_1 + \cdots + r_n + s_1 + \cdots + s_k \end{aligned}$$

From Theorem 3.9.1,

$$1 + 2 + \cdots + \frac{1}{2}(p-1) = (p - r_1) + \cdots + (p - r_n) + s_1 + \cdots + s_k$$

Subtracting this second equation from the previous one, we have

$$(a-1)(1 + 2 + \cdots + \frac{1}{2}(p-1)) = tp - np + 2(r_1 + \cdots + r_n)$$

Since a is odd (given), $tp - np$ is even. Since p is odd, t and n have the same parity. The result now follows from Theorem 3.9.1.

Theorem 3.9.3 (The Law of Quadratic Reciprocity) *If p and q are distinct odd primes, then*

$$\left(\frac{p}{q}\right) = (-1)^{\frac{p-1}{2}\frac{q-1}{2}} \left(\frac{q}{p}\right)$$

Proof: Let S be the set of all ordered pairs (x, y) where x is an integer between 1 and $\frac{1}{2}(p-1)$ inclusive, and y is an integer between 1 and $\frac{1}{2}(q-1)$ inclusive.

If (x, y) is in this set, $qx \neq py$. Thus S consists of two disjoint subsets: S_1 containing pairs (x, y) with $qx > py$, and S_2 containing pairs (x, y) with $qx < py$.

Now S_1 consists of just those pairs of integers (x, y) with $1 \leq x \leq \frac{1}{2}(p-1)$ and $1 \leq y < qx/p$. (If $y < qx/p$ then $y < q(p-1)/(2p) < q/2$ so $y \leq \frac{1}{2}(q-1)$.) Hence S_1 contains

$$A = \left[\frac{q}{p}\right] + \left[\frac{2q}{p}\right] + \cdots + \left[\frac{\frac{1}{2}(p-1)q}{p}\right]$$

elements.

Similarly, S_2 contains

$$B = \left[\frac{p}{q}\right] + \left[\frac{2p}{q}\right] + \cdots + \left[\frac{\frac{1}{2}(q-1)p}{q}\right]$$

elements.

Moreover, $A + B = \frac{1}{2}(p-1)\frac{1}{2}(q-1)$ (from the original definition of S).

By Theorem 3.9.2, $\left(\frac{q}{p}\right) = (-1)^A$, and $\left(\frac{p}{q}\right) = (-1)^B$. Thus $\left(\frac{p}{q}\right)\left(\frac{q}{p}\right) = (-1)^{A+B}$ and the result follows.

For example,

$$\left(\frac{5}{71}\right) = \left(\frac{71}{5}\right) = \left(\frac{1}{5}\right) = 1$$

Exercises 3.9

1. Use the Law of Quadratic Reciprocity to calculate $\left(\frac{7}{97}\right)$.
2. Show that every prime of the form $3m + 1$ can be expressed in the form $a^2 + 3b^2$ and in exactly one way.

3.10 Jacobi Symbol

In this section we study the generalisation of the Legendre symbol which is named after Carl Jacobi (1804–1851). Jacobi's early death was due to smallpox.

Let $Q = 1$ or let Q be a product $q_1 q_2 \dots q_s$ of odd primes (not necessarily distinct). Let P be an integer such that $(P, Q) = 1$. Then the *Jacobi symbol* is defined as follows:

$$\left(\frac{P}{Q}\right) = 1 \text{ if } Q = 1$$

Otherwise

$$\left(\frac{P}{Q}\right) = \left(\frac{P}{q_1}\right)\left(\frac{P}{q_2}\right)\cdots\left(\frac{P}{q_s}\right)$$

Thus if Q is an odd prime, the Jacobi symbol coincides with the Legendre symbol.

If $x^2 \equiv P \pmod{Q}$ has a solution, then so does $x^2 \equiv P \pmod{q_j}$ (for $j = 1, \ldots, s$). Hence $\left(\frac{P}{q_j}\right) = 1$ for each j, and thus $\left(\frac{P}{Q}\right) = 1$. The converse is false: $\left(\frac{-1}{9}\right) = 1$ but $x^2 \equiv -1 \pmod{9}$ has no solution.

Using Theorem 3.8.3, we can prove

Theorem 3.10.1 *Suppose that Q and Q' are odd positive integers, and P and P' are integers such that* $\gcd(PP', QQ') = 1$. *Then*

$$\left(\frac{P}{QQ'}\right) = \left(\frac{P}{Q}\right)\left(\frac{P}{Q'}\right)$$

$$\left(\frac{PP'}{Q}\right) = \left(\frac{P}{Q}\right)\left(\frac{P'}{Q}\right)$$

$$\left(\frac{P^2}{Q}\right) = \left(\frac{P}{Q^2}\right) = 1$$

$$\left(\frac{P'P^2}{Q'Q^2}\right) = \left(\frac{P'}{Q'}\right)$$

$$\left(\frac{P'}{Q}\right) = \left(\frac{P}{Q}\right) \quad \text{if } P' \equiv P \pmod{Q}$$

We also have

Theorem 3.10.2 *Let Q be an odd positive integer. Then*

$$\left(\frac{-1}{Q}\right) = 1 \text{ iff } Q \equiv 1 \pmod 4$$

$$\left(\frac{2}{Q}\right) = 1 \text{ iff } Q \equiv \pm 1 \pmod 8$$

Proof: Let $Q = q_1 \ldots q_s$, where the q's are prime. Then

$$\left(\frac{-1}{Q}\right) = \left(\frac{-1}{q_1}\right)\left(\frac{-1}{q_2}\right)\cdots\left(\frac{-1}{q_s}\right)$$

so that, by Theorem 3.5.2, this is 1 just in case an even number of the q's have the form $4m+3$, and this is true iff $Q \equiv 1 \pmod 4$.

The second result follows in a similar way from Theorem 3.8.4.

We now have Jacobi's generalisation of the Law of Quadratic Reciprocity.

Theorem 3.10.3 *Let P and Q be odd, relatively prime positive integers. Then*

$$\left(\frac{P}{Q}\right) = (-1)^{\frac{P-1}{2}\frac{Q-1}{2}}\left(\frac{Q}{P}\right)$$

Proof: Let $P = p_1 \ldots p_k$ and $Q = q_1 \ldots q_s$ where the p's and q's are prime. Then, by Theorem 3.10.1,

$$\left(\frac{P}{Q}\right) = \left(\frac{p_1}{q_1}\right)\left(\frac{p_2}{q_1}\right)\cdots\left(\frac{p_k}{q_1}\right)\left(\frac{p_1}{q_2}\right)\cdots\left(\frac{p_k}{q_2}\right)\cdots\cdots\left(\frac{p_1}{q_s}\right)\cdots\left(\frac{p_k}{q_s}\right)$$

By the Law of Quadratic Reciprocity (Theorem 3.9.3),

$$\left(\frac{p_i}{q_j}\right) = \pm\left(\frac{q_j}{p_i}\right)$$

with the negative sign just in case both p_i and q_j have the form $4m+3$. Suppose that a of the p's and b of the q's have the form $4m+3$. Then

$$\left(\frac{P}{Q}\right) = -\left(\frac{q_1}{p_1}\right)\cdots\left(\frac{q_1}{p_k}\right)\left(\frac{q_2}{p_1}\right)\cdots\left(\frac{q_2}{p_k}\right)\cdots\left(\frac{q_s}{p_1}\right)\cdots\left(\frac{q_s}{p_k}\right) = -\left(\frac{Q}{P}\right)$$

iff ab is odd. And this is true iff both a and b are odd, and both P and Q have the form $4m+3$.

Suppose we wish to determine whether $x^2 \equiv 105 \pmod{317}$ has a solution. Since 317 is prime, it suffices to compute $\left(\frac{105}{317}\right)$ and see whether it equals 1. To do this, we can use the theorems of this section:

$$\left(\frac{105}{317}\right) = \left(\frac{317}{105}\right) = \left(\frac{2}{105}\right) = 1$$

Hence there is a solution.

Exercises 3.10

1. Show that $\left(\frac{-2}{Q}\right) = 1$ iff $Q \equiv 1$ or $3 \pmod 8$.
2. Show that $x^2 \equiv 599527 \pmod{1000039}$ has a solution.

3.11 More on $x^2 \equiv R \pmod C$ *

We are now in a position to give a fast, direct way of solving the equation $x^2 \equiv R \pmod C$. Having done that, we shall show, in the next section, how this congruence is used to solve certain Diophantine equations. The reader may wish to find the Diophantine equation associated with the following puzzle. Following the next section, the reader will be able to solve it.

> Mrs. Ball baked three equal square cakes and cut them up into equal squares. She gave 10 pieces each to her 6 children and 15 pieces to Mr. Ball, who is a very keen mathematician. The remainder she distributed equally among 14 hungry students, who, although they did not do quite so well as Mr. Ball, thoroughly enjoyed themselves. Assuming that the charming lady did not keep a single crumb for herself, how many pieces did each of the students get?

In this section, we always take C to be a positive integer > 1. If C is small, the best way to solve $x^2 \equiv R \pmod{C}$ is by trial and error. In general, however, the following method is better.

First we prove a theorem that shows that we can reduce the problem to the case where $\gcd(R, C) = 1$.

Theorem 3.11.1 *Let $(R, C) = a^2b$ where b is square-free.*
Then $x^2 \equiv R \pmod{C}$ with $0 \le x < C$ iff
(1) $(b, C/(a^2b)) = 1$, so that b has an inverse b^{-1} mod $C/(a^2b)$, and
(2) there is some integer y such that $y^2 \equiv b^{-1}R/(a^2b) \pmod{C/(a^2b)}$ with $0 \le y < C/(ab)$ and
(3) $x = aby$.

Proof: First note that $(b^{-1}R/(a^2b), C/(a^2b)) = 1$ since

$$bb^{-1} \equiv 1 \pmod{C/(a^2b)}$$

Suppose $x^2 \equiv R \pmod{C}$ with $0 \le x < C$. Then $ab|x$ and we can write $x = aby$ where $0 \le y < C/(ab)$. Hence $a^2b^2y^2 \equiv R \pmod{C}$, so that $by^2 \equiv R/(a^2b) \pmod{C/(a^2b)}$ and, since $(R/(a^2b), C/(a^2b)) = 1$, we have $(b, C/(a^2b)) = 1$. Thus $y^2 \equiv b^{-1}R/(ab^2) \pmod{C/(a^2b)}$.

Conversely, suppose $(b, C/(a^2b)) = 1$, and

$$y^2 \equiv b^{-1}R/(a^2b) \pmod{C/(a^2b)}$$

and $0 \le y < C/(ab)$. If $x = aby$ then $0 \le x < C$ and $x^2 \equiv R \pmod{C}$. This concludes the proof.

For example, if $R = 36$ and $C = 54$, then $\gcd(R, C) = 3^2 \times 2$. The solutions of $y^2 \equiv 4 \pmod{3}$ between 0 and 9 are 1, 2, 4, 5, 7, and 8. Multiplying these by 6, we get the solutions of $x^2 \equiv 90 \pmod{54}$, namely, 6, 12, 24, 30, 42, and 48.

Note that $x^2 \equiv R \pmod{C}$ with $0 \le x < C$ has a times as many solutions as $y^2 \equiv b^{-1}R/(a^2b) \pmod{C/(a^2b)}$ with $0 \le y < C/(ab)$.

It follows from Theorem 3.5.5 and 3.5.6 that, in order to solve $x^2 \equiv R \pmod{C}$ with $(R, C) = 1$, it is enough to know how to solve $x^2 \equiv R \pmod{p^e}$ where p is a prime which does not factor R. Moreover, it follows from Theorem 3.5.3 and 3.5.4 that it is enough to know how to find a *single* solution of this latter congruence, or show there is none. The remaining theorems in this section give fast ways of finding a solution of $x^2 \equiv R \pmod{p^e}$ (if there is one). We begin with the case where the prime $p = 2$. Our next theorem (together with Theorem 3.5.4) gives a fast way of solving $x^2 \equiv R \pmod{2^n}$ where R is odd, and $n \geq 3$.

Theorem 3.11.2 *Suppose R is odd, and $n \geq 3$.*
If $R \equiv 1 \pmod 8$ then $x^2 \equiv R \pmod{2^n}$ has solution a_n where $a_3 = 1$ and $a_{t+1} \equiv a_t + \frac{1}{2}(a_t^2 - R) \pmod{2^{t+1}}$.
Otherwise, $x^2 \equiv R \pmod{2^n}$ has no solution.

Proof: $x^2 \equiv R \pmod{2^n}$ implies that $x^2 \equiv R \pmod 8$. Since 1 is the only odd square modulo 8, the second assertion follows.

Suppose $R \equiv 1 \pmod 8$. The first assertion is true for $n = 3$. Suppose it true for n. Then $a_n^2 \equiv R \pmod{2^n}$. Hence, for some integer y, $a_n^2 = R + 2^n y$.

Case 1. y is even, say, $y = 2m$. Then $a_{n+1} \equiv a_n + 2^n m \pmod{2^{n+1}}$ and

$$a_{n+1}^2 \equiv a_n^2 + 2a_n 2^n m + 2^{2n} m^2 \equiv a_n^2 \equiv R + 2^n 2m \equiv R \pmod{2^{n+1}}$$

as required.

Case 2. y is odd, say, $y = 2m + 1$. Then $a_{n+1} \equiv a_n + 2^n m + 2^{n-1} \pmod{2^{n+1}}$. Since $n \geq 3$ and a_n is odd, we also have

$$a_{n+1}^2 \equiv a_n^2 + 2a_n(2^n m + 2^{n-1}) + (2^n m + 2^{n-1})^2$$

$$\equiv a_n^2 + 2^n a_n \equiv R + 2^n(2m+1) + 2^n a_n \equiv R + 2^n(1 + a_n) \equiv R \pmod{2^{n+1}}$$

as required. The result now follows.

Let p be an odd prime which is not a factor of R. To find a single solution of $x^2 \equiv R \pmod{p^n}$ — or show there is none — it is enough to know how to find a single solution of $x^2 \equiv R \pmod p$ — or show there is none:

Theorem 3.11.3 *Let p be an odd prime not dividing R.*
If $x^2 \equiv R \pmod p$ has no solution, neither has $x^2 \equiv R \pmod{p^n}$.

If $x^2 \equiv R \pmod{p}$ *has a solution* a_1, *then* $x^2 \equiv R \pmod{p^n}$ *has solution* a_n, *where*

$$a_{t+1} \equiv a_t + (R - a_t^2)b_t \pmod{p^{t+1}}$$

with b_t *a solution of* $2a_t y \equiv 1 \pmod{p}$.

Proof: The first statement follows at once.

The second statement is true for $n = 1$. Suppose it true for n. Then, since $p^n|(R - a_n^2)$, we have the following, modulo p^{n+1}:

$$a_{n+1}^2 \equiv (a_n + (R - a_n^2)b_n)^2 \equiv a_n^2 + 2a_n(R - a_n^2)b_n$$

$$\equiv a_n^2 + (kp+1)(R - a_n^2) \equiv R \pmod{p^{n+1}}$$

We have now reduced the problem of solving $x^2 \equiv R \pmod{C}$ to the case in which C is an odd prime. The next theorems handle this case.

Where p is an odd prime not dividing evenly into R, we can tell whether $x^2 \equiv R \pmod{p}$ has a solution by using the Jacobi symbol: there is a solution just in case $\left(\frac{R}{p}\right) = 1$. We have already shown how to compute $\left(\frac{R}{p}\right)$ rapidly (using Jacobi's generalised Law of Quadratic Reciprocity). The following theorems complete our treatment of $x^2 \equiv R \pmod{C}$ by dealing with the case when C is an odd prime, and R is a quadratic residue of C.

We begin by extending the notion of congruence in a way suggested by Lagrange. Let B be a quadratic nonresidue of the odd prime p. Where m, n, r, and s are integers, we define

$$m + n\sqrt{B} \equiv r + s\sqrt{B} \pmod{p}$$

iff

$$m \equiv r \pmod{p} \text{ and } n \equiv s \pmod{p}$$

It is not hard to show that we can add, subtract, and multiply these congruences in the usual way. Furthermore, we have

Theorem 3.11.4 *If* $\left(\frac{A}{p}\right) = 1$ *and* $\left(\frac{B}{p}\right) = -1$ *then*

$$(\sqrt{A} + \sqrt{B})^{p+1} \equiv A - B \pmod{p}$$

Proof: Note that $\sqrt{A}$ just denotes some solution of $x^2 \equiv A \pmod{p}$.

By the Binomial Theorem (which goes back at least to Al-Kashi (1427)),

$$(\sqrt{A} + \sqrt{B})^p \equiv (\sqrt{A})^p + (\sqrt{B})^p \pmod{p}$$

(since the binomial coefficients are multiples of p). Since

$$A^{(p-1)/2} \equiv \left(\frac{A}{p}\right) = 1$$

and

$$B^{(p-1)/2} \equiv \left(\frac{B}{p}\right) = -1$$

(Theorem 3.8.1), this gives

$$(\sqrt{A} + \sqrt{B})^p \equiv \sqrt{A} - \sqrt{B} \pmod{p}$$

Multiplying both sides by $\sqrt{A} + \sqrt{B}$, we obtain the result.

Theorem 3.11.5 *Suppose* $\left(\frac{A}{p}\right) = 1$, $\left(\frac{B}{p}\right) = -1$, *and* $\left(\frac{A-B}{p}\right) = 1$. *Let*

$$(\sqrt{A} + \sqrt{B})^{(p+1)/2} \equiv m + n\sqrt{B} \pmod{p}$$

where m and n are integers. Then $n \equiv 0 \pmod{p}$.

Proof: By Theorem 3.11.4, $A - B \equiv m^2 + n^2B + 2mn\sqrt{B} \pmod{p}$, and hence $2mn \equiv 0 \pmod{p}$. If $m \equiv 0 \pmod{p}$ then

$$1 = \left(\frac{A-B}{p}\right) = \left(\frac{n^2B}{p}\right) = \left(\frac{B}{p}\right) = -1$$

(Theorem 3.8.3). Contradiction.

Theorem 3.11.6 *If If p is an odd prime not dividing R and $\left(\frac{R}{p}\right) = 1$ then there is an integer h such that $\left(\frac{h^2-R}{p}\right) = -1$.*

Proof: The integers

$$-R, \ -2R, \ -3R, \ \ldots, \ -(p-1)R$$

are all distinct and nonzero modulo p. Let $-aR$ be the first quadratic nonresidue on this list. If $a = 1$, we may take $h = 0$. Suppose, however, that $a \neq 1$. Then $-(a-1)R \equiv h^2 \pmod p$ for some integer h, and $h^2 - R \equiv -aR \pmod p$, with $-aR$ a quadratic nonresidue.

To find such a quadratic nonresidue in practice, we try

$$-R, \ 1-R, \ 4-R, \ 9-R, \ \ldots$$

in turn, using the theory of the Jacobi symbol to calculate $\left(\frac{h^2-R}{p}\right)$. Usually very few trials suffice.

Finally, we have

Theorem 3.11.7 *If $\left(\frac{R}{p}\right) = 1$ and $\left(\frac{h^2-R}{p}\right) = -1$ then one solution of $x^2 \equiv R \pmod p$ is $(h + \sqrt{h^2 - R})^{(p+1)/2}$.*

Proof: Take $A = h^2$ and $B = h^2 - R$ in Theorem 3.11.4 and 3.11.5.

Note that the congruence equation has exactly two solutions, one the negative of the other. Hence Theorem 3.11.7 gives a complete solution of it. Note also that if $p \equiv 3 \pmod 4$, we can take $h = 0$, and the solution is simply $R^{(p+1)/4}$ (Theorems 3.5.2, 3.8.3).

To calculate $C^{(p+1)/2}$, we 'factor out squares' in the exponent whenever possible, so that the number of steps is proportionate to $\log_2 \frac{p+1}{2}$. For example, to solve $x^2 \equiv 378 \pmod{991}$, we calculate as follows.

$$\begin{aligned}
378^{(991+1)/4} &\equiv (378^2)^{124} \\
&\equiv 180^{124} \\
&\equiv (180^2)^{62} \\
&\equiv (688^2)^{31} \\
&\equiv 637(637^2)^{15}
\end{aligned}$$

$$\begin{aligned}
&\equiv 637 \times 450(450^2)^7 \\
&\equiv 251(336)^7 \\
&\equiv 251 \times 336(336^2)^3 \\
&\equiv 101(913)^3 \\
&\equiv 954 \equiv -37
\end{aligned}$$

The same 'trick' works when C is of the form $m + n\sqrt{B}$. For example, to solve $x^2 \equiv -1 \pmod{997}$, we can take $h = 1$ and calculate

$$\begin{aligned}
(1+\sqrt{2})^{(997+1)/2} &\equiv (1+\sqrt{2})(3+2\sqrt{2})^{249} \\
&\equiv (1+\sqrt{2})(3+2\sqrt{2})(17+12\sqrt{2})^{124} \\
&\equiv (7+5\sqrt{2})(577+408\sqrt{2})^{62} \\
&\equiv (7+5\sqrt{2})(858+248\sqrt{2})^{31} \\
&\equiv (7+5\sqrt{2})(858+248\sqrt{2})(755+846\sqrt{2})^{15} \\
&\equiv (510+44\sqrt{2})(755+846\sqrt{2})(478+303\sqrt{2})^{7} \\
&\equiv (878+78\sqrt{2})(478+303\sqrt{2})(341+538\sqrt{2})^{3} \\
&\equiv (356+230\sqrt{2})(341+538\sqrt{2})(260+20\sqrt{2}) \\
&\equiv (983+768\sqrt{2})(260+20\sqrt{2}) \\
&\equiv 161+0\sqrt{2}
\end{aligned}$$

Hence the solutions are ± 161, mod 997.

The reader should note that not many other introductory Number Theory books give a fast way to solve $x^2 \equiv R \pmod{p}$ when p has the form $4m+1$. Theorem 3.11.7, with its explicit solution to that equation, was discovered by the author.

Exercises 3.11

1. Solve $x^2 \equiv 1,970,125,838 \pmod{6,895,440,433}$. Hint: 997 is prime.
2. Solve $x^2 \equiv 43,474 \pmod{128,331}$.
3. Solve $x^2 \equiv 3,899,721 \pmod{4,194,304}$.
4. Solve $x^2 \equiv 84,680,902 \pmod{3^{18}}$.
5. Solve $x^2 \equiv 599,527 \pmod{1,000,039}$.
6. Solve $x^2 \equiv 761,234 \pmod{1,000,033}$.

7. Solve $x^2 \equiv 17 \pmod{1,000,004}$.
8. Suppose p is a prime of the form $8m+5$. Then $x^2 \equiv -1 \pmod p$ has solution $(1+\sqrt{2})^{(p+1)/2}$.
9. Show that we can add, subtract, and multiply the new Lagrange congruences in the 'usual way'.

3.12 $Ax^2 + By = C$ *

We can apply the theory of congruence to solve the Diophantine equation $Ax^2 + By = C$. Indeed, we can handle the equation

$$Ax^2 + Bxy + Cy^2 + Dx + Ey = F$$

where $A \neq 0$ and $R = B^2 - 4AC = 0$. Equations of the latter sort are tricky. In Part II of his *Algebra* (on p. 488), Chrystal gives a solution to

$$9x^2 - 12xy + 4y^2 + 3x + 2y = 12$$

but he misses the obvious solution $x = 1$, $y = 0$.

To begin, consider $Ax^2 + By = C$. If A and B have a common factor, either it divides evenly into C or it does not.

If it does, we can divide it out of the equation, to obtain an equivalent equation in which the coefficients of x^2 and y are relatively prime.

If it does not, there is no solution.

Hence, without loss of generality, we may take it that $\gcd(A, B) = 1$. In that case, A has an inverse A^{-1} modulo B, and it is necessary and sufficient for (x, y) to be a solution of the original equation that $x^2 \equiv A^{-1}C \pmod B$, and $y = (C - Ax^2)/B$. Hence

To solve $Ax^2 + By = C$ with $\gcd(A, B) = 1$:
let $z_1, \ldots, z_n$ be the solutions of $x^2 \equiv A^{-1}C \pmod B$ with $0 \le x \le |B|/2$. Then, where K is a variable running over the integers, the solution is

$$\begin{aligned} x &= \pm z_i + BK \\ y &= \frac{C - Az_i^2}{B} \mp 2Az_iK - ABK^2 \end{aligned}$$

For example, to solve $3x^2 + 16000y = 176,147$, we first find the inverse of 3 modulo 16,000: it is 10,667. We then solve

$$x^2 \equiv 10,667 \times 176,147 \pmod{16,000}$$

to get

$$x \equiv \pm 7, \quad \pm 3257, \quad \pm 4743, \quad \pm 7993$$

Thus $3x^2 + 16000y = 176,147$ has the following solutions:

x	y
$\pm 7 + 16000K$	$11 \mp 42K - 48000K^2$
$\pm 3257 + 16000K$	$-1978 \mp 19,542K - 48000K^2$
$\pm 4743 + 16000K$	$-4207 \mp 28,458K - 48000K^2$
$\pm 7993 + 16000K$	$-11,968 \mp 47,958K - 48000K^2$

Note that $x = 7$, and $y = 11$ is the only positive integer solution of the equation.

We can extend this method, using the following theorem.

Theorem 3.12.1 *Let $R = B^2 - 4AC$, $S = BD - 2AE$ and $T = 4AF + D^2$. Suppose $A \neq 0$ and $R = 0$. Then $Ax^2 + Bxy + Cy^2 + Dx + Ey = F$ iff $(2Ax + By + D)^2 - 2Sy = T$.*

Proof:

$$Ax^2 + Bxy + Cy^2 + Dx + Ey = F$$

iff

$$4A^2x^2 + 4ABxy + B^2y^2 + 4ADx + 4AEy = 4AF$$

iff

$$(2Ax + By)^2 + 2(2Ax + By)D + D^2 - 2BDy - D^2 + 4AEy = 4AF$$

iff

$$(2Ax + By + D)^2 - 2Sy = T$$

Suppose we wish to solve the Diophantine equation

$$Ax^2 + Bxy + Cy^2 + Dx + Ey = F$$

where $A \neq 0$ and $R = B^2 - 4AC = 0$. (In the xy-plane, the corresponding curve, if it exists, is a straight line, two parallel straight lines, or a parabola, and what we wish to do is find all the lattice points on this curve.) Using Theorem 3.12.1, we transform the equation into

$$(2Ax + By + D)^2 - 2Sy = T$$

If $S = 0$ the equation is easy to solve. Assume $S \neq 0$. For every solution z of $u^2 \equiv T \pmod{2S}$ we have a solution

$$\begin{aligned} u &= z - 2SK \\ y &= \frac{z^2 - T}{2S} - 2zK + 2SK^2 \end{aligned}$$

of $u^2 - 2Sy = T$, and conversely. If $u = 2Ax + By + D$ then

$$\begin{aligned} 2Ax &= z - 2SK - B(z^2 - T)/2S + 2BzK - 2BSK^2 - D \\ &= z - D - B(z^2 - T)/2S + 2(Bz - S)K - 2BSK^2 \end{aligned}$$

Since $B^2 = 4AC$, $2A$ is a factor of $2BS = 2B(BD - 2AE)$. Thus, for a given solution z of the congruence equation, to get a solution to the original equation, it is necessary and sufficient to have

$$qK \equiv -p \pmod{2A}$$

where $q = 2Bz - 2S$ and $p = z - D - B(z^2 - T)/2S$. If this congruence has no solution (K is the unknown), then there is no solution to the original equation for the z in question. Suppose, however, that it has solution $L \pmod{M}$, where $M = 2A/(2A, q)$ (Theorem 3.1.1, 3.1.2). Then K is restricted to the form $MK' + L$ (for that z). The solution to the original equation (for that z) is thus

$$\begin{aligned} x &= \frac{p + qL - 2BSL^2 + (qM - 4BSLM)K' - 2BSM^2K'^2}{2A} \\ y &= \frac{z^2 - T}{2S} - 2zL + 2SL^2 - 2M(z - 2SL)K' + 2SM^2K'^2 \end{aligned}$$

Exercises 3.12

1. Solve the Mrs. Ball problem from section 3.11.
2. Solve $3x^2 + 5x + 7y = 1$.
3. Solve $45x^2 - 30xy + 5y^2 - 7x + 2y = 2$.
4. Solve Chrystal's equation: $9x^2 - 12xy + 4y^2 + 3x + 2y = 12$.
5. Show that if $A \neq 0$ and $B^2 = 4AC$ then

$$Ax^2 + Bxy + Cy^2 + Dx + Ey = F$$

is a parabola iff $BD \neq 2AE$.

Chapter 4

$x^2 - Ry^2 = C$

In the first three chapters, we presented the Number Theory of Fermat, Lagrange, and Gauss (respectively). In this chapter, we present a new solution of the Diophantine equation $x^2 - Ry^2 = C$, and we present a new solution to a puzzle proposed by Edouard Lucas in 1875. We also establish Lucas's test for perfect numbers, and, finally, look at some recent work of Alan Baker.

4.1 SCF Solution

Throughout this chapter, R is a positive nonsquare integer, and P_1 and Q_1 are integers such that $P_1^2 \equiv R \pmod{Q_1}$. (If R were square, the equation $x^2 - Ry^2 = C$ could be solved simply by factoring $x^2 - Ry^2$.)

If $C = 0$ then, since we are assuming R is nonsquare, $x^2 - Ry^2 = C$ has only one solution: $x = 0$ and $y = 0$.

If x and y have gcd f then f^2 factors x^2 and Ry^2 and hence C. Thus the solution (x, y) of $x^2 - Ry^2 = C$ can be derived by multiplying by f the corresponding relatively prime solution of $x^2 - Ry^2 = C/f^2$. Hence, without loss of generality, we shall take it that $(x, y) = 1$. Such solutions are *primitive*. We shall also take it that x and y are both positive.

If, as we are assuming, x and y are relatively prime, so are y and C (from the equation) and hence (by Theorem 3.1.2) there is a unique integer z such that $x \equiv yz \pmod{C}$ and $-|C|/2 < z \leq |C|/2$. We say that the solution (x, y) *belongs to* z. Note that

$$Ry^2 \equiv Ry^2 + C = x^2 \equiv y^2 z^2 \pmod{C}$$

and hence $z^2 \equiv R \pmod{C}$ (Theorem 3.1.1). To solve $x^2 - Ry^2 = C$, then, it suffices to find all positive, primitive solutions (if any) belonging to each integer z such that $z^2 \equiv R \pmod{C}$ and $-|C|/2 < z \le |C|/2$.

At the beginning of Chapter 2 we noted that Albert Beiler thinks that, when $C > \sqrt{R}$, the equation $x^2 - Ry^2 = C$ is a 'goblin' and a 'monster', and he refers his reader to Chrystal's *Algebra,* where the 'dauntless mathematical Siegfried' will find 'the fragments to forge into a sword to attack this monster'. Many authors do treat the equation in terms of two cases, one in which $C < \sqrt{R}$ and one in which it is not, and they do, indeed, make the second case seem rather terrifying. What we shall do is to treat both cases together, and in a manner that is no more difficult than that used for the first case. Our key theorem is the following.

Theorem 4.1.1 (Siegfried's Sword) *Let R be a positive nonsquare integer, and C a nonzero integer.*
Let z be an integer such that $z^2 \equiv R \pmod{C}$ and $-|C|/2 < z \le |C|/2$.
Let f_m/g_m be the m-th convergent of $\dfrac{-z+\sqrt{R}}{C}$.
Let $\dfrac{P_m+\sqrt{R}}{Q_m}$ *be the m-th complete quotient of* $\dfrac{-z+\sqrt{R}}{C}$.
Then (x, y) is a positive, primitive solution of $x^2 - Ry^2 = C$ belonging to z
iff, for some positive integer n,
$P_{2n+1} \ge 0$, $Q_{2n+1} = 1$, $x = g_{2n}P_{2n+1} + g_{2n-1}$, *and* $y = g_{2n}$.

Proof: First assume the right hand side of the equivalence. Since g_{2n} and g_{2n-1} are positive integers, x and y are positive. Since $(g_{2n-1}, g_{2n}) = 1$, $(x, y) = 1$. By Theorem 2.9.2, $x = Q_1 f_{2n} - P_1 g_{2n}$ and hence, by Theorem 2.9.1,

$$x^2 - Ry^2 = (-1)^{2n+1-1}C = C$$

Finally, $x = Cf_{2n} + zy \equiv yz \pmod{C}$ so that (x, y) does belong to z.

Suppose now that (x, y) is a positive, primitive solution of $x^2 - Ry^2 = C$ belonging to z. Let $t = (x - yz)/C$ so that $x = Ct + zy$. Since

(x, y) belongs to z, t is an integer. Since $x = Ct + zy$ and $\gcd(x, y) = 1$, it follows that $\gcd(t, y) = 1$. We have two cases to consider.
Case 1. $x/y \geq 1$. Then

$$\begin{aligned} \frac{t}{y} - \frac{-z+\sqrt{R}}{C} &= \frac{Ct + zy - \sqrt{R}y}{Cy} \\ &= \frac{x - \sqrt{R}y}{(x^2 - Ry^2)y} \\ &= \frac{1}{(x/y + \sqrt{R})y^2} < \frac{1}{2y^2} \end{aligned}$$

since $R \geq 2$. By Theorem 2.6.2, t/y is a convergent of $(-z + \sqrt{R})/C$. Moreover, it is an even numbered convergent f_{2n}/g_{2n} since they are the even convergents that are greater than the irrational number (Theorem 2.1.11). Also $y = g_{2n}$. By Theorem 2.9.1,

$$Q_{2n+1}C = (Ct + zy)^2 - Ry^2 = x^2 - Ry^2 = C$$

and thus $Q_{2n+1} = 1$. By Theorem 2.9.2, $x = g_{2n}P_{2n+1} + g_{2n-1}$. Since $g_1, g_2, g_3, \ldots$ is an ascending sequence of positive integers, and since x is positive, it follows that $P_{2n+1} \geq 0$.
Case 2. $x/y < 1$. Let $t/y = (a_1, \ldots, a_{2n})$, with an even number of partial quotients. Let $f/g = (a_1, \ldots, a_{2n-1})$ with $g > 0$, and $\gcd(f, g) = 1$. By Plato's Theorem, $tg - fy = 1$, and, by Theorem 2.1.3,

$$(a_1, \ldots, a_{2n}, \sqrt{R}) = \frac{t\sqrt{R} + f}{y\sqrt{R} + g}$$

Let $t' = (Ry - xz)/C$. Since $z^2 \equiv R \pmod{C}$, t' is an integer. Also

$$tx - t'y = 1 = tg - fy$$

and hence $g \equiv x \pmod{y}$ (since $\gcd(t, y) = 1$). Since $1 \leq g \leq y$ and $1 \leq x < y$, it follows that $g = x$, and hence $t' = f$. Hence

$$(a_1, \ldots, a_{2n}, \sqrt{R}) = \frac{t\sqrt{R} + t'}{y\sqrt{R} + x}$$

$$
\begin{aligned}
&= (t\sqrt{R}+t')\frac{x-\sqrt{R}y}{C} \\
&= \frac{t'x - Rty + \sqrt{R}}{C} \\
&= \frac{-z+\sqrt{R}}{C}
\end{aligned}
$$

so that t/y is the $2n$-th convergent of $(-z+\sqrt{R})/C$. Furthermore, since

$$\frac{-z+\sqrt{R}}{C} = (a_1, \ldots, a_{2n}, \sqrt{R})$$

it follows that $\sqrt{R}$ is the $2n+1$-st complete quotient of $(-z+\sqrt{R})/C$, so that $P_{2n+1} = 0$, and $Q_{2n+1} = 1$. Also $y = g_{2n}$, and $x = Ct + zy = g_{2n}P_{2n+1} + g_{2n-1} = g_{2n-1}$ (Theorem 2.9.2).

Corollary: Let n be the least positive integer (if any) such that $P_{2n+1} \geq 0$ and $Q_{2n+1} = 1$. Then $x = g_{2n}P_{2n+1} + g_{2n-1}$ and $y = g_{2n}$ is the least positive, primitive solution of $x^2 - Ry^2 = C$ belonging to z.

As an example, the solution $x = 19$ and $y = 5$ of $x^2 - 7y^2 = 186$ belongs to 41. In the PQ sequence for $(-41+\sqrt{7})/186$, $Q_5 = 1$ and $g_4 = 5$. Indeed, we have the following table.

P	−41	−145	32	−5	3	2	1	1	2
Q	186	−113	9	−2	1	3	2	3	1
	−1	1	3	1	5	1	1	1	
g_n	1	1	4	5	29	34	63	97	

Here $Q_5 = 1$, $g_4 = 5$, and $g_4P_5 + g_3 = 5 \times 3 + 4 = 19$, as required.

According to Theorem 4.1.1, every primitive nonnegative solution of $x^2 - 7y^2 = 186$ belonging to 41 can be obtained from the PQ sequence for $(-41+\sqrt{7})/186$. The solution next higher than $x = 19$ and $y = 5$ is

$$
\begin{aligned}
x &= g_8P_9 + g_7 = 97 \times 2 + 63 = 257 \\
y &= g_8 = 97
\end{aligned}
$$

It is not always the case that if $z^2 \equiv R \pmod{C}$ and $-|C|/2 < z \leq |C|/2$ then $x^2 - Ry^2 = C$ has a primitive solution belonging to z. For example, although $2^2 \equiv 44 \pmod 4$, the equation $x^2 - 44y^2 = 4$ has no primitive solution belonging to 2. This we may conclude from Theorem 4.1.1, and the following PQ sequence for $(-2 + \sqrt{44})/4$.

$$\begin{array}{ccccc} P & -2 & 6 & 6 & 6 \\ Q & 4 & 2 & 4 & 2 \\ & 1 & 6 & 3 & 6 \end{array}$$

(Clearly this expansion contains no $Q_{2n+1} = 1$.) The equation $x^2 - 44y^2 = 4$ does, however, have primitive solutions belonging to 0 (for example, $x = 20$ and $y = 3$) and it also has nonprimitive solutions (for example, $x = 2$ and $y = 0$).

Thanks to Theorem 4.1.1, we are in a position to solve any Diophantine equation of the form $x^2 - Ry^2 = C$. Siegfried's Sword has been forged.

We close this section with a theorem about *positive* solutions belonging to an integer z.

If (u, v) is a solution of $x^2 - Ry^2 = C$ belonging to z, then $(-u, v)$ is a solution of $x^2 - Ry^2 = C$ belonging to $-z$. Moreover, we have

Theorem 4.1.2 *If $x^2 - Ry^2 = C$ has a primitive solution belonging to z then it has a positive primitive solution belonging to z.*

Proof: Suppose (u, v) is a primitive solution belonging to z. Then so is $(-u, -v)$. If neither of these is positive, then $uv \leq 0$. Suppose this is so.

There are infinitely many positive integer solutions of $x^2 - Ry^2 = 1$ (Theorem 2.9.4). Let (s, t) be a solution of $x^2 - Ry^2 = 1$ such that $st > -uv/|C|$.

Let $f = Rvt + us$ and $g = ut + vs$. Then we have

$$\begin{aligned} s^2t^2 &> u^2v^2/C^2 \\ s^2t^2(u^2 - Rv^2)^2 &> u^2v^2 \\ s^2t^2(u^2 + Rv^2)^2 &> u^2v^2 + 4Rs^2t^2u^2v^2 \end{aligned}$$

$$\begin{aligned}
s^2t^2(u^2+Rv^2)^2 &> u^2v^2(1+4Rt^2(Rt^2+1)) \\
s^2t^2(u^2+Rv^2)^2 &> u^2v^2(1+2Rt^2)^2 \\
s^2t^2(u^2+Rv^2)^2 &> u^2v^2(s^2+Rt^2)^2 \\
st(u^2+Rv^2) &> -uv(Rt^2+s^2) \\
Ruvt^2+uvs^2+u^2st+stRv^2 &> 0 \\
(Rvt+us)(ut+vs) &> 0 \\
fg &> 0
\end{aligned}$$

Since $sf - Rtg = u$, and $sg - tf = v$, and since $\gcd(u,v) = 1$, it follows that $\gcd(f,g) = 1$.

By straight calculation, $f^2 - Rg^2 = C$, and $f \equiv gz \pmod{C}$.

Thus one of (e,f) and $(-e,-f)$ is a positive primitive solution belonging to z.

In Section 6 of this chapter, we shall give another solution of $x^2 - Ry^2 = C$, one that is simpler but much more time-consuming.

Exercises 4.1

1. Find 11 consecutive positive integers the sum of whose squares is the square of an integer.
2. Solve $x^2 - 61y^2 = 75$.
3.'Give me advice,' demanded Sheik Noshack. 'I want to be able to arrange my rubies in a square.' One of the servants suggested that the Sheik buy 49 more rubies. 'What!' roared the Sheik, 'as if I could not afford to double my collection to gratify my desire! What are a few hundred rubies to me!' How many rubies did the Sheik in fact have?
4. Prove that every prime of the form $8m \pm 1$ can be written in the form $x^2 - 2y^2$.
5. If x and y are positive integers such that $x^2 - 5y^2 = \pm 4$ then y is a Fibonacci number. Indeed all Fibonacci numbers are found in this way.

4.2 Recursive Formulas for Solutions

The solutions of the Diophantine equation $x^2 - Ry^2 = C$ are linked up in interesting ways that provide a key to Lucas's square pyramid puzzle. Recall that one of the objects of this book is to give a completely elementary proof of the fact that if a square number of cannon-balls are stacked in a square-base pyramid, then there are exactly 4900 of them. The results in this section will help us do just that.

Let R be a positive nonsquare integer, and let C be a nonzero integer. Let $(a_1, b_1) = (a, b)$ be the least positive solution of $x^2 - Ry^2 = 1$. Let the other solutions, in ascending order, be

$$(a_2, b_2), \quad (a_3, b_3), \quad \ldots$$

(Note that x and y increase together.) Let z be an integer such that $z^2 \equiv R \pmod{C}$ and $-|C|/2 < z \leq |C|/2$. Let (u_0, v_0) be the least nonnegative primitive solution of $x^2 - Ry^2 = C$ belonging to z (if such solutions exist). Let the other such solutions, in ascending order, be

$$(u_1, v_1), \quad (u_2, v_2), \quad \ldots$$

Then we have

Theorem 4.2.1 *$u_{m+1} = au_m + bRv_m$ and $v_{m+1} = bu_m + av_m$.*

Proof: Let $s = au_m + bRv_m$ and $t = bu_m + av_m$. A brief calculation shows that $s^2 - Rt^2 = C$. Moreover, since $as - bRt = u_m$ and $at - bs = v_m$, it follows that s and t are relatively prime (since u_m and v_m are). Also

$$s \equiv av_m z + bv_m z^2 \equiv tz \pmod{C}$$

Thus (s, t) is a primitive solution of $x^2 - Ry^2 = C$ belonging to z and (s, t) is greater than (u_m, v_m).

It suffices to show that $v_{m+1} \geq t$. Let

$$e = (u_m u_{m+1} - Rv_m v_{m+1})/C$$

$$f = (u_m v_{m+1} - v_m u_{m+1})/C$$

Since $u_m \equiv v_m z$ and $u_{m+1} \equiv v_{m+1} z \pmod{C}$, it follows that

$$u_m u_{m+1} \equiv v_m z v_{m+1} z \equiv Rv_m v_{m+1} \pmod{C}$$

$$u_m v_{m+1} \equiv v_m z v_{m+1} \equiv v_m u_{m+1} \pmod{C}$$

so that both e and f are integers. Furthermore, $e^2 - Rf^2 = 1$.

Now $v_m = -fu_{m+1} + ev_{m+1}$ and $v_{m+1} = fu_m + ev_m$. Since $v_m \geq 0$, we cannot have $f \geq 0$ and $e < 0$. Since $v_{m+1} \geq 0$, we cannot have $f < 0$ and $e < 0$. So $e > 0$. Hence $e \geq a$.

Since $v_m < v_{m+1}$, the fact that $v_m = -fu_{m+1} + ev_{m+1}$ implies that $f > 0$, and hence $f \geq b$. Hence

$$v_{m+1} = fu_m + ev_m \geq bu_m + av_m = t$$

Corollary: $u_m + v_m\sqrt{R} = (u_0 + v_0\sqrt{R})(a + b\sqrt{R})^m$ and, in particular, $a_n + b_n\sqrt{R} = (a + b\sqrt{R})^n$.

The corollary follows from the theorem by mathematical induction.

For example, $(1, 0)$ is the least nonnegative primitive solution of $x^2 - 2y^2 = 1$, and $(3, 2)$ is the least positive solution of $x^2 - 2y^2 = 1$. Thus $u_1 = 3 \times 1 + 2 \times 2 \times 0 = 3$ and $v_1 = 2 \times 1 + 3 \times 0 = 2$. Similarly,

$$u_2 = 3 \times 3 + 2 \times 2 \times 2 = 17$$

$$v_3 = 2 \times 3 + 3 \times 2 = 12$$

We now generalise Theorem 4.2.1.

Theorem 4.2.2 *$u_{m+n} = a_n u_m + b_n R v_m$ and $v_{m+n} = b_n u_m + a_n v_m$.*

Proof: This follows straightforwardly from the above corollary.

Corollary: $a_{2n} = 2a_n^2 - 1 = 2Rb_n^2 + 1$, and $b_{2n} = 2a_n b_n$.

Corollary: $u_{m+2} = 2au_{m+1} - u_m$ and $v_{m+2} = 2av_{m+1} - v_m$.

For example, when $R = 2$, $u_2 = 17 = 2 \times 3 \times 3 - 1$. Clearly, it is now easy to calculate large solutions of $x^2 - 2y^2 = 1$.

From the corollary to Theorem 4.2.1, we also have

$$u_{m-n} + v_{m-n}\sqrt{R} = (u_m + v_m\sqrt{R})(a_n - b_n\sqrt{R})$$

and this gives the following theorem.

Theorem 4.2.3 *Where m and n are nonnegative integers, and $m-n$ is also nonnegative,*

$$u_{m-n} = a_n u_m - b_n R v_m$$

$$v_{m-n} = -b_n u_m + a_n v_m$$

The next theorem is a rather queer technical result which we shall use in our solution of the square pyramid problem. It links some of the larger solutions of the Pell equation to $x = a_2$.

Theorem 4.2.4 *If m is a positive integer, and r is an odd positive integer, $a_{2rm\pm 2} \equiv -a_2 \pmod{a_m}$.*

Proof: The proof is by mathematical induction on the positive odd integers r. When $r = 1$, we have

$$a_{2rm\pm 2} = a_2 a_{2m} \pm b_2 R b_{2m}$$

(Theorem 4.2.2, 4.2.3). By the first corollary to Theorem 4.2.2, $b_{2m} \equiv 0 \pmod{a_m}$ and $a_{2m} \equiv -1 \pmod{a_m}$. Hence the theorem is true for $r = 1$.

Suppose it true for some odd r. Then

$$a_{2(r+2)m\pm 2} = a_{2rm\pm 2} a_{4m} + b_{2rm\pm 2} R b_{4m}$$

$$\equiv -a_2 a_{4m} \equiv -a_2(2a_{2m}^2 - 1) \equiv -a_2 \pmod{a_m}$$

The material is this section is due to E. Lucas (1842–1891), the French schoolteacher who, in an 1885 Prize Day speech, encouraged the schoolchildren to attack Germany. It is interesting that the first solution to Lucas's Square Pyramid Problem was given only in 1918, the year France defeated Germany, in World War I. The author of this solution was G. N. Watson, a British mathematician.

Exercises 4.2

1. Find the 4 smallest nonnegative solutions of $x^2 - 38y^2 = 1$.
2. Let $A = a + b\sqrt{R}$ and let A' be its conjugate. Let $U = u_0 + v_0\sqrt{R}$ and let U' be its conjugate. Then

$$u_m = \frac{UA^m + U'A'^m}{2}$$

$$v_m = \frac{UA^m - U'A'^m}{2\sqrt{R}}$$

3. Show that a_m is the integer nearest $\frac{1}{2}(a + b\sqrt{R})^m$.
4. Prove that $b_m | b_n \iff m|n$.
5. Find a formula for all triangular numbers which are square.
6. When is the product of 3 consecutive triangular numbers a square?
7. Prove that $b_m^2 | b_n \iff mb_m|n$.

4.3 $Ax^2 + Bxy + Cy^2 + Dx + Ey = F$ *

Let A, B, C, D, E, and F be any integers. In this section we generalise the solution of $x^2 - Ry^2 = C$ to cover the Diophantine equation

$$Ax^2 + Bxy + Cy^2 + Dx + Ey = F$$

where $A \neq 0$ and $R = B^2 - 4AC$ is a positive nonsquare integer.

Let $S = BD - 2AE$ and $T = 4AF + D^2$. By the Conic Transformation Theorem (Theorem 1.7.1), the Diophantine equation is equivalent to

$$(Ry + S)^2 - R(2Ax + By + D)^2 = S^2 - RT$$

If $D = E = 0$, we can use the simpler equivalent equation

$$(2Ax + By)^2 - Ry^2 = 4AF$$

A solution (u, v) of $x^2 - Ry^2 = S^2 - RT$ is *basic* iff there is a positive integer f such that f^2 is a factor of $S^2 - RT$, and there is some integer

z such that $z^2 \equiv R \pmod{\frac{S^2-RT}{f^2}}$, with

$$-\frac{1}{2}\left|\frac{S^2-RT}{f^2}\right| < z \le \frac{1}{2}\left|\frac{S^2-RT}{f^2}\right|$$

and $(u,v) = (fu_0, fv_0)$ where (u_0, v_0) is the least nonnegative primitive solution of $x^2 - Ry^2 = (S^2 - RT)/f^2$ belonging to z. By Theorem 4.4.2 (with $m = 0$), the set of all solutions of $x^2 - Ry^2 = S^2 - RT$ is thus the set of all pairs

$$(\pm(a_n u + b_n vR),\ \ \pm(b_n u + a_n v))$$

— the signs are not linked — such that (u,v) is a basic solution.

If $Ry + S = \pm(a_n u + b_n vR)$ then $y = (\pm(a_n u + b_n vR) - S)/R$ and this is an integer iff $\pm a_n u \equiv S \pmod{R}$.

If $2Ax + By + D = \pm(b_n u + a_n v)$ then

$$x = \frac{\pm(b_n u + a_n v)R - B(\pm(a_n u + b_n vR) - S) - DR}{2AR}$$

and this is an integer iff

$$\pm(b_n u + a_n v)R \equiv B(\pm(a_n u + b_n vR) - S) + DR \pmod{2AR}$$

The following theorem shows that one can easily determine those n for which both the above congruences hold.

Theorem 4.3.1 *The sequence* $(a_0, b_0), (a_1, b_1), (a_2, b_2), \ldots$ *(mod d) is purely periodic.*

Proof: Since there are only d^2 choices for $(a_n, b_n) \pmod{d}$, the sequence eventually repeats. Say $a_q \equiv a_p$ and $b_q \equiv b_p \pmod{d}$ where $q > p > 0$. Since

$$a_{n+1} = aa_n + bb_nR \text{ and } b_{n+1} = ba_n + ab_n$$

(Theorem 4.2.2), it follows that $a_{q+1} \equiv a_{p+1} \pmod{d}$ and $b_{q+1} \equiv b_{p+1} \pmod{d}$. Thus, by Theorem 4.2.2,

$$a_{q-1} = 2aa_q - a_{q+1} \equiv a_{p-1}$$

$$b_{q-1} = 2ab_q - b_{q+1} \equiv b_{p-1}$$

(mod d). Hence the sequence repeats from the beginning.

If L is the length of the period of

$$(a_0, b_0), (a_1, b_1), (a_2, b_2), \ldots \pmod{2AR}$$

then with L trials we can discover those n (mod L) for which x and y are integers.

Note that the double signs require that four separate cases be treated.

As an example, we take an equation given by Gauss (1777–1855) in his *Disquisitiones Arithmeticae:*

$$x^2 + 8xy + y^2 + 2x - 4y = -1$$

Here $R = 60$, $S = 24$, $T = 0$, and $S^2 - RT = 24^2$. There are 2 basic solutions of $x^2 - 60y^2 = 24^2$, namely, $(96, 12)$ with $f = 12$, and $(24, 0)$ with $f = 24$.

In the first case, we need $\pm a_n \times 96 \equiv 24 \pmod{60}$ and

$$\pm(96b_n + 12a_n) \times 60 \equiv 8(\pm(96a_n + 12 \times 60b_n) - 24) + 2 \times 60 \pmod{120}$$

These two conditions simplify to $\pm a_n \equiv 4 \pmod 5$. The first few solutions of $x^2 - 60y^2 = 1$ are

$$(1, 0),\ (31, 4),\ (1921, 248),\ \ldots$$

which, modulo 5, are

$$(1, 0),\ (1, 4),\ (1, 3),\ (1, 2),\ (1, 1),\ (1, 0),\ \ldots$$

Hence it is necessary and sufficient to take the minus sign in $\pm a_n \equiv 4 \pmod 5$. Thus one class of solutions to the original equation is

$$x = \pm(48b_n + 6a_n) + \frac{32a_n + 8}{5} + 48b_n - 1$$

$$y = \frac{-8a_n - 2}{5} - 12b_n$$

With $n = 0$ we get $x = 13$ or 1, and $y = -2$. With $n = 1$, we get $x = 769$ or 13 and $y = -98$.

In the second case, we need $\pm a_n \times 24 \equiv 24 \pmod{60}$ and

$$\pm b_n \times 24 \times 60 \equiv 8(\pm 24a_n - 24) + 2 \times 60 \pmod{120}$$

These two conditions simplify to $\pm a_n \equiv 1 \pmod 5$. Thus this time we take the plus sign. Hence the second class of solution to Gauss's equation is

$$x = \pm 12b_n - \frac{8a_n - 8}{5} - 1$$

$$y = \frac{2a_n - 2}{5}$$

With $n = 0$ we get $x = -1$ and $y = 0$. With $n = 1$, we get $x = -1$ or -97, and $y = 12$.

Exercises 4.3

1. Find all integer solutions of $3x^2 + 5xy + y^2 - 5x - 10y = 2$ which contain fewer than 10 digits.
2. Solve $5x^2 - 14xy + 7y^2 = -1$.
3. Consider $Ax^2 + Bxy + Cy^2 + Dx + Ey = F$ (with $A \geq 0$). Let

$$L = BDE - 4ACF - AE^2 - CD^2 + FB^2$$

Show that if $B^2 > 4AC$ and $L \neq 0$ then the original equation represents a hyperbola. This was first proved by Descartes, in an Appendix to a philosophy book.

4.4 Square Pyramid Problem

As we noted in Section 1.1, Edouard Lucas challenged the readers of the *Nouvelles Annales de Mathématiques* to prove that

> A square pyramid of cannon-balls contains a square number of cannon-balls only when it has 24 cannon-balls along its base.

In other words, the only nontrivial solution of

$$1^2 + 2^2 + 3^2 + \cdots + x^2 = y^2$$

is $x = 24$ and $y = 70$.

For over a hundred years, no one found a simple, elementary proof of this fact. Then, in 1988, W. S. Anglin simplified a proof given by D. G. Ma, and produced the proof given in this section.

We begin by considering the following table for the solutions $x = a_n$, and $y = b_n$ of the Pell equation $x^2 - 3y^2 = 1$.

n	0	1	2	3	4	5	6	7
a_n	1	2	7	26	97	362	1351	5042
$a_n \pmod 5$	1	2	2	1	2	2	1	2
$a_n \pmod 8$	1	2	-1	2	1	2	-1	2

By Theorem 4.2.2, the 'mod rows' are periodic. From the Law of Quadratic Reciprocity (Theorem 3.10.3), and the row for $a_n \pmod 5$, we obtain the following.

Theorem 4.4.1 *Suppose n is even. Then* $\gcd(a_n, 10) = 1$. *Also*

$$\left(\frac{5}{a_n}\right) = 1 \text{ iff } 3|n$$

By Exercise 3.10, number 1, and the row for $a_n \pmod 8$, we also obtain

Theorem 4.4.2 *Suppose n is even. Then* $\gcd(a_n, 2) = 1$. *Also*

$$\left(\frac{-2}{a_n}\right) = 1 \text{ iff } 4|n$$

The key lemma is the following.

Theorem 4.4.3 *The solution $x = a_n$ of $x^2 - 3y^2 = 1$ has the form $m^2 + 3$ only when $n = 2$ (and $a_n = 7$).*

Proof: Suppose $a_n = m^2 + 3$ for some integer m and suppose $n > 2$. Then $a_n \equiv 3$, 4 or $-1 \pmod 8$ and, from the above table, n has the form $8k \pm 2$. Since $n > 2$, we can write n in the form $2r2^s \pm 2$ where r is odd and s is an integer ≥ 2. By Theorem 4.2.4,

$$a_n = a_{2r2^s \pm 2} \equiv -a_2 \equiv -7 \pmod{a_{2^s}}$$

From this it follows that

$$m^2 = a_n - 3 = -7 - 3 \equiv -10 \pmod{a_{2^s}}$$

and hence

$$\left(\frac{-2}{a_{2^s}}\right)\left(\frac{5}{a_{2^s}}\right) = \left(\frac{-10}{a_{2^s}}\right) = 1$$

by Theorem 3.10.1. By Theorem 4.4.2 and the fact that $s \geq 2$, it follows that the first factor is 1. By Theorem 4.4.1, it follows that the second factor is -1. Contradiction. Thus $a_n \neq m^2 + 3$.

Theorem 4.4.4 (The Square Pyramid Theorem) *If* $1^2+2^2+3^2+\cdots+x^2 = y^2$ *with* x *an integer* > 1, *then* $x = 24$ *and* $y = 70$.

Proof: The equation is equivalent to $x(x+1)(2x+1) = 6y^2$.

We divide the proof into two cases, the first with x odd, which we handle using the previous theorem, and the second with x even, which we handle using some theorems we proved in Chapter 1, using classical divisibility theory.

Suppose that x is odd. Then, since x, $x+1$, and $2x+1$ are pairwise relatively prime, x is either a square or a triple of a square, and hence x is not congruent to 2, mod 3. Moreover, $x+1$ is either double a square or six times a square, and hence $x+1$ is not congruent to 1, mod 3. Thus $x \equiv 1 \pmod 3$ and thus $x+1 \equiv 2 \pmod 3$, and, finally, $2x+1 \equiv 0 \pmod 3$. Hence, for some nonnegative integers u, v, and w, we have

$$\begin{aligned} x &= u^2 \\ x+1 &= 2v^2 \\ 2x+1 &= 3w^2 \end{aligned}$$

From this we have $6w^2 + 1 = 4x + 3 = (2u)^2 + 3$, which is a number of the form $m^2 + 3$. Also

$$(6w^2 + 1)^2 - 3(4vw)^2 = 12w^2(3w^2 + 1 - 4v^2) + 1 = 1$$

Hence, by Theorem 4.4.3, $6w^2 + 1 = 7$. Thus $w = 1$ and $x = 1$.

Suppose now that x is even. Then $x + 1$ is odd and is a square, or a triple of a square. Thus $x + 1$ is not congruent to 2, mod 3. Similarly, $2x + 1$ is not congruent to 2, mod 3. Hence x is congruent to 0, mod 3, and, for some nonnegative integers p, q, r, we have

$$\begin{aligned} x &= 6q^2 \\ x + 1 &= p^2 \\ 2x + 1 &= r^2 \end{aligned}$$

Now

$$6q^2 = (2x + 1) - (x + 1) = (r - p)(r + p)$$

Since p and r are both odd, q is even. Say $q = 2q'$. This gives

$$6q'^2 = \frac{r - p}{2} \frac{r + p}{2}$$

and, since $\frac{r-p}{2}$ and $\frac{r+p}{2}$ are relatively prime, we obtain one of the following cases.

Case 1. For some nonnegative integers A and B, $\frac{r-p}{2} = 3A^2$ and $\frac{r+p}{2} = 2B^2$, or vice versa. Then $p = \pm(3A^2 - 2B^2)$ and $q = 2q' = 2AB$. Since $6q^2 + 1 = x + 1 = p^2$, this gives

$$24A^2B^2 + 1 = (3A^2 - 2B^2)^2$$

and hence $(3A^2 - 6B^2)^2 - 2(2B)^4 = 1$. By Theorem 1.7.2, $B = 0$, and hence $x = 6q^2 = 6(2AB)^2 = 0$.

Case 2. Here $\frac{r-p}{2} = 6A^2$ and $\frac{r+p}{2} = B^2$, or vice versa. Then $p = \pm(6A^2 - B^2)$ and $q = 2AB$. Since $6q^2 + 1 = p^2$, we have

$$24A^2B^2 + 1 = (6A^2 - B^2)^2$$

and hence $(6A^2 - 3B^2)^2 - 8B^4 = 1$. By Theorem 1.7.4, $B = 0$ or 1. Thus $x = 6q^2 = 6(2AB)^2 = 0$ or 24.

We may conclude that if a square number of cannon-balls are stacked in a square pyramid then there are exactly 4900 of them.

What if the square number of cannon-balls are stacked in a pyramid whose base is, not a square, but an equilateral triangle of side x ? In such a pyramid, the n-th level of cannon-balls contains $n(n+1)/2$ cannon-balls. Thus the whole pyramid contains

$$1 + 3 + 6 + 10 + \cdots + x(x+1)/2$$

cannon-balls. The question we now wish to answer is: when is this sum a square?

Theorem 4.4.5 (The Tetrahedron Theorem) *If* $1 + 3 + 6 + 10 + \cdots + x(x+1)/2 = y^2$, *with x an integer* > 2, *then* $x = 48$ *and* $y = 140$.

Proof: The equation is equivalent to $x(x+1)(x+2) = 6y^2$.

First suppose x is even. Then $x+2$ is even, and $6y^2$ is divisible by 4. Hence y is even. Let $x = 2x'$ and $y = 2y'$. Then the equation is equivalent to $x'(x'+1)(2x'+1) = 6y'^2$. By the previous theorem, $x' = 0$, 1, or 24 — so that $x = 48$.

Second suppose x is odd. Then x is a square or a triple of a square. Hence x is not congruent to 2, mod 3. Similarly, $x+2$ is not congruent to 2, mod 3. Hence $x \equiv 1 \pmod 3$. Thus $x = 6m+1$. Also there are positive integers a, b, and c such that

$$\begin{aligned} 6m+1 &= a^2 \\ 6m+2 &= 2b^2 \\ 6m+3 &= 3c^2 \end{aligned}$$

Since a is odd, and $\gcd(a,b) = 1$, it follows that $2b - a$ and $2b + a$ are relatively prime. Since

$$3c^2 = 4b^2 - a^2 = (2b-a)(2b+a)$$

there are odd positive integers d and e such that

$$\begin{aligned} 2b \pm a &= 3d^2 \\ 2b \mp a &= e^2 \\ c &= de \end{aligned}$$

Thus

$$\begin{aligned} 9(d^2 - e^2)^2 &= (3d^2 - e^2)^2 - 12d^2e^2 + 8e^4 \\ &= (2a)^2 - 12c^2 + 8e^4 \\ &= -8 + 8e^4 \\ &= 64\frac{e^2 - 1}{4}\frac{e^2 + 1}{2} \end{aligned}$$

Since $\frac{e^2-1}{4}$ and $\frac{e^2+1}{2}$ are relatively prime integers, it follows that $\frac{e^2-1}{4}$ is a square. This means that $e^2 - 1$ is a square, and hence $e = 1$. Hence $d = e = 1$, and we get $c = 1$, $m = 0$, and $x = 1$.

Thus if a square number of cannon-balls are stacked in a tetrahedron, there are 19,600 of them.

Exercises 4.4

1. Show that $1^5 + 2^5 + \cdots + x^5 = y^2$ iff $x = (3f - 1)/2$ and $y = g(9f^2 - 1)/8$ where f/g is an odd numbered convergent of $\sqrt{6}/3$.

4.5 Lucas's Test for Perfect Numbers *

The SCF expansion of $\sqrt{3}$ is as follows:

P_n	0	1	1	1	1
Q_n	1	2	1	2	1
a_n	1	1	2	1	2
f_n	1	2	5	7	
g_n	1	1	3	4	

— where f_n/g_n is the n-th convergent of $\sqrt{3}$.

The object of this section is to prove that if n is an odd prime, then $2^n - 1$ is prime iff $2^n - 1 | f_{2^{n-1}}$. (Note that $2^n - 1$ is not prime unless n is prime.) As we shall show, it is not hard to calculate $f_{2^{n-1}}$, so this theorem gives us a practical way of finding Mersenne primes, and hence even perfect numbers. Indeed, it is this very theorem that has been used to find the largest known perfect numbers.

If i is even, $x = f_i = a_{i/2}$ gives a solution of the Pell equation $x^2 - 3y^2 = 1$. Since, by Theorem 4.2.2, $a_{2t} = 2a_t^2 - 1$, we have

$$2f_{2^{n+1}} = (2f_{2^n})^2 - 2$$

Thus, defining

$$\begin{aligned} s_1 &= 4 \\ s_{n+1} &= s_n^2 - 2 \end{aligned}$$

we have $s_n = 2f_{2^n}$, and the theorem we are going to prove in this section is equivalent to the statement

$$\boxed{2^n - 1 \text{ is prime iff } (2^n - 1) | s_{n-1}}$$

— assuming n is an odd prime.

To prove this, let $a = 1 + \sqrt{3}$ and $b = 1 - \sqrt{3}$, and, for all positive integers n, define

$$\begin{aligned} u_n &= \frac{a^n - b^n}{a - b} \\ v_n &= a^n + b^n \end{aligned}$$

For example, $u_1 = 1$, $u_2 = 2$, $v_1 = 2$, and $v_2 = 8$.

Theorem 4.5.1 *If n is a positive integer,*

$$v_{2^n} = 2^{2^{n-1}} s_n$$

Proof: This is true when $n = 1$. Suppose it true for n. Then

$$2^{2^n} s_{n+1} = 2^{2^n}(s_n^2 - 2) = \left(2^{2^{n-1}} s_n\right)^2 - 2^{2^n+1} = v_{2^n}^2 - 2^{2^n+1}$$

From the definition of v_n,

$$v_{2k} = v_k^2 + (-2)^{k+1}$$

and hence

$$v_{2^{n+1}} = v_{2^n}^2 - 2^{2^n+1}$$

Thus the theorem is true for $n+1$ and the result follows by mathematical induction.

Theorem 4.5.2 *If q is a prime > 3 then q is a factor of both $u_q - 3^{(q-1)/2}$ and $v_q - 2$.*

Proof: By the definition of u_n, and the Binomial Theorem,

$$u_q = \sum_{k=0}^{(q-1)/2} \binom{q}{2k+1} 3^k$$

Since all the binomial coefficients are divisible by q, except the one with $k = (q-1)/2$, the result for u_n follows. The proof for v_n is similar.

We are already in a position to prove half of our main theorem.

Theorem 4.5.3 *If p and $2^p - 1$ are odd primes then $2^p - 1 | s_{p-1}$.*

Proof: Let $q = 2^p - 1$. Since 3 divides $2^{p-1} - 1$, and $2^{p-1} - 1$ is half of $q - 1$, it follows that 3 divides $q - 1$. Since $q - 1$ has the form $8t + 6$, it follows that q has the form $24k + 7$.

As noted above, it follows from the definition of v_n that

$$v_{2k} = v_k^2 + (-2)^{k+1}$$

and hence

$$v_{2^p} = v_{2^{p-1}}^2 - 4 \times 2^{(q-1)/2}$$

By Theorems 3.8.2 and 3.8.4, $q \mid 2^{(q-1)/2} - 1$, and hence

$$q \mid v_{2^p} - v_{2^{p-1}}^2 + 4 \quad (*)$$

By Theorem 3.8.2, modulo q we have

$$3^{(q-1)/2} \equiv \left(\frac{3}{q}\right) \equiv -\left(\frac{q}{3}\right) \equiv -\left(\frac{7}{3}\right) \equiv -1$$

(using the Law of Quadratic Reciprocity). From Theorem 4.5.2 it now follows that q is a factor of $u_q + 1$. From the definitions of u_n and v_n we have

$$v_{q+1} = v_q + 6u_q$$

and hence

$$v_{2^p} = (v_q - 2) + 6(u_q + 1) - 4$$

Thus, using Theorem 4.5.2 again, q is a factor of $v_{2^p} + 4$, whence by $(*)$, q factors $v_{2^{p-1}}$. By Theorem 4.5.1 it now follows that q (being odd) factors s_{p-1}, as required.

To prove the converse, we need two preliminary theorems. These presuppose a definition: if q is an odd prime, we define $\omega(q)$ as the least natural number n (if there is one) such that $q \mid u_n$.

Theorem 4.5.4 *The odd prime q factors u_n iff $\omega(q)$ factors n.*

Proof: Let S be the set of natural numbers n such that $q \mid u_n$. From the definitions of u_n and v_n we have

$$2u_{k+l} = u_k v_l + v_k u_l$$

$$(-2)^{l+1} u_{k-l} = u_l v_k - u_k v_l$$

Thus if any two natural numbers are in S, so are their sum and (positive) difference. Hence if S is nonempty, it consists of multiples of S's least member d. (If n is in S then $n = qd + t$ with $0 \le t < d$ and hence t is either 0 or a member of S.)

Theorem 4.5.5 *Let q be a prime > 3. Then, if $\omega(q)$ exists, it is less than or equal to $q+1$.*

Proof: From the definitions of u_n and v_n we have

$$2u_{q+1} = 2u_q + v_q$$

$$-4u_{q-1} = 2u_q - v_q$$

whence

$$-8u_{q+1}u_{q-1} = 4u_q^2 - v_q^2$$

By Theorem 4.5.2, q factors $u_q^2 - 3^{q-1}$. By Fermat's Little Theorem, q factors $3^{q-1} - 1$. Thus q factors $u_q^2 - 1$ and also $4u_q^2 - 4$. By Theorem 4.5.2, q factors $v_q^2 - 4$, and hence q factors $4u_q^2 - v_q^2 = -8u_{q+1}u_{q-1}$.

Hence q factors one of u_{q+1} and u_{q-1}, so that, by Theorem 4.5.4, $\omega(q) \leq q+1$.

Finally, we have

Theorem 4.5.6 *Suppose p is an odd prime and $2^p - 1$ divides s_{p-1}. Then $2^p - 1$ is prime.*

Proof: By Theorem 4.5.1, $2^p - 1$ divides $v_{2^{p-1}}$. Now, from the definitions of u_n and v_n we have

$$u_{2k} = u_k v_k$$

and hence $2^p - 1$ divides u_{2^p}.

Let q be any prime divisor of $2^p - 1$. Then $q \neq 3$. Since q is a factor of u_{2^p}, Theorem 4.5.4 implies that $\omega(q) \mid 2^p$.

Moreover, $\omega(q)$ is not a factor of 2^{p-1} lest, by Theorem 4.5.4, q be a factor of $u_{2^{p-1}}$, and hence, from the fact that

$$v_k^2 - 12u_k^2 = (-2)^{k+2}$$

— with $k = 2^{p-1}$ — q be a factor of 2. (Since $2^p - 1$ divides $v_{2^{p-1}}$, so does q.)

Hence $\omega(q) = 2^p$. By Theorem 4.5.5 it now follows that $2^p \leq q+1$, so that $2^p - 1 \leq q$. Since q is a factor of $2^p - 1$, it follows that $2^p - 1 = q$.

But q is prime.

We may therefore conclude that, if p is an odd prime, $2^p - 1$ is prime just in case

$$2^p - 1 \mid s_{p-1}$$

In 1994, D. Slowinski and P. Gage used this result to show that $2^{858433} - 1$ is prime. At the moment (1994), this is the largest known prime number.

Exercises 4.5

1. What is s_4?
2. Use the Lucas-Lehmer theorem to show 8128 is perfect.

4.6 Simultaneous Fermat Equations *

Let R and S be positive nonsquare integers, with $S > R$. Suppose RS is not a square. Let C and D be nonzero integers. Then the following simultaneous Diophantine equations are *simultaneous Fermat equations:*

$$x^2 - Ry^2 = C$$
$$z^2 - Sy^2 = D$$

For example, we might have

$$x^2 - 3y^2 = -2$$

$$z^2 - 8y^2 = -7$$

This system was solved by A. Baker and H. Davenport (see the *Quarterly J. Math.* Oxford (2), 20, 129–37). They showed it has solutions (1, 1, 1), (19, 11, 31), and no others.

The object of this section is to give a practical way of solving such systems when R, S, $|C|$, and $|D|$ are all less than, say, 1000. This practical method is based on a theorem of Michel Waldschmidt, proved pages 257 to 283 of volume 37 of *Acta Arithmetica.* We shall not give the proof here.

Note that if $RS = U^2$ (with U a positive integer), then the above equations imply that

$$(Sx - Uz)(Sx + Uz) = S(SC - RD)$$

and the problem can be solved by factoring. Let us suppose, then, that RS is nonsquare.

Let (x, y, z) be a nonnegative integer solution of

$$x^2 - Ry^2 = C$$

$$z^2 - Sy^2 = D$$

Let $j = \gcd(x, y)$ and $k = \gcd(y, z)$. Then $j^2|C$ and $k^2|D$. Moreover, $(x/j, y/j)$ is a primitive solution of $x^2 - Ry^2 = C/j^2$ and, as such, it belongs to some integer z' with

$$z'^2 \equiv R \pmod{C/j^2} \text{ and } -\frac{|C|}{2j^2} < z' \le \frac{|C|}{2j^2}$$

Similarly, $(z/k, y/k)$ is a primitive solution of $z^2 - Sy^2 = D/k^2$, belonging to some integer z'' with

$$z''^2 \equiv S \pmod{D/k^2} \text{ and } -\frac{|D|}{2k^2} < z'' \le \frac{|D|}{2k^2}$$

Let us say that solution (x, y, z) *belongs to* (j, k, z', z'').

To solve the simultaneous Fermat equations, it suffices to find all possible quadruples (j, k, z', z'') subject to the above conditions, and, for each one, find all pairs of nonnegative integers (X, Y) and (Z, Y') such that (X, Y) is a primitive solution of $x^2 - Ry^2 = C/j^2$ belonging to z', and (Z, Y') is a primitive solution of $z^2 - Sy^2 = D/k2$ belonging to z'', and $Yj = Y'k$. A typical solution to the simultaneous Fermat equations is (Xj, Yj, Zk). For what follows, we fix a particular quadruple (j, k, z', z'').

Let (a, b) be the least positive integer solution of $x^2 - Ry^2 = 1$. To apply Waldschmidt's result, we need to know something about $A =$

$a + b\sqrt{R}$. A's minimal polynomial is $x^2 - 2ax + 1$ If the *height* of an algebraic number is the maximum of the absolute values of the coefficients of its minimal polynomial in $\mathbf{Z}[x]$ — with the gcd of these coefficients being 1 — then the height of A is $2a$.

If $R < 1000$ then $a < 2 \times 10^{37}$. (The largest a for any $R < 1000$ is

$$a = 16,421,658,242,965,910,275,055,840,472,270,471,049$$

for $R = 661$.) Since $a^2 - Rb^2 = 1$, it follows that $b\sqrt{R} < a$. Hence

$$A = a + b\sqrt{R} < 4 \times 10^{37}$$

$$H = \text{height}(A) < 4 \times 10^{37}$$

Similarly, if (a', b') is the least positive integer solution of $x^2 - Sy^2 = 1$, and $A' = a' + b'\sqrt{S}$, then, assuming $S < 1000$,

$$A' = a' + b'\sqrt{S} < 4 \times 10^{37}$$

$$H' = \text{height}(A') < 4 \times 10^{37}$$

We also need a bound for (u_0, v_0).

Theorem 4.6.1 *Suppose that (u_0, v_0) is the least nonnegative primitive solution of $x^2 - Ry^2 = C$ belonging to some number z. Then if (a, b) is the least positive solution of $x^2 - Ry^2 = 1$,*

$$\begin{aligned} v_0 &\leq a\sqrt{|C|/R} \\ u_0 &\leq a\sqrt{|C|} \\ u_0 + v_0\sqrt{R} &\leq 2a\sqrt{|C|} \end{aligned}$$

Proof: We must consider two cases.
Case 1. $C > 0$. Suppose $x^2 - Ry^2 = C$ with $x,\ y \geq 0$, and $\gcd(x, y) = 1$, and $x \equiv yz \pmod{C}$.

Also suppose $y > b\sqrt{C}$. Then

$$\frac{1}{Rb^2} + 1 > \frac{C}{Ry^2} + 1 > \frac{1}{\frac{1}{Rb^2} + 1}$$

and hence

$$\frac{a}{b\sqrt{R}} > \frac{x}{y\sqrt{R}} > \frac{b\sqrt{R}}{a}$$

so that

$$\frac{a}{b} > \frac{x}{y} > \frac{bR}{a}$$

and hence

$$\begin{aligned} ax - bRy &> 0 \\ -bx + ay &> 0 \end{aligned}$$

Thus

$$x = a(ax - bRy) + bR(-bx + ay) > a(ax - bRy) > ax - bRy$$

$$y = b(ax - bRy) + a(-bx + ay) > a(-bx + ay) > -bx + ay$$

Moreover, from this it follows that $\gcd(ax - bRy, -bx + ay) = 1$, since $\gcd(x, y) = 1$. Since

$$(ax - bRy)^2 - R(-bx + ay)^2 = C$$

and

$$ax - bRy \equiv ayz - bz^2y \equiv ayz - bzx \equiv (-bx + ay)z \pmod{C}$$

it follows that $(ax - bRy,\ -bx + ay)$ is a primitive, nonnegative solution of $x^2 - Ry^2 = C$ belonging to z. Thus, still on the assumption that $y > b\sqrt{C}$, it follows that (x, y) is not the smallest nonnegative primitive solution of $x^2 - Ry^2 = C$ belonging to z. Hence $v_0 \leq b\sqrt{C}$.

From this it follows that

$$u_0 = \sqrt{Rv_0^2 + C} \leq \sqrt{RCb^2 + C} = a\sqrt{C}$$

and $u_0 + v_0\sqrt{R} \leq 2a\sqrt{C}$.

Case 2. $C < 0$. Suppose x and y are as above, and suppose $y > a\sqrt{-C/R}$. Then $y^2 > -C(1 + Rb^2)/R$ and, again,

$$\frac{1}{Rb^2} + 1 > \frac{C}{Ry^2} + 1 > \frac{1}{\frac{1}{Rb^2} + 1}$$

Hence, as in Case 1, $v_0 \leq a\sqrt{-C/R}$. Since $u_0^2 - Rv_0^2 < 0$, we also have $u_0 \leq a\sqrt{|C|}$.

Thus, whether C is positive or negative, the result follows.

Note that, with Exercise 4.2 # 2, Theorem 4.6.1 gives another solution of $x^2 - Ry^2 = C$.

From the previous theorem, it follows that if (u_0, v_0) is the smallest nonnegative primitive solution of $x^2 - Ry^2 = C/j^2$ belonging to z', then $u_0 + v_0\sqrt{R} \leq 2a\sqrt{|C|}$. If $R,\ |C| < 1000$ this implies that $u_0 + v_0\sqrt{R} \leq 2 \times 10^{39}$.

Similarly, if (w_0, t_0) is the smallest nonnegative primitive solution of $z^2 - Sy'^2 = D/k^2$ belonging to z'' (with $z = w_0$, $y' = t_0$), then $w_0 + t_0\sqrt{S} \leq 2 \times 10^{39}$ — provided $S,\ |D| < 1000$.

At this stage, we are almost ready to assemble a 'linear form in logarithms' and apply Waldschmidt's result. First, however, we need to think about the algebraic number

$$E = \frac{j\sqrt{S}}{k\sqrt{R}} \frac{u_0 + v_0\sqrt{R}}{w_0 + t_0\sqrt{S}}$$

Then $E < 10^{42}$ and $1/E < 10^{42}$ also. Let

$$\begin{aligned} E_2 &= -\frac{j\sqrt{S}}{k\sqrt{R}} \frac{u_0 - v_0\sqrt{R}}{w_0 + t_0\sqrt{S}} \\ E_3 &= -\frac{j\sqrt{S}}{k\sqrt{R}} \frac{u_0 + v_0\sqrt{R}}{w_0 - t_0\sqrt{S}} \\ E_4 &= \frac{j\sqrt{S}}{k\sqrt{R}} \frac{u_0 - v_0\sqrt{R}}{w_0 - t_0\sqrt{S}} \end{aligned}$$

Then

$$p(x) = (x - E)(x - E_2)(x - E_3)(x - E_4)$$

$$= \frac{1}{D^2R^2}(D^2R^2x^4 + 4DjkR^2St_0v_0x^3 - 2RS(CD + 2Ck^2St_0^2 + 2Dj^2Rv_0^2)x^2$$

$$+4CjkRS^2t_0v_0x + C^2S^2)$$

No single one of the linear polynomial factors of $p(x)$ is in $\mathbf{Q}[x]$, so E does not have degree 1 or 3. If it has degree 2 then its minimal polynomial has the form $(x - E')(x - E'')$ and its height is $\leq \max(E' + E'', E'E'') < 10^{84}$. If E has degree 4, then $p(x)$ is its minimal polynomial. Thus the height of E is bound by the maximum of the absolute values of the coefficients of $p(x)$. From what we said above, it follows that

$$\begin{aligned} j^2 v_0^2 &\leq a^2|C|/R \\ k^2 t_0^2 &\leq a'^2|D|/S \\ jkt_0 v_0 &\leq aa'\sqrt{|CD|/(RS)} \end{aligned}$$

Hence H'', the height of E, is no greater than 10^{86}.

From the above, we may conclude that the six numbers $1 + \ln H$, $1 + \ln H'$, $1 + \ln H''$, $\ln A$, $\ln A'$, $|\ln E|$ are bound by $V = 200$.

From Section 4.2, we know that all the primitive nonnegative solutions of $x^2 - Ry^2 = C/j^2$ belonging to z' are given by

$$u_m = \frac{(u_0 + v_0\sqrt{R})(a + b\sqrt{R})^m + (u_0 - v_0\sqrt{R})(a - b\sqrt{R})^m}{2}$$

$$v_m = \frac{(u_0 + v_0\sqrt{R})(a + b\sqrt{R})^m - (u_0 - v_0\sqrt{R})(a - b\sqrt{R})^m}{2\sqrt{R}}$$

for $m = 0, 1, 2, \ldots$. Furthermore, all the primitive, nonnegative solutions of $z^2 - Sy'^2 = D/k^2$ belonging to z'' are given by

$$w_n = \frac{(w_0 + t_0\sqrt{S})(a' + b'\sqrt{S})^n + (w_0 - t_0\sqrt{S})(a' - b'\sqrt{S})^n}{2}$$

$$t_n = \frac{(w_0 + t_0\sqrt{S})(a' + b'\sqrt{S})^n - (w_0 - t_0\sqrt{S})(a' - b'\sqrt{S})^n}{2\sqrt{S}}$$

for $n = 0, 1, 2, \ldots$. To solve the simultaneous Fermat equations, it suffices to find all (m, n) such that $v_m j = t_n k$.

From Section 4.2, we know that $v_{m+2} = 2av_{m+1} - v_m$ and a similar relation holds for t_{n+2}. Thus, using a computer with a multiprecision arithmetic package, it is not hard to check to see if the simultaneous Fermat equations have a solution with one of m and n less than, say, 100.

Let

$$P = \frac{(u_0 + v_0\sqrt{R})(a + b\sqrt{R})^m}{\sqrt{R}}$$

Then

$$\frac{1}{P} = \frac{u_0 - v_0\sqrt{R}}{C/j^2}(a - b\sqrt{R})^m\sqrt{R}$$

Let

$$Q = \frac{(w_0 + t_0\sqrt{S})(a' + b'\sqrt{S})^n}{\sqrt{S}}$$

Then

$$\frac{1}{Q} = \frac{w_0 - t_0\sqrt{S}}{D/k^2}(a' - b'\sqrt{S})^n\sqrt{S}$$

Note that $P > (a + b\sqrt{R})^{m-1}$ and $Q > (a' + b'\sqrt{S})^{n-1}$. Since the smallest value for $a + b\sqrt{R}$ is $K = 2 + \sqrt{3}$, it follows that $P > K^{m-1}$ and $Q > K^{n-1}$. Note also that

$$\frac{Pj}{Qk} = \frac{EA^m}{A'^n}$$

Furthermore, $v_m j = t_n k$ just in case

$$(P - \frac{C}{Rj^2P})j = (Q - \frac{D}{Sk^2Q})k$$

or

$$Pj - \frac{C}{PRj} = Qk - \frac{D}{QSk}$$

Note that if $m, n \geq 10$, then $Pj \neq Qk$, lest we have

$$e + f\sqrt{R} + g\sqrt{S} + h\sqrt{RS} = 0$$

for some integers e, f, g, and h — which is impossible.

In what follows, suppose $m, n \geq 10$.

Suppose $Pj > Qk$. Then, in the case of a solution,

$$\begin{aligned}
\frac{Pj}{Qk} - 1 &= \frac{C}{PQjkR} - \frac{D}{Q^2k^2S} \\
&< \frac{1000}{PQ} \\
&< \frac{1000}{K^{m-1}K^{n-1}} \\
&\leq \frac{1000K^2}{K^{\max(m,n)}K^{10}} \\
&< K^{-\max(m,n)}
\end{aligned}$$

Since the slope of the log function is < 1 when $x > 1$, it follows that

$$0 < |\ln \frac{Pj}{Qk}| < K^{-\max(m,n)}$$

Now suppose $Pj < Qk$. Then, in the case of a solution,

$$\begin{aligned}
1 - \frac{Pj}{Qk} &= \frac{D}{Q^2k^2S} - \frac{C}{PQjkR} \\
&< \frac{D}{Q^2k^2S} \\
&< \frac{D}{PQjkS} \\
&< \frac{500}{PQ} \\
&< \frac{1}{2}K^{-\max(m,n)}
\end{aligned}$$

Since the slope of the log function is < 2 when $x > 1/2$, it follows that

$$0 < |\ln \frac{Pj}{Qk}| < K^{-\max(m,n)}$$

Hence, assuming $m,\ \ n \geq 10$, we have

$$0 < |\ln \frac{EA^m}{A'^n}| < K^{-\max(m,n)}$$

or

$$0 < |m \ln A - n \ln A' + \ln E| < K^{-\max(m,n)}$$

This brings us to Waldschmidt's Theorem. Actually, we do not need that theorem in its full generality, but only a corollary of it, say, the following.

Theorem 4.6.2 (Corollary to Waldschmidt) *Let A, A', and E be nonzero, nonnegative algebraic numbers, each of degree ≥ 2 and ≤ 4. Let H, H' and H'' be their heights. Suppose*

$$V \geq \max(1 + \ln H,\ \ 1 + \ln H',\ \ 1 + \ln H'',\ \ |\ln A|,\ \ |\ln A'|,\ \ |\ln E|)$$

Let m and n be positive integers, and let $W = \max(\ln m,\ \ln n)$. Let

$$L = m \ln A - n \ln A' + \ln E$$

Then if $L \neq 0$,

$$|L| > \exp(-2^{101} V^3 (W + \ln(64eV)) \ln(64eV))$$

Proof: See *Acta Arithmetica* 32, pages 257–83 or *New Advances in Transcendence Theory,* ed. A. Baker, pages 280-81.

In our case, we can take $V = 200$ (see above), giving us

$$K^{-\max(m,n)} > \exp(-2^{101} 200^3 (W + 11) 11)$$

so that

$$2^{101} 200^3 (W + 11) 11 > e^W \ln K$$

If $W > 55$ this gives

$$2^{101} 200^3 \times (6/5) \times 11 > (e^W \ln K)/W$$

or

$$88.5 > 101 \ln 2 + 3 \ln 200 + \ln(6/5) + \ln 11 - \ln(\ln K) > W - \ln W$$

so that $W \leq 93.5$. Of course, if $W \leq 55$, we reach the same conclusion. Hence $\max(m, n) \leq e^{93.5} < 10^{41}$.

The above inequality was reached on the assumption that m, $n \geq 10$. What if one of them is < 10? Since $v_m j = t_n k$ it follows that the other certainly cannot exceed 10^{41}.

Needless to say, we cannot check all the possibilities less than 10^{41}, so a further theorem is needed. This is Davenport's Lemma.

Theorem 4.6.3 (Davenport's Lemma) *Suppose x_1 and x_2 are reals. Suppose M is a positive integer with $K > (10^6 M)^{1/M}$, where $K = 2 + \sqrt{3}$. Suppose p and q are integers with $1 \le q \le 1000M$ and*

$$|x_1 q - p| \le \frac{2}{1000M}$$

Suppose m and n are positive integers such that

$$|mx_1 - n - x_2| \le \frac{1}{K^m}$$

Then, where $\|r\|$ denotes the distance of a real number r from the nearest integer,
(1) $m < (\ln 10^6 M)/\ln K$
or
(2) $m > M$
or
(3) $\|qx_2\| < 0.003$ and $pm - qn = [qx_2 + 1/2]$.

Proof: Let $w = qx_1 - p$. Then $|w| \le 2/1000M$. Now

$$|mx_1 q - nq - x_2 q| \le q/K^m \le 1000M/K^m$$

so that

$$|mp - nq + mw - x_2 q| \le 1000M/K^m$$

Suppose $\|qx_2\| \ge 3/1000$. To obtain a contradiction, also suppose

$$(\ln 10^6 M)/\ln K \le m \le M$$

Then $10^6 M \le K^m$ and $1000M/K^m \le 1/1000$. Also, since $m \le M$, we have $|mw| \le 2/1000$. Since $\|qx_2\| \ge 3/1000$, it follows that $\|mw - qx_2\| > 1/1000$. Since $mp - nq$ is an integer,

$$1/1000 < \|mp - nq + mw - x_2 q\| \le 1000M/K^m \le 1/1000$$

Contradiction. Thus if $\|qx_2\| \ge 3/1000$ then (1) or (2) holds.

Suppose $\|qx_2\| < 3/1000$. Again suppose that

$$(\ln 10^6 M)/\ln K \le m \le M$$

Then, as above, $1000M/K^m \leq 1/1000$ and $|mw| \leq 2/1000$. Since $x_2 q = [qx_2 + 1/2] \pm \|qx_2\|$, we have

$$\begin{aligned} |mp - nq - [qx_2 + 1/2]| &\leq |mp - nq - (x_2 q - mw)| \\ &\quad + |(x_2 q - mw) - [qx_2 + 1/2]| \\ &\leq 1000M/K^m + |-mw \pm \|qx_2\|| \\ &< 1/1000 + 2/1000 + 3/1000 \leq 6/1000 \end{aligned}$$

Hence $mp - nq = [qx_2 + 1/2]$. Thus if $\|qx_2\| < 3/1000$ then (1) or (2) or $mp - nq = [qx_2 + 1/2]$.

We apply Davenport's Lemma to our problem as follows. Let

$$x_1 = \frac{\ln A}{\ln A'} \text{ and } x_2 = -\frac{\ln E}{\ln A'}$$

Let $M = 10^{41}$. Let x_1' be a rational (e.g. decimal) approximation to x_1, so that $|x_1 - x_1'| < 10^{-90}$. Let x_2' be a rational (e.g. decimal) approximation to x_2, so that $|x_2 - x_2'| < 10^{-50}$. Let f/g and f'/g' be consecutive simple continued fraction convergents of x_1' such that $g \leq 10^{44}$ but $g' > 10^{44}$. Then

$$\begin{aligned} |x_1 - f/g| &\leq |x_1 - x_1'| + |x_1' - f/g| \\ &< 10^{-90} + 1/gg' \\ &\leq 10^{-44}/g + 10^{-44}/g \end{aligned}$$

Thus $|x_1 g - f| < 2/10^{44}$.

Now if $m,\ n \geq 10$, we have $|m \ln A - n \ln A' + \ln E| < K^{-\max(m,n)}$ so that

$$|mx_1 - n - x_2| < K^{-m}$$

Also

$$K > (10^6 10^{41})^{\frac{1}{10^{41}}}$$

All the conditions of Davenport's Lemma are met.

Let r be a rational approximation to gx_2', so that $|r - gx_2'| < 10^{-5}$. Then

$$|r - gx_2| < |r - gx_2'| + g|x_2' - x_2| < 10^{-5} + 10^{44}10^{-50} < 2 \times 10^{-5}$$

so r gives gx_2 accurate to 4 decimal places. Thus if $\|r\| > 4/1000$ then $\|gx_2\| > 3/1000$.

Suppose this is so (as is probable). Then we are in case (1) or (2) in Davenport's Lemma. Moreover, Waldschmidt's Theorem assures us that we are not in case (2). Hence $m < \ln 10^{47}/\ln K < 83$. This many m's can be checked one by one (by, say, using the fact that, if $m,\ n \geq 10$ then $|mx_1 - x_2|$ has to be close to an integer). Indeed, we could even use a second application of Davenport's Lemma to reduce the bound further.

If $\|gx_2\| \leq 3/1000$ then we also have to check solutions of $fm - gn = [gx_2 + 1/2]$ with $m < 10^{41}$. Usually, there will not be more than one of these.

Given the above, the only remaining practical problem in solving simultaneous Fermat equations is that of calculating the logarithms to sufficient accuracy. In this connection, it is useful to note the following. If $-1 < x < 1$ then

$$\ln\frac{1+x}{1-x} = 2x(1 + \frac{x^2}{3} + \frac{x^4}{5} + \frac{x^6}{7} + \cdots)$$

If we truncate this series just after the term $x^{2n}/(2n+1)$, the error is bound by

$$\frac{x^{2n+3}}{n(1-x^2)}$$

Moreover,

$$\ln(a + b\sqrt{R}) = \frac{1}{2}\ln\left(\frac{1 + b\sqrt{R}/a}{1 - b\sqrt{R}/a}\right)$$

If $C > 0$ we have

$$\ln j(u_0 + v_0\sqrt{R}) = \frac{1}{2}\ln\left(\frac{1 + v_0\sqrt{R}/u_0}{1 - v_0\sqrt{R}/u_0}\right) + \frac{1}{2}\ln C$$

If $C < 0$ we have

$$\ln j(u_0 + v_0\sqrt{R}) = \frac{1}{2}\ln\left(\frac{1 + u_0/(v_0\sqrt{R})}{1 - u_0/(v_0\sqrt{R})}\right) + \frac{1}{2}\ln(-C)$$

Also, to calculate logs of positive integers, we have

$$\ln(N+1) = \ln N + \ln\left(\frac{1+1/(2N+1)}{1-1/(2N+1)}\right)$$

To compute $\ln 15$ note that $\ln 15 = \ln 3 + \ln 5$. To compute $\ln 79$, note that

$$\ln 79 = \ln\left(\frac{1+9/(2\times 70+9)}{1-9/(2\times 70+9)}\right) + \ln 7 + \ln 10$$

In general, if $x = c + 2d$, then $\ln(1 + c/d) = \ln((1+x)/(1-x))$.

Exercises 4.6

1. Solve the system

$$x^2 - 2y^2 = 1$$
$$z^2 - 2312y^2 = 1$$

(Hint: RS is a square.)

2. Prove that the only positive integer solution of

$$\begin{aligned} x^2 - 11y^2 &= 1 \\ z^2 - 56y^2 &= 1 \end{aligned}$$

is (199, 60, 449).

3. Show that if (u, v) is the smallest nonnegative solution of $x^2 - Ry^2 = C$ belonging to either z or $-z$ then

$$v \le \sqrt{\frac{(a+1)|C|}{2R}}$$

(Hint: See T. Nagell's *Introduction to Number Theory,* page 206.)

Chapter 5

Classical Construction Problems

The ancient Greeks searched for a way of using straightedge and compass to trisect an arbitrary angle, and to draw a segment of length $\sqrt[3]{2}$. They also tried to 'square the circle', that is, construct a segment of length $\sqrt{\pi}$. Finally, they struggled to find straightedge and compass constructions for regular polygons with 7, 9, 11, 13, and 17 sides. In all this they failed, but it was not proved until the nineteenth century that the reason for their failure was that all these problems are insoluble — except one. In 1796 Gauss discovered a straightedge and compass construction for the regular 17–sided polygon. It was this discovery, the first advance on construction problems in 2000 years, that motivated Gauss to devote himself to mathematics.

In this chapter we give an explicit construction for the regular 'heptadecagon', and show why the other problems are, indeed, insoluble.

5.1 Euclidean Constructions

Sadly, it is now possible to obtain a PhD in mathematics and not know that Euclid lived in Alexandria, Egypt, about 300 BC, and wrote a book called the *Elements.* When we today do geometry, we usually start with a plane which already contains a point corresponding to every ordered pair of reals. Euclid was more parsimonious. He started with just two points (corresponding to (0, 0) and (1, 0)), and then constructed, one by one, just enough extra points, lines and circles to meet his immediate needs.

The rules for construction were strict.

(1) If A and B are previously given or constructed points, you can 'join AB', constructing the line segment AB; if this segment intersects any previously constructed line segments or circles, you have thereby constructed the points of intersection.
(2) If AB is a previously constructed segment, and O is a previously given or constructed point, you can draw a circle with centre O and radius AB; if this circle intersects any previously constructed line segments or circles, you have thereby constructed the points of intersection.
(3) If AB is a previously constructed segment, you can lengthen, or 'produce', it in either direction to meet a previously constructed segment or circle (assuming that segment or circle lies 'in its way'), and thereby construct a point.
(4) The only way to construct anything is to apply the above rules a finite number of times.

As examples, we give the following 8 straightedge and compass constructions.

C1. To bisect an angle

Let ABC be an angle, with previously constructed 'arms' AB and BC. With centre B and radius BA, cut BC in E. (That is, construct a circle with centre B and radius BA. If the circumference meets BC in a point, call that point E. Otherwise, produce BC, in the direction going from B to C, until it meets the circumference in a point, which we shall call E.) With centres A and E, construct two circles each with radius AE. These circles meet in two points. Let F be the meeting point which is on the side of AE away from B. (Note that AEF is an equilateral triangle.) Join BF. Then BF is the required bisector. This can be proved using the 'side-side-side' congruence theorem to show that triangles BAF and BEF are congruent.

If $\angle ABC = 180°$ then BF is perpendicular to AC. Thus construction C1 is also a construction for drawing a perpendicular to a given segment through a given point in that segment.

C2. To construct the right bisector of a segment
Let AB be a previously constructed segment. With centres A and B, draw two circles, each with radius AB. These circles meet in exactly two points C and D. Join CD. Then CD is the required right bisector.

Note that CD meets AB in its midpoint, and hence this construction also works as a construction of the midpoint of a given segment.

C3. To construct a segment through a given point and parallel to a given segment
Let A be the point, and BC the segment. It is assumed that A is not on the line BC. With centre C and radius AB, draw a circle. With centre A and radius BC, draw a second circle to cut the first circle in point D, where D and B are on opposite sides of AC. Then AD is the required parallel.

C4. To add two segments
Let AB and CD be two previously constructed segments. With centre B and radius CD, draw a circle. Produce AB (in the direction from A to B) so that it meets this circle at E. The segment AE is the required sum.

C5. To multiply two segments
Let AB and CD be previously constructed segments. With centres C and D, and radius CD, construct two circles meeting in E and E'. With centre C and radius AB, cut CE (or CE produced in the direction from C to E) in F. If O and X are the two points with which Euclid starts, so that OX is a unit segment, then, with centre C and radius OX, cut CD (or CD produced in the direction from C to D) in G. Join FG. Using C3, draw a segment through D parallel to FG, to meet CE (or CE produced) in H. Then CH is the required product.

This is proved using the theory of similar triangles. Since $CH : CF :: CD : 1$, it follows that $CH = CF \times CD = AB \times CD$.

C6. To draw the multiplicative inverse of a segment
Let AB be a previously constructed segment. With centres A and B, construct circles with radius AB, to meet in C and C'. With centre

A and radius OX (the unit segment), cut AC (or AC produced in the direction from A to C) in D. With centre A and radius OX, cut AB (or AB produced in the direction from A to B) in E. Draw a line through E which is parallel to BD to meet AC in F. Then AF is the required segment.

C7. To construct the square root of a segment

Let AB be a previously constructed segment. Add the unit segment OX to it, drawing a segment $AC = AB + 1$, with B between A and C. Using C1, erect a perpendicular to AC through B. Using C2, construct the midpoint D of AC. With centre D and radius DC, draw a circle to cut the perpendicular at E. Then BE is the required square root.

This is proved by noting that $\angle AEC$, being an angle in a semicircle, is right. Hence triangles ABE and EBC are similar. This gives $AB : BE :: BE : BC$, so that $AB \times BC = BE^2$. But $BC = OX = 1$.

C8. To construct a Pythagorean star

With centre O and radius OX draw a circle. Join XO and produce it to meet the circle in Y. Construct the midpoint C of OX. Construct the right bisector of YX, meeting the circle in E. With centre C and radius CE, cut OY in F. With centre E and radius EF, cut the original circle in G and H. With centre G, and the same radius, cut the original circle again at J. With centre H, and the same radius, cut the original circle again at K. Join EJ, EK, GK, GH, and HJ.

From the above, it is clear that, starting with the unit segment OX, Euclid could construct segments of any positive rational length. He could also construct segments with length equal to numbers like

$$3 + \sqrt{5\sqrt{3} + \frac{7}{\sqrt{10} + 2}}$$

The reason that the Greeks failed to 'duplicate the cube' is simply that $\sqrt[3]{2}$ is not a number of this type. This we shall prove below.

Exercises 5.1

1. Get a straightedge and compass, and construct a regular hexagon.
2. Give a straightedge and compass construction for a line through a given point not on a given line, and perpendicular to the given line.
3. Give a Euclidean construction for an angle of 3°.
4. Prove that the above construction for the five-pointed star works.
5. Prove that if a regular polygon with n sides is constructible, then so is a regular polygon with $2n$ sides.
6. Construct a common tangent to two given circles. You must apply Euclid's rules, and not just 'move the ruler round until it touches both circles'.
7. Construct an isosceles triangle given the base and a bisector of a base angle.

5.2 Fields and Vector Spaces

In the previous section, we saw that we can add, subtract, multiply and divide segments in Euclidean geometry. This means that the segments form a 'field'. Knowing just which field they form will help us answer questions about what figures are constructible with straightedge and compass.

We also saw that we can take square roots in Euclidean geometry. This means that we can find roots of polynomials such as the quadratic polynomial

$$ax^2 + bx + c$$

If a root of a polynomial is added to the field of rationals, we get a vector space over the rationals, and, again, in order to understand just what figures are constructible, we need to say something about vector spaces.

In this section, then, we review fields and vectors spaces.

A *field* is a set containing at least two elements, 0 and 1, which is closed under two unary operations, $-$ and $^{-1}$, and two binary operations, + and . such that

$$\begin{array}{ll} (a+b)+c = a+(b+c) & (a.b).c = a.(b.c) \\ a+b = b+a & a.b = b.a \\ a+0 = a & a.1 = a \\ a+(-a) = 0 & a.a^{-1} = 1 \\ a.(b+c) = a.b+a.c & \end{array}$$

— with one exception, namely, there is no 0^{-1}.

For example the set $\mathbf{Q}$ of rationals forms a field, and so do the residue classes modulo p, if p is prime.

We next turn our attention to the interaction between fields and polynomials.

Let F be a field, and let $F[x]$ be the set of polynomials with coefficients in F.

Note that if $g(x)$ and $h(x)$ are members of $F[x]$ and $g(x)h(x) = 0$ (i.e. is identical to the constant polynomial 0) then either $g(x) = 0$ or $h(x) = 0$.

A polynomial $p(x)$ in $F[x]$ is *irreducible* if it cannot be written as a product of two lower degree polynomials in $F[x]$. We say $p(x)$ is *monic* if the coefficient of its highest power of x is 1.

Let $f(x)$ and $p(x)$ be any polynomials in $F[x]$. By polynomial division, we can obtain a series of equations as follows.

$$\begin{array}{rcll} f(x) & = & q_1(x)p(x) + r_1(x) & \text{with } \deg r_1 < \deg p \\ p(x) & = & q_2(x)r_1(x) + r_2(x) & \text{with } \deg r_2 < \deg r_1 \\ r_1(x) & = & q_3(x)r_2(x) + r_3(x) & \text{with } \deg r_3 < \deg r_2 \\ & \vdots & & \end{array}$$

Since the degrees of the polynomials cannot decrease forever, we eventually get

$$\begin{array}{rcll} r_{n-1}(x) & = & q_{n+1}(x)r_n(x) + r_{n+1}(x) & \text{with } \deg r_{n+1} < \deg r_n \\ r_n(x) & = & q_{n+2}(x)r_{n+1}(x) & \end{array}$$

If $t(x)$ is any polynomial dividing evenly into $p(x)$ and $f(x)$, then $t(x)$ also divides evenly into $r_1(x)$, and hence also into $r_2(x)$, ..., and hence also into $r_{n+1}(x)$. Conversely, $r_{n+1}(x)$ divides evenly into $r_n(x)$ and hence also into $r_{n-1}(x)$, ..., and hence also into $p(x)$ and hence also into $f(x)$. Thus $r_{n+1}(x)$ is a gcd of $f(x)$ and $p(x)$.

If $p(x)$ is irreducible and not a factor of $f(x)$, this gcd has degree 0: it is some constant (e.g. 1) in F. And hence there are polynomials $m(x)$ and $n(x)$ in $F[x]$ such that

$$m(x)f(x) + n(x)p(x) = 1$$

— for

$$1 = c \times r_{n+1}(x) = cr_{n-1}(x) - cq_{n+1}(x)r_n(x) = \ldots$$

and so on (with c in F). Thus, if $p(x)$ is irreducible, and divides evenly into $f(x)g(x)$, but not into $f(x)$, then, since

$$m(x)f(x)g(x) + n(x)p(x)g(x) = g(x)$$

it follows that $p(x)$ divides $g(x)$. Hence, just as in the case of integers, we have a unique factorisation theorem for members of $F[x]$:

Theorem 5.2.1 *If F is a field, and $f(x) \epsilon F[x]$ then $f(x)$ can be written as a product of a member c of F and some monic irreducible polynomials, in essentially one way.*

We also have

Theorem 5.2.2 (Gauss's Lemma) *Let $f(x)$ be a polynomial with only integer coefficients. Suppose it is the product of two lower degree polynomials $g(x)$ and $h(x)$ which have rational coefficients. Then $f(x)$ is also the product of two lower degree polynomials $g'(x)$ and $h'(x)$ which have only integer coefficients.*

Proof: Every polynomial with rational coefficients can be written uniquely in the form $\pm ak(x)$ where a is a fraction, and $k(x)$ is a *primitive* polynomial, that is, one with relatively prime integer coefficients. To prove the result, it suffices to show that the product of two primitive polynomials is also primitive. For suppose this is the case and suppose $f(x) = g(x)h(x)$ with f primitive. Let $g(x) = ag'(x)$ where a is a fraction and $g'(x)$ is primitive. Let $h(x) = bh'(x)$ where b is a fraction and $h'(x)$ is primitive. Then $1f(x) = abg'(x)h'(x)$ with $f(x)$ and $g'(x)h'(x)$

both primitive. From the uniqueness statement given at the beginning of this proof, it follows that $ab = \pm 1$. And $f(x) = \pm g'(x)h'(x)$ is a product of polynomials with only integer coefficients. The result is now easily extended to the case in which $f(x)$ has only integer coefficients, but is not primitive.

To show that the product of two primitive polynomials is primitive, we reason as follows. Let the primitive polynomials be

$$f(x) = a_0 + a_1 x + \cdots + a_n x^n$$

$$g(x) = b_0 + b_1 x + \cdots + b_m x^m$$

To obtain a contradiction suppose the coefficients of their product have some prime common factor p. Let a_i be the first coefficient of $f(x)$ (starting from the left) not divisible by p, and let b_j be the first coefficient of $g(x)$ not divisible by p. (Since these polynomials are primitive, those coefficients exist.) Now consider the coefficient of x^{i+j} in the product $f(x)g(x)$:

$$c_{i+j} = \sum_{i+j-m \leq k < i} a_k b_{i+j-k} + a_i b_j + \sum_{i+j-n \leq l < j} a_{i+j-l} b_l$$

The prime p is a factor of c_{i+j} and also of the two sums expressed with a $\sum$. Hence $p \mid a_i b_j$. Contradiction.

We say that two polynomials with integer coefficients are *congruent* modulo p (where p is a prime) iff, for all powers i, the coefficient of x^i in the first is congruent, modulo p, to the coefficient of x^i in the second.

Note that if p is an odd prime, then p factors $\binom{p}{j}$ for $j = 1, 2, \ldots, p-1$. Hence $(x-1)^p \equiv x^p - 1 \pmod p$

The notion of polynomial congruence is important in the proofs of the next two theorems. These theorems will help us set limits on the sort of regular polygons one can construct with ruler and compass.

Theorem 5.2.3 *If p is an odd prime, then $x^{p-1} + x^{p-2} + \cdots + x + 1$ is irreducible in* $\mathbf{Q}[x]$.

Proof: Suppose not. By Theorem 5.2.2, the polynomial factors into two lower degree polynomials $f(x)$ and $g(x)$ with only integer coefficients. Moreover, we can take it that both $f(x)$ and $g(x)$ are monic. Letting $x = 1$, we get $p = f(1)g(1)$. Without loss of generality, let $g(1) = \pm 1$.

Since

$$(x^{p-1} + x^{p-2} + \cdots + x + 1)(x - 1) = x^p - 1 \equiv (x - 1)^p \pmod p$$

we have

$$x^{p-1} + x^{p-2} + \cdots + x + 1 \equiv (x - 1)^{p-1} \pmod p$$

The residue classes mod p form a field $\mathbf{Z}_p$, and $\mathbf{Z}_p[x]$ has unique factorisation (Theorem 5.2.1). Since $f(x)$ and $g(x)$ are monic, they have the same degree when they are considered as elements of $\mathbf{Z}_p[x]$. Since

$$f(x)g(x) = x^{p-1} + \cdots + x + 1 \equiv (x - 1)^{p-1} \pmod p$$

it follows that $g(x) \equiv (x - 1)^s \pmod p$ for some integer s with $1 \leq s \leq p - 1$. Hence $\pm 1 = g(1) \equiv 0 \pmod p$. Contradiction.

The next theorem is similar.

Theorem 5.2.4 *If p is an odd prime then $x^{(p-1)p} + x^{(p-2)p} + \cdots + x^{2p} + x^p + 1$ is irreducible in $\mathbf{Q}[x]$.*

Proof: Suppose not. Then it factors into two lower degree monic polynomials, $f(x)$ and $g(x)$, with only integer coefficients (Theorem 5.2.2). Since $p = f(1)g(1)$, we can take it that $g(1) = \pm 1$. Now

$$f(x)g(x)(x^p - 1) = x^{p^2} - 1 \equiv (x^p - 1)^p \equiv ((x-1)^p)^p \equiv (x-1)^{p^2} \pmod p$$

so that

$$f(x)g(x) \equiv (x - 1)^{p^2-1} \pmod p$$

and $g(x) \equiv (x - 1)^s \pmod p$ for some integer s with $1 \leq s < p^2 - 1$. Hence $\pm 1 = g(1) \equiv 0 \pmod p$. Contradiction.

The key relationship among roots, polynomials and fields is given in the next theorem.

Theorem 5.2.5 *Let a be a root of a monic irreducible polynomial $f(x)$ in $F[x]$. Then $f(x)$ is the only monic irreducible polynomial in $F[x]$ of which it is a root. Moreover, the set of polynomials in $F[a]$ form a field.*

Proof: Suppose $g(x)$ is a monic irreducible polynomial in $F[x]$ with $g(a) = 0$. If $g(x)$ is not $f(x)$, then they have gcd 1, since they are both monic and irreducible. Thus there are polynomials $m(x)$ and $n(x)$ such that

$$m(x)f(x) + n(x)g(x) = 1$$

Hence $m(a)f(a) + n(a)g(a) = 1$. But $f(a) = g(a) = 0$. Contradiction.

Suppose $h(x)$ is any polynomial with coefficients in F. If $h(a) \neq 0$ then $f(x)$ and $h(x)$ are relatively prime, and, for some $m(x)$ and $n(x)$ in $F[x]$, we have

$$m(x)f(x) + n(x)h(x) = 1$$

Hence $n(a)h(a) = 1$, that is, $h(a)$ has a multiplicative inverse in $F[a]$. Since the members of $F[a]$ satisfy the other field requirements, the result follows.

The polynomial $f(x)$ is the *minimal* polynomial of a over F. Its degree is the *degree of a* over F.

The field $F[a]$ of the previous theorem is best understood as a vector space. Recall that a *vector space* over a field F is a set V such that V is an abelian group under addition, and there is a mapping $(f, v) \longmapsto fv$ from $F \times V$ into V which satisfies

$$\begin{aligned} f(v + v') &= fv + fv' \\ (f + f')v &= fv + f'v \\ (ff')v &= f(f'v) \\ 1v &= v \end{aligned}$$

for all v, v' in V and all f, f' in F.

Vectors $v_1, \ldots, v_n$ are *linearly independent* iff

$$f_1v_1 + \cdots + f_nv_n = 0$$

implies that $f_1 = \cdots = f_n = 0$.

Vectors $v_1, \ldots, v_n$ *span* or *generate* vector space V iff every member of V has the form $f_1 v_1 + \cdots + f_n v_n$.

Vectors $v_1, \ldots, v_n$ are a *basis* for V iff they are linearly independent and span V.

It can be proved that if $v_1, \ldots, v_n$ is a basis for V then any other basis also has n elements. This number, denoted by $[V : F]$, is the *dimension* of the vector space (over F). Note that $[F : F] = 1$.

Theorem 5.2.6 *Let F be a field and a the root of a monic irreducible polynomial $f(x)$ of degree d in $F[x]$. Then the field $F[a]$ is a vector space over F with basis 1, a, a^2, ..., a^{d-1}. And $[F[a] : F] = \deg(a)$.*

Proof: Suppose $g(x) = f_d x^{d-1} + f_{d-1} x^{d-2} + \cdots + f_2 x + f_1$, with the f's in F. If $g(a) = 0$ then $g(x)$ has an irreducible factor $h(x)$ such that $h(a) = 0$. But this is impossible, unless $g(x)$ is the 0 polynomial, since $g(x)$ has degree less than d. Thus 1, a, a^2, ..., a^{d-1} are linearly independent.

Since a^d can be expressed as a linear combination of 1, a, a^2, ..., a^{d-1} (over F) — since $f(a) = 0$ and $f(x)$ is monic — it follows that 1, a, a^2, ..., a^{d-1} span $F[a]$.

The next theorem shows that vector spaces can be built up in 'towers'. This corresponds to the geometrical fact that we can construct square roots of expressions already containing square roots.

Theorem 5.2.7 *Let F be a field and a the root of a monic irreducible polynomial $f(x)$ of degree d in $F[x]$. Then $F[a]$ is a field. Let a' be the root of a monic irreducible polynomial $f'(x)$ of degree d' in $(F[a])[x]$, so that $F[a][a']$ (i.e. $(F[a])[a']$) is a vector space of dimension d' over $F[a]$.*
Then $F[a][a]$ is a vector space of dimension dd' over F. Indeed, it has basis $a^i a'^j$ with $i = 0, \ldots, d-1$, and $j = 0, \ldots d'-1$.

Proof: If w is in $F[a][a']$, then

$$w = f'_1 + f'_2 a' + \cdots + f'_{d'} a'^{d'-1}$$

where each f'_k has the form $f_{k1} + f_{k2}a + \cdots + f_{kd}a^{d-1}$ with the f's in F. Thus the vectors $a^i a'^j$ span $F[a][a']$.

Their linear independence can be established in an equally straightforward manner.

Corollary: $[F[a][a'] : F] = [F[a][a'] : F[a]] \times [F[a] : F]$

Similarly, we have

Theorem 5.2.8 *If a_1 has degree d_1 over F, a_2 has degree d_2 over $F[a_1]$, $\ldots$, and a_t has degree d_t over $F[a_1][a_2]\ldots[a_{t-1}]$ then $F[a_1][a_2]\ldots[a_t]$ has degree $d_t \ldots d_2 d_1$ over F.*

The final theorem in this section gives important information about the degrees of the 'towers' of vector spaces. As we shall see, this information implies that there are certain limits on what is constructible using only straightedge and compass.

Theorem 5.2.9 *Suppose a' is an element of the field $F[a_1][a_2]\ldots[a_t]$ where each a_k is the root of a monic irreducible polynomial with all its coefficients in $F[a_1]\ldots[a_{k-1}]$. Then a' is the root of a monic irreducible polynomial in F, and $[F[a'] : F]$ is a factor of $[F[a_1]\ldots[a_t] : F]$.*

Proof: Consider $1, a', a'^2, \ldots, a'^d$ where d is the degree of $F[a_1]\ldots[a_t]$ over F. If these $d+1$ numbers are linearly independent over F then they generate a vector subspace of $F[a_1]\ldots F[a_t]$ with dimension $d+1$. But this is impossible. Hence, for some f's in F, we have

$$f_1 + f_2 a' + \cdots + f_{d+1} a'^d = 0$$

with not all the f's equal to 0. Thus a' is the root of a polynomial in $F[x]$, and hence of some monic irreducible polynomial in $F[x]$.

Since a' is in the field $F[a_1]\ldots[a_t]$, it follows that $F[a'][a_1]\ldots[a_t] \subseteq F[a_1]\ldots F[a_t]$. Since $F \subseteq F[a']$, it follows that

$$F[a_1]\ldots[a_t] \subseteq F[a'][a_1]\ldots[a_t]$$

Hence

$$d = [F[a_1]\ldots[a_t] : F] = [F[a'][a_1]\ldots[a_t] : F]$$

$$\begin{aligned} &= [F[a'][a_1]\dots[a_t] : F[a'][a_1]\dots[a_{t-1}]] \\ &\quad \times[F[a'][a_1]\dots[a_{t-1}] : F[a'][a_1]\dots[a_{t-2}]] \\ &\quad \times\cdots \\ &\quad \times[F[a'][a_1] : F[a']] \\ &\quad \times[F[a'] : F] \end{aligned}$$

so that $[F[a'] : F]$ is indeed a factor of d.

It is in the next section that we shall use the above field theory to establish limits on Euclidean constructions.

Exercises 5.2

1. Prove that, in a vector space, $0v = 0$.
2. Prove that a vector space with a finite basis cannot also have an infinite basis.
3. Show that if field H contains field G and field G contains field F then $[H : F] = [H : G][G : F]$.
4. Suppose a field G contains the field $\mathbf{Q}$, and $[G : \mathbf{Q}]$ is a power of 2. Does G contain any numbers with degree 3 over $\mathbf{Q}$? Why not?
5. Let c be an element of field F. Then $F[\sqrt{c}]$ is the set of all numbers of the form $a + b\sqrt{c}$ with a and b in F. Why? How should one express the inverse $1/(d + e\sqrt{c})$ in the form $a + b\sqrt{c}$?
6. Let p be a prime. Suppose b and d are integers such that $0 \le b < p$ and $0 \le d < p$. Then

$$\binom{ap+b}{cp+d} \equiv \binom{a}{c}\binom{b}{d} \pmod{p}$$

(If $d > b$ then the number of ways of choosing d out of b things is 0, but if $b = d = 0$ then it is 1.)

Hint: in $\mathbf{Z}_p[x]$, $(1+x)^{ap+b} = (1+x^p)^a(1+x)^b$. Look at the coefficients of x^{cp+d}.

7. If p is prime and r_i, s_i the base p digits of r and s, then

$$\binom{s}{r} \equiv \binom{s_n}{r_n}\cdots\binom{s_1}{r_1}\binom{s_0}{r_0} \pmod{p}$$

8. Each binary digit of positive integer r is less than or equal to the corresponding binary digit of s iff $\binom{s}{r} \equiv 1 \pmod 2$.

This, and the previous result, were first proved by E. Lucas, in 1878.

5.3 Limits of Ruler and Compass Construction

The Delians were told by their oracle that, to avert a plague, they should double the size of Apollo's cubical altar. This, it is said, led to the ancient Greek attempt to 'double the cube' — that is, construct a segment of $\sqrt[3]{2}$ using only straightedge and compass. As we show in this section, this is not possible. Nor is it possible to trisect an arbitrary angle.

Let r and s be reals. We call the point, or complex number, $(r,s) = r + s\sqrt{-1}$ *constructible* iff, starting with the points $(0,0)$ and $(1,0)$, and using only straightedge and compass (in the way explained above), it is possible to construct the point (r,s). We say that a real, r, is constructible iff $(r,0)$ is constructible. Thus if r is a positive real, a segment of length r can be constructed with straightedge and compass iff r is constructible.

Let $\mathbf{Q}$ be the field of rationals. Let $c_1, \ldots, c_n$ be reals such that

$$\begin{aligned} c_1 &\in \mathbf{Q} \\ c_2 &\in \mathbf{Q}[\sqrt{c_1}] \\ c_3 &\in \mathbf{Q}[\sqrt{c_1}][\sqrt{c_2}] \\ &\vdots \\ c_n &\in \mathbf{Q}[\sqrt{c_1}][\sqrt{c_2}]\ldots[\sqrt{c_{n-1}}] \end{aligned}$$

We define

$$G(c_1, \ldots c_n) = \mathbf{Q}[\sqrt{c_1}][\sqrt{c_2}]\ldots[\sqrt{c_{n-1}}][\sqrt{c_n}]$$

Then $G(c_1, \ldots, c_n)$ is a field whose degree over $\mathbf{Q}$ is a power of 2 (Theorem 5.2.8). Let us call a complex number a *G-number* (or *geometry number*) if it is an element of a field of the above form. The G-numbers are thus the smallest field which contains the rationals and is closed under the $\sqrt{\ }$ operation. From Theorem 5.2.9 it follows that every G-number has degree 2^t over $\mathbf{Q}$ for some nonnegative integer t. Hence we have

Theorem 5.3.1 *If a is a root of a monic irreducible polynomial in $\mathbf{Q}[x]$ with degree not a power of 2, then a is not a G-number.*

The next theorem, and its converse, form the core of this section.

Theorem 5.3.2 *All the G-numbers are constructible.*

Proof: For (1) all the rationals are constructible; (2) if a complex number (r, s) is constructible, so is its additive inverse, and its multiplicative inverse; (3) if (r, s) and (r', s') are both constructible, so is their sum and their (complex number) product; (4) if (r, s) is constructible, so are its square roots (using De Moivre's Theorem, and angle bisection).

Note that if the sum and product of two numbers are both constructible, then each of the two numbers is constructible. For if the sum is s and the product p, then the numbers are

$$\frac{s \pm \sqrt{s^2 - 4p}}{2}$$

Note also that an angle can be constructed just in case its cosine is constructible.

To show that every constructible number is a G-number, we prove the following theorems.

Theorem 5.3.3 *If (a, b) is a G-number, so are $(a, -b)$, $(a, 0)$ and $(0, b)$.*

Proof: The fact that $(a, -b)$ is a G-number follows using mathematical induction on the degree of (a, b). The rest of the theorem follows from the fact that the sum and difference of any two G-numbers are G-numbers.

Theorem 5.3.4 *If (a, b) and (c, d) are G-numbers, so are the coefficients in the Cartesian equation for the straight line joining them.*

Proof: The line joining (a, b) and (c, d) is

$$(b-d)x + (c-a)y + ad - bc = 0$$

Theorem 5.3.5 *If (a, b) is a G-number, and if the positive real r is a G-number, then so are the coefficients in the Cartesian equation for the circle with centre (a, b) and radius r.*

Proof: That equation is $(x-a)^2 + (y-b)^2 = r^2$.

Theorem 5.3.6 *If d, e, f, d', e', and f' are all G-numbers, and if the lines $dx + ey + f = 0$ and $d'x + e'y + f' = 0$ meet, then they do so at a point which is a G-number.*

Proof: The lines meet in the point

$$\left(\frac{fe' - ef'}{d'e - e'd}, \frac{df' - d'f}{d'e - e'd} \right)$$

Theorem 5.3.7 *If d, e, f, d', e', and f' are G-numbers, and if the line $dx + ey + f = 0$ meets the circle $(x - d')^2 + (y - e')^2 = f'^2$ in some point, then that point is a G-number.*

Proof: If $e \neq 0$, $dx + ey + f = 0$ is equivalent to $y = -(dx + f)/e$. Substituting this into

$$(x - d')^2 + (y - e')^2 = f'^2$$

we obtain a quadratic equation for x. This can be solved using rational operations ($+$, $-$, $\times$, $/$) and one square root operation. Hence x is a G-number. And so is y.

The result follows in a similar fashion if $e = 0$.

Theorem 5.3.8 *If d, e, f, d', e', and f' are G-numbers, and if the circles $(x-d)^2+(y-e)^2 = f^2$ and $(x-d')^2+(y-e')^2 = f'^2$ meet in some point, then that point is a G-number.*

Proof: If

$$x^2 - 2dx + d^2 + y^2 - 2ey + e^2 = f^2$$

and

$$x^2 - 2d'x + d'^2 + y^2 - 2e'y + e'^2 = f'^2$$

then

$$2(d-d')x + 2(e-e')y + d'^2 - d^2 + e'^2 - e^2 + f^2 - f'^2 = 0$$

The circles intersect where this line meets one of them (and hence the other). By Theorem 5.3.7, the meeting points are G-numbers.

We now have

Theorem 5.3.9 *All constructible numbers are G-numbers.*

Proof: To construct a point, we start with $(0,0)$ and $(1,0)$ and join points, extend lines (to meet lines or circles), and draw circles (to meet lines or circles). We do nothing else. Hence, by the above theorems, any point so constructed is a G-number.

Corollary: If a is a root of a monic irreducible polynomial in $\mathbf{Q}[x]$ with degree which is not a power of 2, then no segment of length a is constructible with straightedge and compass (Theorem 5.3.1).

Pierre Wantzel (1814–1848) gave the above corollary in 1837, and used it to establish some of the limits on straightedge and compass constructions. In particular, he showed that the ancient Greeks had laboured in vain:

Theorem 5.3.10 *No constructible segment has length $\sqrt[3]{2}$.*

Proof: The cube root of 2 is a root of the monic irreducible polynomial $x^3 - 2$, and its degree, 3, is not a power of 2.

Theorem 5.3.11 *You cannot trisect a 60° angle using only straightedge and compass.*

Proof: Let $x = 2\cos 20°$ Since

$$\cos 3A = 4\cos^3 A - 3\cos A$$

and since $\cos 60° = 1/2$, we have $1/2 = x^3/2 - 3x/2$, or

$$x^3 - 3x - 1 = 0$$

By Gauss's Lemma and the Remainder Theorem, this polynomial is irreducible over $\mathbf{Q}[x]$. Hence x is not constructible. And neither is $\cos 20°$. Thus an angle of 60° cannot be trisected using only straightedge and compass.

Theorem 5.3.12 *Let p be an odd prime not of the form $2^n + 1$. Then one cannot construct a regular p-sided polygon using only straightedge and compass.*

Proof: If one can construct a regular p-gon then one can construct the complex number

$$z = (\cos \frac{360°}{p}, \ \sin \frac{360°}{p})$$

Moreover z is the root of a monic irreducible polynomial in $\mathbf{Q}[x]$ of degree 2^s.

By Theorem 5.2.3,

$$f(x) = x^{p-1} + x^{p-2} + \cdots + x + 1$$

is irreducible in $\mathbf{Q}][x]$. Since $(x-1)f(x) = x^p - 1$, and $z^p = 1$, it follows that $f(z) = 0$. Hence z has degree $p - 1$ Thus, if one can construct a regular p-gon, then $p - 1$ is a power of 2.

Thus, for example, it is not possible to give a straightedge and compass construction for a regular 7 or 11 sided polygon.

Theorem 5.3.13 *If p is an odd prime, there is no straightedge and compass construction for a regular polygon with p^2 sides.*

Proof: If there is such a construction, then

$$z = \left(\cos \frac{360°}{p^2}, \ \sin \frac{360°}{p^2} \right)$$

is a G-number, and hence has a degree which is a power of 2.

By Theorem 5.2.4,

$$f(x) = x^{(p-1)p} + x^{(p-2)p} + \cdots + x^{2p} + x^p + 1$$

is irreducible in $\mathbf{Q}[x]$. Since $(x^p - 1)f(x) = x^{p^2} - 1$, it follows that $f(z) = 0$, and hence z has degree $(p-1)p$. But this is never a power of 2.

Hence, for example, there is no straightedge and compass construction for a regular polygon with 9 sides.

Finally, we have

Theorem 5.3.14 *If there is a straightedge and compass construction for an n-sided regular polygon, then $\phi(n)$ is a power of 2.*

Proof: If m is a factor of n, and we can construct a regular n-gon then we can construct a regular m-gon. (Take every (n/m)-th vertex of the regular n-gon.) Thus, from the above, if we can construct a regular n-gon, then n has the form $2^s p_1 \ldots p_r$ where the p's are distinct odd primes of the form $2^k + 1$. But $\phi(2^s p_1 \ldots p_r)$ is then a power of 2 (Theorem 3.2.2).

Exercises 5.3

1. Express $\sin 72°$ in terms of rationals and square roots.
2. Odd primes of the form $2^m + 1$ are *Fermat primes.* Find the first 5 Fermat primes.
3. Show that $2^m + 1$ is prime only if m is 0 or a power of 2.
4. Is every algebraic number of degree 4 constructible?

5.4 Gauss's Constructions

In 1796 Gauss discovered a construction that had eluded the Greeks. For the first time in history, someone constructed a regular 17-sided polygon using only straightedge and compass. In this section we show how Gauss did it.

The key concept we shall use is that of the 'p-character'. We shall develop its basic properties in the first four theorems in this section. Then we shall define the 'gauss sum' belonging to a p-character, and the J function associated with any two given p-characters. This apparatus will lead us to straightedge and compass constructions for all the regular polygons which have them.

Let p be an odd prime. A *p-character* is a function $X : \mathbf{Z} \to \mathbf{C}$ such that

(1) $a \equiv b \pmod p$ implies $X(a) = X(b)$
(2) $X(ab) = X(a)X(b)$
(3) $X(a) = 0$ iff $p|a$.

Theorem 5.4.1 *If X is a p-character, then*
$X(1) = 1$
$(X(a))^{p-1} = 1$ *unless* $p|a$
$(X(a))^{-1} = X(a^{-1})$ *where* a^{-1} *is an inverse of* a*, modulo* p
$(X(a))^{-1} = \overline{X(a)}$ *(the complex conjugate of* $X(a)$*).*

Proof: The first statement follows from the fact that $X(1)X(1) = X(1)$. The second statement follows from Fermat's Little Theorem. The second statement implies that $|X(a)| = 1$, and this leads to the fourth statement.

Theorem 5.4.2 *Let q be a primitive root of p. Then the typical p-character is given by*

$$\begin{aligned} X_k(0) &= 0 \\ X_k(q^t) &= \cos\left(\frac{360^\circ kt}{p-1}\right) + i\sin\left(\frac{360^\circ kt}{p-1}\right) \end{aligned}$$

for $k = 0, 1, \ldots, p-2$. Thus there are exactly $p-1$ p-characters.

Defining $X_k * X_{k'}(a) = X_k(a)X_{k'}(a)$, the p-characters form a group with identity X_0. In this group, $X_k^{-1} = X_{p-1-k}$.

Theorem 5.4.3 *If $X_k \neq X_0$ then*

$$X_k(0) + X_k(1) + \cdots + X_k(p-1) = 0$$

Proof: Let a be an integer such that $0 < a < p$ and $X_k(a) \neq 1$. Then

$$X_k(a)\left(X_k(0) + X_k(1) + \cdots + X_k(p-1)\right)$$

$$= X_k(a \times 0) + X_k(a \times 1) + \cdots + X_k(a(p-1))$$

But $a \times 0$, $a \times 1$, $\ldots$, $a(p-1)$ is a complete set of residues mod p, and hence this latter sum equals the original sum. Call it S. Then $X_k(a)S = S$. Since $X_k(a) \neq 1$, we have $S = 0$.

Theorem 5.4.4 *If $a \not\equiv 0,\ 1$ (mod p) then*

$$S = X_0(a) + X_1(a) + \cdots + X_{p-2}(a) = 0$$

Proof: Let q be a primitive root of p. Since $a = q^t$ for some integer t with $0 < t < p-1$, it follows that $X_1(a) \neq 1$. Now

$$X_1(a)\, S = X_1 * X_0(a) + X_1 * X_1(a) + \cdots + X_1 * X_{p-2}(a)$$

Moreover, the p-characters $X_1 * X_k$ with $k = 0, \ldots, p-2$ just are the $p-1$ p-characters. Thus $X_1(a)\, S = S$, and hence $S = 0$.

Note that $X_0(1) + X_1(1) + \cdots + X_{p-2}(1) = p - 1$.

In what follows we shall use N to denote the complex number

$$e^{2\pi i/p} = \cos\left(\frac{360°}{p}\right) + i\sin\left(\frac{360°}{p}\right)$$

Note that $N^p = 1$.

The *gauss sum* belonging to the p-character X is

$$g(X) = \sum_{t=0}^{p-1} X(t)N^t$$

Theorem 5.4.5 $g(X_0) = -1$.

Proof:

$$\begin{aligned} g(X_0) &= X_0(0)N^0 + X_0(1)N^1 + \cdots + X_0(p-1)N^{p-1} \\ &= 1 + N + N^2 + \cdots + N^{p-1} - 1 \\ &= \frac{N^p - 1}{N - 1} - 1 \;=\; -1 \end{aligned}$$

since $N^p = 1$.

Theorem 5.4.6 *Let X and Y be any two p-characters. Then*

$$g(X)g(Y) = \sum_{t=0}^{p-1}\left(\sum_{u=0}^{p-1} X(u)Y(t-u)\right)N^t$$

The proof is left to the reader.

If X and Y are any p-characters, we define

$$J(X,Y) = \sum_{t=0}^{p-1} X(t)Y(1-t) = \sum_{t+u\equiv 1} X(t)Y(u)$$

Note that if p is a prime of the form $2^n - 1$, then the values of X and Y are all constructible (Theorem 5.4.2), and hence the complex number $J(X,Y)$ is constructible.

Theorem 5.4.7

$$\begin{aligned} J(Y,X) &= J(X,Y) \\ J(X_0,X_0) &= p-2 \end{aligned}$$

If $X \neq X_0$ then $J(X_0,X) = -1$ (by Theorem 5.4.3).
If $X \neq X_0$ then $J(X,X^{-1}) = -X(-1)$ (where $X^{-1} * X = X_0$).

Proof: The last assertion is proved as follows. $X^{-1}(a) = X(a^{-1})$ if $a \neq 0$. Also if $a = 2, \ldots, p-1$, then $a^{-1} - 1$ takes as values all the residues mod p except 0 and $p-1$. Thus

$$\begin{aligned} J(X,X^{-1}) &= J(X^{-1},X) \\ &= X^{-1}(2)X(1-2) + \cdots + X^{-1}(p-1)X(1-(p-1)) \\ &= X(2^{-1})X(1-2) + \cdots + X((p-1)^{-1})X(1-(p-1)) \\ &= X(2^{-1}-1) + \cdots + X((p-1)^{-1}-1) \\ &= -X(p-1) \;=\; -X(-1) \end{aligned}$$

by Theorem 5.4.3.

Theorem 5.4.8 *If X is a p-character other than X_0 then*

$$g(X)g(X^{-1}) = \pm p$$

Proof: If p does not divide t then tx goes through all the residues mod p as x does. Thus, if $Y = X^{-1}$,

$$\begin{aligned} & X(0)Y(t-0) + X(1)Y(t-1) + \cdots + X(p-1)Y(t-(p-1)) \\ = \;& X(t\times 0)Y(t-t\times 0) + \cdots + X(t(p-1))Y(t-t(p-1)) \\ = \;& X*Y(t)(\,X(0)Y(1-0) + \cdots + X(p-1)Y(1-(p-1))\,) \\ = \;& X*Y(t)J(X,Y) \;=\; -X(-1) \end{aligned}$$

If p does divide t then

$$\begin{aligned} & X(0)Y(t-0) + X(1)Y(t-1) + \cdots + X(p-1)Y(t-(p-1)) \\ = \;& Y(-1)(X*Y(0) + X*Y(1) + \cdots + X*Y(p-1)) \\ = \;& Y(-1)(p-1) \end{aligned}$$

since $X * Y = X_0$.

Hence, by Theorem 5.4.6,

$$g(X)g(X^{-1}) = Y(-1)(p-1) - X(-1)(N + N^2 + \cdots + N^{p-1})$$

$$= X(-1)(p-1) - X(-1)(-1) = pX(-1)$$

and $X(-1) = \pm 1$.

Theorem 5.4.9 *If $X * Y \neq X_0$ then $g(X)g(Y) = J(X,Y)g(X * Y)$.*

Proof: The proof is similar to that of Theorem 5.4.8.

Theorem 5.4.10 *Let n be an integer > 2 such that $p \equiv 1 \pmod{n}$ Let X be a p-character with order n in the group of p-characters. (That is, $X^n = X_0$ and $X^k \neq X_0$ if $k < n$.) Then*

$$(g(X))^n = pX(-1)J(X,X)J(X,X^2)\ldots J(X,X^{n-2})$$

Proof: Since $X^2 = X * X \neq X_0$, it follows from Theorem 5.4.9 that $(g(X))^2 = J(X,X)g(X^2)$. Thus

$$(g(X))^3 = J(X,X)g(X)g(X^2) = J(X,X)J(X,X^2)g(X^3)$$

(Theorem 5.4.9). Indeed,

$$(g(X))^{n-1} = J(X,X)J(X,X^2)\ldots J(X,X^{n-2})g(X^{n-1})$$

Since $X^n = X_0$, it follows that $g(X)(g(X^{n-1})) = X(-1)p$ (see Theorem 5.4.8).

Theorem 5.4.11 *If $p = 2^n + 1$ (with $n > 1$) then the regular p-gon is constructible using only straightedge and compass.*

Proof: Let X be any p-character. Since it is one of $p-1=2^n$ characters, which form a group, the order of X is 2^m for some nonnegative integer m. If $m=0$, then $X=X_0$ and $g(X)=-1$ (Theorem 5.4.5). If $m=1$, then $g(X)g(X)=\pm p$ (Theorem 5.4.8). If $m>1$, then $2^m>2$, so, by Theorem 5.4.10,

$$(g(X))^{2^m} = \pm pJ(X,X)\dots J(X,X^{2^m-2})$$

Now X, X^2, ..., X^{2^m-2} all map integers to 2^n-th roots of unity, so, by the definition of J,

$$J(X,X^t) \in \mathbf{Q}[\cos\left(\frac{360^\circ}{2^n}\right) + i\sin\left(\frac{360^\circ}{2^n}\right)]$$

Thus $g(X)$ is in a field of the form $G(c_1,\dots,c_s)$.

Furthermore, using Theorem 5.4.4,

$$g(X_0)+g(X_1)+\cdots+g(X_{p-1}) = \sum_{t=0}^{p-1}\left(\sum_{k=0}^{p-1} X_k(t)\right) N^t = (p-1)N$$

Thus N is also a member of a field of the form $G(c_1,\dots,c_s)$. Hence N is constructible. Thus $\cos\left(\frac{360^\circ}{p}\right)$ is constructible.

Combining the previous result with that of the preceding section, we obtain

Theorem 5.4.12 *If n is an integer ≥ 3, the regular n-sided polygon is constructible using only straightedge and compass iff $\phi(n)$ is a power of 2.*

Exercises 5.4

1. If $p=5$, describe all the p-characters.
2. If $p=5$, and $N=\cos 72^\circ + i\sin 72^\circ$, show that

$$g(X_0)+g(X_1)+\cdots+g(X_4) = 4N$$

3. If $p = 5$ and $m = 3$, verify, for each of the 4 p-characters, that

$$(g(X))^{2^m} = \pm pJ(X, X) \ldots J(X, X^{2^m - 2})$$

4. Prove Theorem 5.4.6.
5. Show that it is possible to construct a regular 771-sided polygon using only straightedge and compass.
6. How many constructible polygons are there with fewer than 1000 sides?

5.5 Fermat Primes

In the previous sections we proved that a regular polygon with prime number p sides is constructible with straightedge and compass iff p has the form $2^m + 1$, with m a positive integer. It is not hard to show that in any such prime, m has the form 2^k, with k a nonnegative integer. These primes are named after Fermat, who thought that, for all k, $2^{2^k} + 1$ is prime. In 1732 his belief was refuted by Euler, who found that 641 is a factor of $2^{2^5} + 1$. We know now that $2^{2^k} + 1$ is composite for $k = 5, 6, 7, \ldots, 21$, but we do not know if there are more than 5 Fermat primes. There might be infinitely many, or there might be only the 5 corresponding to $k = 0, 1, 2, 3$, and 4.

Euler proved the following.

Theorem 5.5.1 *If $n \geq 2$ any factor of $2^{2^n} + 1$ has the form $2^{n+2}k + 1$.*

Proof: If p is a prime factor of $2^{2^n} + 1$ then

$$2^{2^n} \equiv -1 \pmod{p}$$

so that

$$2^{2^{n+1}} \equiv 1 \pmod{p}$$

Thus the order of 2 mod p is a factor of 2^{n+1}. Indeed, it is 2^{n+1}.

Since $p \equiv 1 \pmod{8}$, Theorem 3.8.2 implies that $2^{(p-1)/2} \equiv 1 \pmod{p}$ and hence 2^{n+1} divides $(p-1)/2$. Thus $p - 1 = 2^{n+2}k$.

A priest, Jean François Théophile Pépin (1826–1904), gave the following theorem in 1877.

Theorem 5.5.2 *If k is an integer > 1 then $2^{2^k}+1$ is prime iff*

$$5^{2^{2^k-1}} \equiv -1 \pmod{2^{2^k}+1}$$

Proof: Suppose that $p = 2^{2^k}+1$ is prime. Then, by Theorem 3.8.3,

$$5^{(p-1)/2} \equiv \left(\frac{5}{p}\right) \pmod{p}$$

By Quadratic Reciprocity, $\left(\frac{5}{p}\right) = \left(\frac{p}{5}\right)$, and

$$p = (2^4)^{2^{k-2}} + 1 \equiv 2 \pmod 5$$

Hence $\left(\frac{5}{p}\right) = -1$.

Conversely, suppose the congruence holds. By squaring it, we see that the order of 5 mod $2^{2^k}+1$ is a power of 2. Since 2^{2^k-1} is not large enough, the order of 5 is 2^{2^k}. Thus 2^{2^k} is a factor of $\phi(2^{2^k}+1)$, and hence

$$2^{2^k} \leq \phi(2^{2^k}+1)$$

this being possible only if $2^{2^k}+1$ is prime.

Exercises 5.5

1. Factor $2^{2^6}+1$. Hint: it has just two nontrivial factors.
2. Assuming there are only 5 Fermat primes, how many odd-sided regular polygons are constructible?

5.6 The Transcendence of π *

A complex number is *algebraic* (over **Q**) iff it is a root of a polynomial with integer coefficients. Otherwise, it is *transcendental.* The object of this section is to prove that π is transcendental, and hence there is no straightedge and compass construction for a square with area π. This was first proved by C. L. F. Lindemann in 1882.

A complex number is *algebraic over field* F (with $\mathbf{Q} \subseteq F \subseteq \mathbf{C}$) iff it is a root of a polynomial with coefficients in F.

If a is algebraic over F then a is a root of a monic polynomial $f(x) \in F[x]$, which is irreducible and unique (Theorem 5.2.5). Moreover, if a is algebraic over F then $F(a) = F[a]$, and $F(a)$ has a finite dimension over F. Indeed, $[F(a) : F] = \deg(a)$ (Theorem 5.2.5, 5.2.6). Conversely, if $[F(a) : F]$ is finite, say, equal to n, then

$$1 \quad a \quad a^2 \quad \ldots \quad a^n$$

are linearly dependent (lest $F(a)$ have a vector subspace with higher dimension than it has). Thus there are elements $k_0, \ldots, k_n$ in F such that

$$k_0 + k_1 a + k_2 a^2 + \cdots + k_n a^n = 0$$

Thus a is algebraic over F. Hence we have

Theorem 5.6.1 *a is algebraic over F iff $[F(a) : F]$ is finite.*

Now suppose a and b are algebraic over $\mathbf{Q}$. Then, *a fortiori*, b is a root of a polynomial in $(\mathbf{Q}(a))[x]$. Hence $[(\mathbf{Q}(a))(b) : \mathbf{Q}(a)]$ is finite. By Theorem 5.2.7,

$$[(\mathbf{Q}(a))(b) : \mathbf{Q}] = [(\mathbf{Q}(a))(b) : \mathbf{Q}(a)] \times [\mathbf{Q}(a) : \mathbf{Q}]$$

is thus also finite. Now $\mathbf{Q}(a + b) \subseteq (\mathbf{Q}(a))(b)$. Since a subspace of a finite dimensional vector space is also finite, it follows that $[\mathbf{Q}(a+b) : \mathbf{Q}]$ is finite. Hence $a + b$ is algebraic over $\mathbf{Q}$. Similarly, ab is algebraic over $\mathbf{Q}$.

Furthermore, if $a \neq 0$ is algebraic, with minimal polynomial

$$f(x) = x^n + k_{n-1}x^{n-1} + \cdots + k_1 x + k_0 \in \mathbf{Q}[x]$$

let $g(x) = k_0 x^n + \cdots + k_{n-1}x + 1$. Since

$$\begin{aligned} g(a^{-1}) &= a^{-n}(k_0 + k_1 a + \cdots + k_n a^{n-1} + a^n) \\ &= a^{-n} f(a) = 0 \end{aligned}$$

it follows that a^{-1} is also algebraic. Hence we have the following theorem:

Theorem 5.6.2 *The algebraic numbers form a field.*

Thus if π were algebraic, $\pi\sqrt{-1}$ would be algebraic too.

In order to show the transcendence of π (by way of showing the transcendence of $i\pi$), we need the 'Fundamental Theorem of Symmetric Polynomials'. What is a 'symmetric polynomial'?

A *symmetric polynomial* in $\mathbf{Z}[a_1, \ldots, a_n]$ is a polynomial which remains the same under any permutation of the a's (which might be variables or else complex number constants). Included among the symmetric polynomials in $\mathbf{Z}[a_1, \ldots, a_n]$ are the *elementary symmetric polynomials:*

$$\begin{aligned} s_1 &= a_1 + a_2 + \cdots + a_n \\ s_2 &= a_1a_2 + \cdots + a_{n-1}a_n \\ s_3 &= a_1a_2a_3 + \cdots + a_{n-2}a_{n-1}a_n \\ &\vdots \\ s_n &= a_1a_2 \ldots a_n \end{aligned}$$

Note that s_k has $\binom{n}{k}$ terms. Note also that these elementary symmetric polynomials are just $\pm$ the coefficients of

$$(x - a_1)(x - a_2) \ldots (x - a_n)$$

The Fundamental Theorem of Symmetric Polynomials is the following.

Theorem 5.6.3 *Suppose $F(a_1, \ldots, a_n)$ is a symmetric polynomial in $\mathbf{Z}[a_1, \ldots, a_n]$ with degree $\leq s$ (with the a's in $\mathbf{C}$). Then it can be written as a polynomial in $\mathbf{Z}[s_1, \ldots, s_n]$.*
Moreover, if $f(x) = k(x - a_1) \ldots (x - a_n)$, with k an integer, and if $f(x)$ has integer coefficients, then $k^sF(a_1, \ldots, a_n)$ is an integer.

Proof: To prove this, we need a few definitions. Assuming c and d are nonzero integers, and the j's and k's are nonnegative integers, define

$$ca_1^{j_1} \ldots a_n^{j_n} > da_1^{k_1} \ldots a_n^{k_n}$$

to mean that there is a nonzero term in the sequence $j_1 - k_1$, $j_2 - k_2$, ..., and the first nonzero term is positive.

The *leading term* of $F(a_1, \ldots, a_n)$ is the monomial summand which is larger in the above sense. (We assume that like terms in $F(a_1, \ldots, a_n)$ have been collected.)

If $G_1(a_1, \ldots, a_n)$, $G_2(a_1, \ldots, a_n) \in \mathbf{Z}[a_1, \ldots, a_n]$ then

$$G_1 > G_2$$

just in case G_1's leading terms is $>$ than G_2's leading term. Note that $>$ is transitive.

Note also that, if $b_1, \ldots, b_n$ are nonnegative integers, then the leading term of

$$s_1^{b_1} s_2^{b_2} \ldots s_n^{b^n}$$

is

$$a_1^{b_1+b_2+\cdots+b_n} a_2^{b_2+\cdots+b_n} \ldots a_n^{b_n}$$

— maximising the exponent of a_1, then that of a_2, and so on.

Now suppose $F(a_1, \ldots, a_n)$ has leading term

$$c_1 a_1^{m_1} \ldots a_n^{m_n}$$

with c_1 an integer, and $m_1 + \cdots + m_n \leq s$. Since F is symmetric, $m_1 \geq m_2 \geq \ldots \geq m_n$. Let $b_1 = m_1 - m_2$, $b_2 = m_2 - m_3$, ..., $b_{n-1} = m_{n-1} - m_n$, and $b_n = m_n$. Let

$$G(a_1, \ldots, a_n) = s_1^{b_1} s_2^{b_2} \ldots s_n^{b_n}$$

Then $b_1 + \cdots + b_n \leq s$. As noted above, the leading term of G_1 is

$$a_1^{m_1} a_2^{m_2} \ldots a_n^{m_n}$$

(since $m_1 = b_1 + b_2 + \cdots + b_n$ and so on). Let $F_1 = F - c_1 G_1$. Since the leading terms cancel, $F > F_1$. Note that F_1 is also symmetric.

Repeating the process, we get a symmetric polynomial $F_2 = F_1 - c_2 G_2$ such that $F_1 > F_2$ (with c_2 an integer, and G_2 of the same form as G_1).

Eventually, we get to an F_i which is the 0 polynomial. For suppose

$$c_j a_1^{q_1} \ldots a_n^{q_n}$$

is the leading term of F_j. Since F_j is symmetric, $q_1 \geq \ldots \geq q_n$. As $F > F_j$, we have $m_1 \geq q_1$, so there are only finitely many possibilities for the nonnegative integers q. (For example, if

$$F_j = a_1 a_2 \ldots a_n + d$$

then $F_{j+1} = d$ and F_{j+2} is the 0 polynomial.) Thus, for some i, we have

$$F_i = F - c_1 G_1 - c_2 G_2 - \cdots - c_i G_i = 0$$

so that

$$F = c_1 G_1 + c_2 G_2 + \cdots + c_i G_i$$

with all the c's integers, and $\deg(G_j) \leq s$. Thus F can indeed be written as a polynomial in $\mathbf{Z}[s_1, \ldots, s_n]$ of degree $\leq s$.

Now suppose $k(x - a_1) \ldots (x - a_n)$, with k an integer, has integer coefficients. Then ks_1, ks_2, ..., ks_n are all integers. Since F is a polynomial in $\mathbf{Z}[s_1, \ldots, s_n]$ of degree $\leq s$, it follows that $k^s F$ is a polynomial in $\mathbf{Z}[ks_1, \ldots, ks_n]$, and hence an integer.

We are now in a position to begin our final approach to the proof of the fact that π is transcendental. We start with a polynomial $g(x)$, with integer coefficients, and a positive integer k. We then define a large prime p, and, in terms of p, a polynomial $f(x)$, with rational coefficients. In the next theorem, we use continuity to get a bound on an expression involving $f(x)$. This is in preparation for the final theorem in this section, where, on the assumption that π is *not* transcendental, we show that this bound is violated. This gives us the desired reductio ad absurdum.

Theorem 5.6.4 *Suppose $g(x) = cx^r + c_1 x^{r-1} + \cdots + c_{r-1} x + c_r$ is a polynomial with integer coefficients (and $c \neq 0$). Suppose the roots of $g(x)$ are b_1, b_2, ..., b_r. Let k be a given positive integer. Then there is a prime p such that*

$$p > k,\ |c|,\ |c_r|$$

and, moreover, if t is a real between 0 and 1 (inclusive), and j is one of 1, 2, ..., r, then

$$|(c^r b_j g(tb_j))^p e^{(1-t)b}| < \frac{(p-1)!}{2r}$$

Furthermore, if

$$f(x) = \frac{c^{rp-1}x^{p-1}(g(x))^p}{(p-1)!}$$

then

$$\left|\sum_{j=1}^{r} b_j \int_0^1 e^{(1-t)b_j} f(tb_j)dt\right| \leq \frac{1}{2}$$

Proof: Note that we are using functions of a complex variable.

Since a continuous function on a closed region is bounded in that region, there is a positive integer M_j such that, for all reals t between 0 and 1 (inclusive),

$$|c^r b_j g(tb_j)| < M_j$$

and also

$$|e^{(1-t)b_j}| < M_j$$

Since

$$\lim_{q\to\infty} \frac{M_j^{q+1}}{(q-1)!} = 0$$

(q being a positive integer), there is a positive integer p_j such that

$$\frac{M_j^{p_j+1}}{(p_j-1)!} < \frac{1}{2r}$$

and, moreover, if $p \geq p_j$ then

$$\frac{M_j^{p+1}}{(p-1)!} < \frac{1}{2r}$$

Let p be a prime greater than k, $|c|$, $|c_r|$, p_1, p_2, $\ldots$, p_r. Then if t is a real between 0 and 1 (inclusive),

$$|(c^r b_j g(tb_j))^p e^{(1-t)b_j}| < M_j^{p+1} < \frac{(p-1)!}{2r}$$

Moreover, if f is defined as above,

$$\left|\sum_{j=1}^{r} b_j \int_0^1 e^{(1-t)b_j} f(tb_j)dt\right|$$

$$= \left| \sum_{j=1}^{r} \frac{1}{c} \int_0^1 \frac{(c^r b_j g(tb_j))^p}{(p-1)!} t^{p-1} e^{(1-t)b_j} dt \right|$$

$$\leq \sum_{j=1}^{r} \frac{1}{|c|} \int_0^1 \frac{|(c^r b_j g(tb_j))^p \, e^{(1-t)b_j}|}{(p-1)!} |t|^{p-1} dt$$

$$\leq \sum_{j=1}^{r} \frac{1}{|c|} \int_0^1 \frac{1}{2r} dt \leq \sum_{j=1}^{r} \frac{1}{2r} \leq \frac{1}{2}$$

Note that p is given in terms of $g(x)$ and k, and that $f(x)$ is given in terms of p and $g(x)$.

In the next four theorems, we study the higher derivatives of $f(x)$.

Theorem 5.6.5 *Suppose g, k, p and f are as in Theorem 5.6.4. Let z be an nonnegative integer $< p$. Then the z-th derivative $f^{(z)}(b_j) = 0$ for $j = 1, \ldots, r$.*

Proof: If $z = 0$ then $f^{(z)}(x)$ has the form

$$\frac{c^{rp-1}}{(p-1)!}(g(x))^{p-z} h_z(x)$$

for some $h_z(x) \in \mathbf{Z}[x]$. (Indeed, $h_0(x) = x^{p-1}$.)

Suppose this true for some nonnegative integer z. Then

$$f^{(z+1)}(x) = \frac{c^{rp-1}}{(p-1)!}\{(g(x))^{p-z} h_z'(x) + h_z(x)(p-z)(g(x))^{p-z-1} g'(x)\}$$

$$= \frac{c^{rp-1}}{(p-1)!}(g(x))^{p-(z+1)}\{g(x)h_z'(x) + h_z(x)(p-z)g'(x)\}$$

By mathematical induction, it follows that, for any nonnegative integer $z < p$, $f^{(z)}(x)$ has the form

$$\frac{c^{rp-1}}{(p-1)!}(g(x))^{p-z} h_z(x)$$

Since $g(b_j) = 0$, the result follows. (Recall that the b's were defined as the roots of $g(x)$.)

Theorem 5.6.6 *Suppose g, k, p and f are as in Theorem 5.6.4. Suppose $p \le z \le rp + p - 1$, with z an integer. Then*

$$(p-1)!f^{(z)}(x) \in \mathbf{Z}[x]$$

Also $(p-1)!f^{(z)}(x)$ has degree $\le rp + p - 1 - z$ and all its (integer) coefficients are divisible by $p!c^{rp-1}$.

Proof:

$$(p-1)!f(x) = c^{rp-1}\{c^p x^{rp+p-1} + d_{rp+p-2}x^{rp+p-2} + \cdots + d_p x^p + c_r^p x^{p-1}\}$$

with all the d's integers. Thus

$$(p-1)!f^{(p)}(x) = c^{rp-1}\{c^p e_{rp-1}x^{rp-1} + e_{rp-2}x^{rp-2} + \cdots + d_p p!\}$$

where each of the coefficients e_i is divisible by a product of p consecutive positive integers. Since $\binom{u+p}{p}$ is an integer (if u is a positive integer), it follows that

$$(u+p)(u+p-1)\dots(u+1)$$

is divisible by $p!$ Thus each e_i is a multiple of $p!$ Hence

$$(p-1)!f^{(p)}(x) = c^{rp-1}p!h(x)$$

for some $h(x) \in \mathbf{Z}[x]$, with $\deg(h(x)) = rp - 1$.

Finally, if $z > p$ then

$$(p-1)!f^{(z)}(x) = c^{rp-1}p!h^{(z-p)}(x)$$

with the degree $\le rp - 1 - (z - p)$.

Theorem 5.6.7 *Suppose g, k, p and f are as in Theorem 5.6.4. Then*

$$\begin{aligned} f^{(z)}(0) &= 0 && \text{if } z = 0, 1, \dots, p-2 \\ f^{(p-1)}(0) &= c^{rp-1}c_r^p && \text{and finally,} \\ f^{(z)}(0) &= pK_z \end{aligned}$$

for some integer K_z, if $z = p, p+1, \dots, rp+p-1$.

Proof: As noted at the beginning of the proof of Theorem 5.6.6,

$$\frac{(p-1)!}{c^{rp-1}}f(x) = c^p x^{rp+p-1} + d_{rp+p-2}x^{rp+p-2} + \cdots + d_p x^p + c_r^p x^{(p-1)}$$

with the d's integers. Taking the first $p-2$ derivatives, we find there is still an x in every term. Hence $f^{(z)}(0) = 0$ if $z \le p-2$. Furthermore,

$$\frac{(p-1)!}{c^{rp-1}}f^{(p-1)}(x)$$

$$= c^p e_{rp}x^{rp} + e_{rp-1}x^{rp-1} + \cdots + (d_p p(p-1)\ldots 2)x + c_r^p(p-1)!$$

so that $f^{(p-1)}(0) = c^{rp-1}c_r^p$ as required. If $z \ge p$, then, as shown in the proof of Theorem 5.6.6,

$$\frac{(p-1)!}{c^{rp-1}}f^{(z)}(x)$$

has the form $p!h(x)$ for some $h(x) \in \mathbf{Z}[x]$. Thus $f^{(z)}(x)$ has the form $pc^{rp-1}h(x)$ and so $f^{(z)}(0)$ is a multiple of p.

Theorem 5.6.8 *Suppose g, k, p and f are as in Theorem 5.6.4. Suppose z is an integer such that $p \le z \le rp+p-1$. Then, for some integer k_z,*

$$\sum_{j=1}^{r} f^{(z)}(b_j) = pk_z$$

Proof: Let $s = rp-1$. By Theorem 5.6.6,

$$\frac{f^{(z)}(x)}{pc^s}$$

is a polynomial with integer coefficients only and degree $\le s$. Thus

$$\sum_{j=1}^{r}\frac{f^{(z)}(b_j)}{pc^s}$$

is a symmetric polynomial in $\mathbf{Z}[b_1, \ldots, b_r]$ with degree at most s. Also

$$g(x) = c(x-b_1)\ldots(x-b_r)$$

with c an integer, has only integer coefficients. Hence, by the Fundamental Theorem of Symmetric Polynomials,

$$c^s \sum_{j=1}^{r} \frac{f^{(z)}(b_j)}{pc^s}$$

is an integer.

The Fundamental Theorem of Symmetric Polynomials is also invoked in our penultimate lemma:

Theorem 5.6.9 *Suppose q_1 is an integer, and suppose*

$$g_1(x) = q_1(x - a_1)(x - a_2)\dots(x - a_n)$$

is a polynomial with integer coefficients only. For $j = 2, 3, \dots, n$, let $g_j(x)$ be the polynomial with degree $\binom{n}{j}$ defined as follows:

$$g_j(x) = (x - (a_1 + a_2 + \cdots + a_j))\dots(x - (a_{n-(j-1)} + \cdots + a_n))$$

(There are $\binom{n}{j}$ factors corresponding to the $\binom{n}{j}$ ways of picking j summands from the numbers $a_1, \dots, a_n$.)
Then, for each j there is an integer $q_j \neq 0$ such that $q_j g_j(x)$ has integer coefficients only.

Proof: Consider

$$g_2(x) = (x - (a_1 + a_2))\dots(x - (a_{n-1} + a_n))$$

For each nonnegative integer i, the coefficient of x^i is a symmetric polynomial in $\mathbf{Z}[a_1, \dots, a_n]$. By the Fundamental Theorem of Symmetric Polynomials, the coefficient of x^i can be written as a polynomial in $\mathbf{Z}[s_1, \dots, s_n]$. Since $g_1(x)$ has integer coefficients only, each of $s_1, \dots, s_n$ is rational. Hence the coefficient of x^i in $g_2(x)$ is rational. Thus, for some integer q_2, $q_2 g_2(x)$ has integer coefficients only.

The same sort of reasoning applies to g_j for any j.

Finally, we have Lindemann's result:

Theorem 5.6.10 *π is transcendental.*

Proof: It suffices to show that $i\pi$ is transcendental, since the algebraic numbers form a field.

To obtain a contradiction, suppose $i\pi$ is algebraic, with minimal polynomial $(x-a_1)\ldots(x-a_n)$ — where $a_1 = i\pi$. Let

$$g_1(x) = q_1(x-a_1)\ldots(x-a_n)$$

be a polynomial with integer coefficients only, q_1 being an integer. Let $g_2(x), \ldots, g_n(x)$ be as in Theorem 5.6.9, and let $q_2, \ldots, q_n$ be as in Theorem 5.6.9. Let

$$g^*(x) = g_1(x)q_2g_2(x)\ldots q_ng_n(x)$$

Then $g^*(x)$ can be written in the form

$$g^*(x) = (cx^r + c_1x^{r-1} + \cdots + c_{r-1}x + c_r)x^{k-1}$$

where c, $c_r \neq 0$, and all the c's are integers, and k is a positive integer.

Let $b_1, b_2, \ldots, b_r$ be the nonzero roots of $g^*(x)$, that is, the roots of $g(x) = cx^r + c_1x^{r-1} + \cdots + c_r$.

Now consider

$$(e^{a_1}+1)(e^{a_2}+1)\ldots(e^{a_n}+1)$$

Since $a_1 = i\pi$, this product is 0. That is,

$$\begin{aligned} 1 &+ e^{a_1} + e^{a_2} + \cdots + e^{a_n} \\ &+ e^{a_1+a_2} + \cdots + e^{a_{n-1}+a_n} \\ &+ e^{a_1+a_2+a_3} + \cdots + e^{a_{n-2}+a_{n-1}+a_n} \\ &+ \ldots \\ &+ e^{a_1+\cdots+a_n} \\ &= 0 \end{aligned}$$

The exponents of the e's are just the roots of $g^*(x)$. Suppose $k-1$ of these roots are 0 (as above). The complex numbers $b_1, \ldots, b_r$ are, precisely, the nonzero roots of $g^*(x)$, that is, the roots of $g(x)$. Thus

$$1 + (k-1) + \sum_{j=1}^{r} e^{b_j} = 0 \quad (*)$$

with k a positive integer.

Let p be a prime as in Theorem 5.6.4. Let f be defined as in Theorem 5.6.4. If $f^{(j)}(x)$ is the j-th derivative of f with respect to x — so that $f^{(rp+p)}(x) = 0$ — define

$$F(x) = f(x) + f^{(1)}(x) + \cdots + f^{(rp+p-1)}(x)$$

Note that

$$\begin{aligned}
(e^{-x}F(x))' &= e^{-x}F'(x) - e^{-x}F(x) \\
&= -e^{-x}(-F'(x) + F(x)) \\
&= -e^{-x}(-f^{(1)}(x) - f^{(2)}(x) - \cdots - f^{(rp+p)}(x) \\
&\qquad + f(x) + f^{(1)}(x) + \cdots + f^{(rp+p-1)}(x)) \\
&= -e^{-x}f(x)
\end{aligned}$$

Thus, by the Fundamental Theorem of Calculus (for complex variables — since x can be any complex number),

$$e^{-x}F(x) - e^{0}F(0) = \int_0^x -e^{-s}f(s)ds$$

Let $t = s/x$ so that

$$\frac{dt}{ds} = \frac{1}{x}$$

(since x is considered as a constant in relation to s). Then

$$e^{-x}F(x) - F(0) = \int_0^1 -e^{-tx}f(tx)xdt$$

and

$$F(x) - e^{x}F(0) = -x\int_0^1 e^{(1-t)x}f(tx)dt$$

Letting x take values $b_1, \ldots, b_r$, and adding up the r resulting equations, we get

$$\sum_{j=1}^{r} F(b_j) - \sum_{j=1}^{r} e^{b_j}F(0) = -\sum_{j=1}^{r} b_j \int_0^1 e^{(1-t)b_j}f(tb_j)dt$$

Hence, by (*) above,

$$\sum_{j=1}^{r} F(b_j) - kF(0) = -\sum_{j=1}^{r} b_j \int_0^1 e^{(1-t)b_j} f(tb_j)dt$$

Note that the expression on the right is just the one we had in Theorem 5.6.4, where we showed that its absolute value is bounded by 1/2. By Theorem 5.6.5,

$$\begin{aligned}\sum_{j=1}^{r} F(b_j) &= f(b_1) + f^{(1)}(b_1) + \cdots + f^{(rp+p-1)}(b_1) \\ &\qquad +f(b_2) + f^{(1)}(b_2) + \cdots + f^{(rp+p-1)}(b_2) \\ &\qquad +\cdots \\ &\qquad +f(b_r) + f^{(1)}(b_r) + \cdots + f^{(rp+p-1)}(b_r) \\ &= f^{(p)}(b_1) + \cdots + f^{(rp+p-1)}(b_1) \\ &\qquad +f^{(p)}(b_2) + \cdots + f^{(rp+p-1)}(b_2) \\ &\qquad +\cdots \\ &\qquad +f^{(p)}(b_r) + \cdots + f^{(rp+p-1)}(b_r)\end{aligned}$$

By Theorem 5.6.8, it now follows that

$$\sum_{j=1}^{r} F(b_j)$$

is an integer, and a multiple of p. By Theorem 5.6.7,

$$\begin{aligned}F(0) &= f(0) + f^{(1)}(0) + \cdots + f^{(p)}(0) + \cdots + f^{(rp+p-1)}(0) \\ &= c^{rp-1}c_r^p + pM\end{aligned}$$

for some integer M. Thus, since $p > k$, $|c|$, $|c_r|$ and c, $c_r \neq 0$,

$$\sum_{j=1}^{r} F(b_j) - kF(0)$$

is an integer which is not a multiple of p. Hence it is a nonzero integer. Yet, by Theorem 5.6.4,

$$\left|\sum_{j=1}^{r} b_j \int_0^1 e^{(1-t)b_j} f(tb_j)dt\right| \leq \frac{1}{2}$$

Contradiction. Hence π is transcendental — and the Greeks worked in vain to square the circle.

Exercises 5.6

1. Prove that $\pi^2 - 3\pi + 1$ is transcendental.
2. Let ABC be a right triangle, with right angle at A. Construct semicircles outwardly on AB and AC. Let the *lunes* be the two areas enclosed by these semicircles, and also the semicircle through B, A, and C. Hippocrates (440 BC) showed that one *can* 'square the lunes'. Do the same.
3. Let ABC be an equilateral triangle of side 1. The three circles with centres A, B, and C, and the same radius r, overlap to form a familiar Venn diagram (provided $r > 1/\sqrt{3}$). Let x be the area of the part of the circle with centre A that is outside the other two circles, and let y be the area of the region which is common to the circles with centres A and B but is not shared by the circle with centre C. (In set theory terms, x is the area of the region representing $A - (B \cup C)$, while y is the area of the region representing $(A \cap B) - C$.) Prove that, for any $r > 1/\sqrt{3}$, we have $x - y = \sqrt{3}/2$.

Chapter 6

The Polygonal Number Theorem

A *polygonal number* is a nonnegative integer of the form

$$m\frac{t^2-t}{2}+t$$

where m is a positive integer, and t is a nonnegative integer. For example, when $m = 1$, we have the *triangular numbers* 0, 1, 3, 6, 10, 15, and so on. These are called triangular numbers because n pebbles can be arranged in the form of an isosceles right triangle just in case n has the form $(t^2-t)/2+t$.

When $m = 2$ we have the *square numbers* 0, 1, 4, 9, 16, and so on. When $m = 3$ we have the *pentagonal numbers* 0, 1, 5, 12, 22, and so on.

On account of their natural geometric representations, these polygonal numbers were studied as long ago as Pythagoras (525 BC). Nicomachus of Gerasa (near Jerusalem) mentions them in his *Introductio Arithmeticae* (100 AD), and Diophantus (250 AD) wrote a treatise on them, in which he proves that

$$m\frac{t^2-t}{2}+t=1+(1+m)+(1+2m)+\cdots+(1+(t-1)m)$$

Pierre de Fermat (1601–1665) conjectured that every positive integer is a sum of 3 triangular numbers, 4 square numbers, 5 pentagonal numbers, and so on. For example,

$$19 = 1 + 3 + 15 = 1 + 1 + 1 + 16 = 0 + 1 + 1 + 5 + 12$$

In his *Disquisitiones Arithmeticae* (1801), Carl Friedrich Gauss gave the first proof of this conjecture for the case of the triangular numbers, and, in 1813, Augustin Cauchy gave the first proof of the whole conjecture. The purpose of this chapter is to give a relatively short, completely elementary proof of Fermat's conjecture. This proof is an abridgement of the work of Gauss and Cauchy. We begin by discussing matrices.

Exercises

1. Is 153 triangular?
2. In how many ways is 100 a polygonal number?
3. If $f(m,t) = m(t^2 - t)/2 + t$, show that

$$f(m,t+1) - f(m,t) - (f(m,t) - f(m,t-1)) = m$$

6.1 Gaussian Forms

Let D be an integer which is negative and congruent to 1 mod 4. Let a, b, and c be relatively prime integers such that a, $c > 0$ and $b^2 - 4ac = D$. For example, if $D = 4n + 1$, we might have $a = -n$, $b = 1$, and $c = 1$. The polynomial $ax^2 + bxy + cy^2$ is a *gaussian form,* and is denoted by $\begin{bmatrix} a & b & c \end{bmatrix}$. With this gaussian form we associate the matrix

$$\begin{bmatrix} a & \frac{b}{2} \\ \frac{b}{2} & c \end{bmatrix}$$

Note that

$$\begin{bmatrix} a & b & c \end{bmatrix} = (x \;\; y) M (x \;\; y)^T$$

where A^T is the transpose of the matrix A. The number D is the *discriminant* of the form $\begin{bmatrix} a & b & c \end{bmatrix}$ and its corresponding matrix.

Theorem 6.1.1 *If D has r distinct prime factors then the number of gaussian forms $\begin{bmatrix} a & a & c \end{bmatrix}$ (with $b = a$) is 2^r.*

Proof: $ax^2 + axy + cy^2$ is one of the required gaussian forms iff $a(a - 4c) = D$ with a positive and odd, $a - 4c$ negative and odd, and $\gcd(a, a - 4c) = 1$. There are r choices for prime divisors of D factoring a. Whatever the choice, $c = (a^2 - D)/4a$ is a positive integer relatively prime to a. Thus there are 2^r possibilities.

For example, suppose $D = -315$. Then three distinct primes divide D (namely, 3, 5, and 7), and there are 8 choices for a (namely, 1, 3^2, 5, 7, $3^2 5$, $3^2 7$, 5×7, and 315). If, say, $a = 3^2$ then $a - 4c = -35$, and $c = (81 + 315)/36 = 11$.

The multiplicative group of 2 by 2 matrices with integer entries and determinant 1 is given the awkward name $SL_2(\mathbf{Z})$. If

$$G = \begin{bmatrix} r & s \\ t & u \end{bmatrix} \in SL_2(\mathbf{Z})$$

we define $G * \begin{bmatrix} a & b & c \end{bmatrix}$ as

$$(x \;\; y) G M G^T (x \;\; y)^T$$

$$= \begin{bmatrix} ar^2 + brs + cs^2 & 2art + b(ru + st) + 2csu & at^2 + btu + cu^2 \end{bmatrix}$$

where M is the matrix associated with $\begin{bmatrix} a & b & c \end{bmatrix}$.

Theorem 6.1.2 *If* $\begin{bmatrix} a & b & c \end{bmatrix}$ *is a gaussian form, and* $G \in SL_2(\mathbf{Z})$, *then* $G * \begin{bmatrix} a & b & c \end{bmatrix}$ *is a gaussian form (with the same discriminant D).*

Proof: Let M be the matrix associated with $\begin{bmatrix} a & b & c \end{bmatrix}$, and let M' be the matrix associated with

$$G * \begin{bmatrix} a & b & c \end{bmatrix} = \begin{bmatrix} a' & b' & c' \end{bmatrix}$$

Let

$$G = \begin{bmatrix} r & s \\ t & u \end{bmatrix}$$

Any prime which divides a, b, and c also divides

$$\begin{aligned} a' &= ar^2 + brs + cs^2 \\ b' &= 2art + b(ru + st) + 2csu \\ c' &= at^2 + btu + cu^2 \end{aligned}$$

Since

$$(G^{-1}) * \begin{bmatrix} a' & b' & c' \end{bmatrix} = (G^{-1}) * (G * \begin{bmatrix} a & b & c \end{bmatrix}) = \begin{bmatrix} a & b & c \end{bmatrix}$$

it follows, in the same way, that any prime which divides a', b', and c' also divides a, b, and c. Hence a', b', and c' are relatively prime just in case a, b, and c are relatively prime — which they are.

The discriminant of $\begin{bmatrix} a' & b' & c' \end{bmatrix}$ is -4 times the determinant of $M' = GMG^T$. Hence, since the determinant of G is 1, $\begin{bmatrix} a' & b' & c' \end{bmatrix}$ has the same discriminant D as $\begin{bmatrix} a & b & c \end{bmatrix}$.

Since $D < 0$, it follows that $|b| < 2\sqrt{ac}$. (Recall that a and c are positive.) Since

$$\left(\sqrt{a}|r| - \sqrt{c}|s|\right)^2 \geq 0$$

we have

$$ar^2 - |brs| + cs^2 > 0$$

and hence $a' = ar^2 + bsr + cs^2 > 0$. Since $b'^2 - 4a'c' = D < 0$, it follows that $c' > 0$.

The next theorem gives some important examples.

Theorem 6.1.3

$$\begin{bmatrix} 0 & 1 \\ -1 & 0 \end{bmatrix} * \begin{bmatrix} a & b & c \end{bmatrix} = \begin{bmatrix} c & -b & a \end{bmatrix}$$

$$\begin{bmatrix} 0 & 1 \\ -1 & 1 \end{bmatrix} * \begin{bmatrix} a & a & c \end{bmatrix} = \begin{bmatrix} c & 2c-a & c \end{bmatrix}$$

$$\begin{bmatrix} -1 & 2 \\ -1 & 1 \end{bmatrix} * \begin{bmatrix} a & a & c \end{bmatrix} = \begin{bmatrix} 4c-a & 4c-a & c \end{bmatrix}$$

$$\begin{bmatrix} 1 & 0 \\ n & 1 \end{bmatrix} * \begin{bmatrix} a & b & c \end{bmatrix} = \begin{bmatrix} a & b+2an & an^2+bn+c \end{bmatrix}$$

Note that if $a \neq 0$ and n is the integer nearest $-b/2a$, then $|b+2an| \leq a$.

A gaussian form $\begin{bmatrix} a & b & c \end{bmatrix}$ (or corresponding matrix) is *reduced* just in case (1) $|b| \leq a \leq c$ and (2) $b \geq 0$ if $|b| = a$ or $c = a$.

Theorem 6.1.4 *If* $\begin{bmatrix} a & b & c \end{bmatrix}$ *is reduced then* $a \leq \sqrt{-D/3}$.

Proof: $4a^2 \leq 4ac = b^2 - D \leq a^2 - D$, so that $3a^2 \leq -D$.

For example, there is only one reduced gaussian form with discriminant -3, namely, $\begin{bmatrix} 1 & 1 & 1 \end{bmatrix}$, since a has to be 1, and hence b, which must be odd, has to be 1 also.

Theorem 6.1.5 *Suppose a and c are two relatively prime positive integers.*
If $a \leq c$ *then* $\begin{bmatrix} a & a & c \end{bmatrix}$ *is reduced.*
If $c < a < 2c$ *then* $\begin{bmatrix} c & 2c-a & c \end{bmatrix}$ *is reduced.*
If $2c < a \leq 3c$ *then* $\begin{bmatrix} c & -(2c-a) & c \end{bmatrix}$ *is reduced .*
If $3c < a < 4c$ *then* $\begin{bmatrix} 4c-a & 4c-a & c \end{bmatrix}$ *is reduced.*

(All the above forms have discriminant $D = a^2 - 4ac$.)

Two gaussian forms F and F' (or their corresponding matrices) are *properly equivalent* iff for some G in $SL_2(\mathbf{Z})$, $F' = G * F$ (and $M' = GMG^T$). It is not hard to prove that proper equivalence is an equivalence relation (using the fact that $(G^T)^{-1} = (G^{-1})^T$ and the fact that $G^T G'^T = (G'G)^T$).

Theorem 6.1.6 *Every gaussian form is properly equivalent to a reduced gaussian form.*

Proof: By Theorem 6.1.3, $\begin{bmatrix} a & b & c \end{bmatrix}$ is equivalent to $\begin{bmatrix} c & -b & a \end{bmatrix}$ and also to $\begin{bmatrix} a & b+2an & an^2+bn+c \end{bmatrix}$. Moreover, if n is the integer nearest $-b/2a$, then $-a \leq b + 2an \leq a$.

Using these facts, we can construct a sequence of properly equivalent gaussian forms, whose first member is the given form, and which is such that the first coefficient a of the forms steadily decreases. For example, if the given form is $\begin{bmatrix} 10 & 14 & 5 \end{bmatrix}$, we have

$$\begin{bmatrix} 10 & 14 & 5 \end{bmatrix} \begin{bmatrix} 5 & -14 & 10 \end{bmatrix} \begin{bmatrix} 5 & -4 & 1 \end{bmatrix} \begin{bmatrix} 1 & 4 & 5 \end{bmatrix} \begin{bmatrix} 1 & 0 & 1 \end{bmatrix}$$

Since the coefficients a are positive integers, this sequence cannot continue forever without arriving at a form in which $|b| \leq a \leq c$. If $b = -a$, then, using Theorem 6.1.3, with $n = 1$, we can obtain a properly equivalent form with $b = a \leq c$. If $b < 0$ and $c = a$, then using the first statement of Theorem 6.1.3, we can obtain a properly equivalent form with $0 \leq b \leq a = c$.

Theorem 6.1.7 *No two reduced gaussian forms are properly equivalent.*

Proof: Suppose $\begin{bmatrix} a & b & c \end{bmatrix}$ and $\begin{bmatrix} a' & b' & c' \end{bmatrix}$ are reduced and properly equivalent. Then there is a matrix

$$G = \begin{bmatrix} r & s \\ t & u \end{bmatrix} \in SL_2(\mathbf{Z})$$

such that

$$G * \begin{bmatrix} a & b & c \end{bmatrix} = \begin{bmatrix} a' & b' & c' \end{bmatrix}$$

and $a' = ar^2 + brs + cs^2$. Without loss of generality, suppose $a' \leq a$. Then

$$\begin{aligned} & a(r + bs/2a)^2 + (-D/4a)s^2 \\ = & \; ar^2 + bsr + ab^2s^2/4a^2 + (4ac - b^2)s^2/4a = a' \leq a \end{aligned}$$

and hence $(-D/4a)s^2 \leq a$. Thus $-Ds^2 \leq 4a^2 \leq -4D/3$ (Theorem 6.1.4), so that $s = 0$ or ± 1.

Suppose $s = 0$. Then $a(r + bs/2a)^2 \leq a$ implies that $r^2 = 1$, so that $a' = a$. (If $r = 0$ then, since $s = 0$, G is not in $SL_2(\mathbf{Z})$.) Furthermore, $b' = 2art + bru$. Since G is in $SL_2(\mathbf{Z})$, it has determinant 1, and hence $ru = 1$. Thus $b' = 2art + b$. Since $-a \leq b$, $b' \leq a$, and $b' - b = 2art$, it follows that $b' = b$. (Recall that if $b = a$ then $b \geq 0$.) Hence

$$c' = \frac{b'^2 - D}{4a'} = \frac{b^2 - D}{4a} = c$$

Suppose $s = \pm 1$. Then $ar^2 \pm br + c \leq a$ and hence $ar^2 \pm br \leq 0$. Thus $r = 0$ or $a|r| \leq |b|$. Since $|b| \leq a$, it follows that $r = 0$ or ± 1.

If $r = 0$ then $a' = c$. Since $a' \leq a$ this implies $a = c$ and hence $b \geq 0$. Also if $r = 0$ then $st = -1$ and $b' = -b + 2csu$. Since $|b' + b| < 2c$, it follows that $su = 0$ and hence $u = 0$. Thus $c' = a = a'$, so that $b' \geq 0$. Since $b' = -b$ this implies that $b' = b = 0$.

If $r = \pm 1$ then $b = a$ (since $a|r| \leq |b| \leq a$). Since $ar^2 \pm br + c = a' \leq a$, we have $a \pm a + c \leq a$ and hence $a = c$. Since a, b, and c are relatively prime, it follows that they all equal 1, and the discriminant D is therefore -3. We saw above that there is only one reduced gaussian form with discriminant -3.

PROBLEM: Find all reduced forms with $D = -23$.
SOLUTION: If $\begin{bmatrix} a & b & c \end{bmatrix}$ is one of these forms then, by Theorem 6.1.4, $a < 3$. If $a = 1$ then $b = 1$ and $c = (b^2 + 23)/4a = 6$. Indeed, $\begin{bmatrix} 1 & 1 & 6 \end{bmatrix}$ is reduced.

If $a = 2$ then, again, $b = \pm 1$ (b cannot be even) and $c = 3$. The forms $\begin{bmatrix} 2 & 1 & 3 \end{bmatrix}$ and $\begin{bmatrix} 2 & -1 & 3 \end{bmatrix}$ are both reduced.

There are only these 3 possibilities, and we say the *class number* for -23 is 3.

Gaussian forms of the form $\begin{bmatrix} a & a & c \end{bmatrix}$ (all with discriminant D) are called *special ambiguous forms.* In Theorem 6.1.1 we saw that if r is the number of distinct prime divisors of D then the number of special ambiguous forms is 2^r. By Theorem 6.1.3 the two special ambiguous forms $\begin{bmatrix} a & a & c \end{bmatrix}$ and $\begin{bmatrix} 4c-a & 4c-a & c \end{bmatrix}$ are properly equivalent. (These cannot be the same form, since a and c are relatively prime, so that $4c - a = a$ implies $c = 1$, $a = 2$, and hence $b = 2$ — against the fact that $b^2 - 4ac$ is odd.) In the special ambiguous form $\begin{bmatrix} a & a & c \end{bmatrix}$, $a \neq 2c$ and $a < 4c$ (the latter since $a^2 - 4ac = D < 0$). By Theorem 6.1.3 and 6.1.5, $\begin{bmatrix} a & a & c \end{bmatrix}$ is properly equivalent to a reduced form $\begin{bmatrix} * & * & c \end{bmatrix}$ with final coefficient c. Thus if $\begin{bmatrix} a & a & c \end{bmatrix}$ and $\begin{bmatrix} a' & a' & c' \end{bmatrix}$ are properly equivalent, so that they are properly equivalent to the same reduced form, then $c' = c$, and thus, since

$$a^2 - 4ac = D = a'^2 - 4a'c'$$

we have $(a'-2c)^2 = (a-2c)^2$ whence $a' = a$ or $4c-a$. Thus each special ambiguous form $\begin{bmatrix} a & a & c \end{bmatrix}$ is properly equivalent to exactly one other special ambiguous form, namely, $\begin{bmatrix} 4c-a & 4c-a & c \end{bmatrix}$. For example, $\begin{bmatrix} 1 & 1 & 1 \end{bmatrix}$ is properly equivalent to $\begin{bmatrix} 3 & 3 & 1 \end{bmatrix}$, and to no other special ambiguous form.

The 'properly equivalent' equivalence relation partitions the gaussian forms into pairwise disjoint equivalence classes. What the above tells us is that the 2^r special ambiguous forms are found in exactly 2^{r-1} of these equivalence classes. If an equivalence class contains $\begin{bmatrix} a & a & c \end{bmatrix}$ then it also contains $\begin{bmatrix} 4c-a & 4c-a & c \end{bmatrix}$, and no other special ambiguous form.

For example, if $D = -23$ there are three equivalence classes, corresponding to the three reduced forms $\begin{bmatrix} 1 & 1 & 6 \end{bmatrix}$ and $\begin{bmatrix} 2 & \pm 1 & 3 \end{bmatrix}$. Here the number of distinct primes $r = 1$, and the special ambiguous forms are $\begin{bmatrix} 1 & 1 & 6 \end{bmatrix}$ and $\begin{bmatrix} 23 & 23 & 6 \end{bmatrix}$, both found in the equivalence class of forms properly equivalent to $\begin{bmatrix} 1 & 1 & 6 \end{bmatrix}$.

Let $[\begin{bmatrix} a & b & c \end{bmatrix}]$ be the equivalence class containing the gaussian form $\begin{bmatrix} a & b & c \end{bmatrix}$. An equivalence class containing a special ambiguous form — one which therefore can be written $[\begin{bmatrix} a & a & c \end{bmatrix}]$ — is a *special ambiguous class.* From the above it follows that there are 2^{r-1} special ambiguous classes (where r is the number of distinct prime divisors of D).

Exercises 6.1

1. For $D = -39$, find the special ambiguous forms, and the reduced forms to which they are equivalent.
2. Find all the reduced gaussian forms with $D = -163$.
3. Show that the class number for -15 is 2.
4. What reduced gaussian form is properly equivalent to $\begin{bmatrix} 12 & 5 & 13 \end{bmatrix}$ and what matrix G in $SL_2(\mathbf{Z})$ reduces it ?

6.2 Ternary Quadratic Form Matrices

In order to show that every natural number is a sum of three triangular numbers, we need to study three by three matrices.

$$\text{Let } M = \begin{bmatrix} a & b & c \\ d & e & f \\ h & i & j \end{bmatrix}$$

be an invertible 3 by 3 matrix. Define

$$\overline{M} = \begin{bmatrix} ej - fi & fh - dj & di - eh \\ ci - bj & aj - ch & bh - ai \\ bf - ce & cd - af & ae - bd \end{bmatrix}$$

Then $M\overline{M}^T = (\det M)I$, where I is the 3 by 3 identity matrix. Hence $\overline{M}^T = (\det M)M^{-1}$, and

$$\det(\overline{M}) = (\det M)^3(\det M^{-1}) = (\det M)^2.$$

Also

$$\overline{\overline{M}} = (\det\overline{M})(\overline{M}^{-1})^T = (\det M)^2(\det M)^{-1}M = (\det M)M.$$

Furthermore, $\overline{MM'} = \overline{M}\,\overline{M'}$ and $\overline{M^T} = \overline{M}^T$. Note also that, where s is any real number, $\overline{sM} = s^2\overline{M}$.

For example, if

$$M = \begin{bmatrix} 1 & 0 & 0 \\ 0 & e & f \\ 0 & f & j \end{bmatrix} \text{ then } \overline{M} = \begin{bmatrix} 1 & 0 & 0 \\ 0 & j & -f \\ 0 & -f & e \end{bmatrix}$$

Again, if

$$F = \begin{bmatrix} a & u/2 & w/2 \\ u/2 & b & v/2 \\ w/2 & v/2 & c \end{bmatrix}$$

with $\det F \neq 0$, then

$$\overline{F} = \begin{bmatrix} bc - v^2/4 & vw/4 - cu/2 & uv/4 - bw/2 \\ vw/4 - cu/2 & ac - w^2/4 & uw/4 - av/2 \\ uv/4 - bw/2 & uw/4 - av/2 & ab - u^2/4 \end{bmatrix}$$

If a, b, c, u, v, and w are integers then a matrix of the form F (above) is an *integral ternary quadratic form matrix.* Let's call that a *ternary* for short.

The name $GL_3(\mathbf{Z})$ is given to the multiplicative group of 3 by 3 matrices with integer entries and determinant ± 1. If G is a member of $GL_3(\mathbf{Z})$, then so is $\overline{G}$. (Recall from above that $\det(\overline{M}) = (\det \overline{M})^2$.) If F is a ternary, so is GFG^T. Two ternaries F and F' are *equivalent* just in case there is some G in $GL_3(\mathbf{Z})$ such that $F' = GFG^T$. 'Equivalent' is an equivalence relation.

Let G be a matrix in $GL_3(\mathbf{Z})$ of the form

$$\begin{bmatrix} r & s & 0 \\ t & q & 0 \\ 0 & 0 & 1 \end{bmatrix}$$

with the 'upper left determinant', $rq - st$, equal to 1. Then G is *top left heavy.* The set of top left heavy matrices is a subgroup of $GL_3(Z)$. Note that if G is top left heavy, so is $\overline{G}$. Similarly, the *bottom right heavy* matrices

$$\begin{bmatrix} 1 & 0 & 0 \\ 0 & r & s \\ 0 & t & q \end{bmatrix}$$

with 'lower left determinant' $rq - st$ equal to 1 also form a subgroup of $GL_3(\mathbf{Z})$. Again, if G is bottom right heavy, so is $\overline{G}$.

If F is a ternary, and G is top left heavy, then GFG^T has the form

$$\begin{bmatrix} \begin{bmatrix} r & s \\ t & q \end{bmatrix} \begin{bmatrix} a & u/2 \\ u/2 & a \end{bmatrix} \begin{bmatrix} r & t \\ s & q \end{bmatrix} & * \\ * & c \end{bmatrix}$$

— with the bottom right entry c the same in F and GFG^T. And note also that the upper left determinant is invariant too, since $rq - st = 1$. If G is bottom right heavy then GFG^T has the form

$$\begin{bmatrix} a & * \\ * & \begin{bmatrix} r & s \\ t & q \end{bmatrix} \begin{bmatrix} b & v/2 \\ v/2 & c \end{bmatrix} \begin{bmatrix} r & t \\ s & q \end{bmatrix} \end{bmatrix}$$

— with the top left entry a the same in F and GFG^T. And note also that the lower right determinant is invariant too, since $rq - st = 1$.

Thanks to the next two theorems we can use top left heavy and bottom right heavy matrices to 'reduce' ternaries — much the way we found a 'reduced' gaussian form equivalent to a given gaussian form.

Theorem 6.2.1 *If*

$$F = \begin{bmatrix} a & u/2 & w/2 \\ u/2 & b & v/2 \\ w/2 & v/2 & c \end{bmatrix}$$

is a ternary and

$$G = \begin{bmatrix} 0 & 1 & 0 \\ -1 & 0 & 0 \\ 0 & 0 & 1 \end{bmatrix}$$

then

$$GFG^T = \begin{bmatrix} b & -u/2 & * \\ -u/2 & a & * \\ * & * & c \end{bmatrix}$$

where the $*$'s represent integers or half integers. Moreover, if

$$G = \begin{bmatrix} 1 & 0 & 0 \\ n & 1 & 0 \\ 0 & 0 & 1 \end{bmatrix}$$

then

$$GFG^T = \begin{bmatrix} a & an+u/2 & * \\ an+u/2 & an^2+un+b & * \\ * & * & c \end{bmatrix}$$

Note that $a(an^2+un+b)-(an+u/2)^2 = ab-(u/2)^2$. Note also that if $a \neq 0$ and n is the integer nearest $-u/2a$ then $|2an+u| \leq |a|$.

Theorem 6.2.2 *If*

$$F = \begin{bmatrix} a & u/2 & w/2 \\ u/2 & b & v/2 \\ w/2 & v/2 & c \end{bmatrix}$$

is a ternary and

$$G = \begin{bmatrix} 1 & 0 & 0 \\ 0 & 0 & 1 \\ 0 & -1 & 0 \end{bmatrix}$$

then

$$GFG^T = \begin{bmatrix} a & * & * \\ * & c & -v/2 \\ * & -v/2 & b \end{bmatrix}$$

Moreover, if

$$G = \begin{bmatrix} 1 & 0 & 0 \\ 0 & 1 & n \\ 0 & 0 & 1 \end{bmatrix}$$

then

$$GFG^T = \begin{bmatrix} a & * & * \\ * & cn^2+vn+b & cn+v/2 \\ * & cn+v/2 & c \end{bmatrix}$$

Theorem 6.2.3 *Let*

$$F = \begin{bmatrix} a & u/2 & w/2 \\ u/2 & b & v/2 \\ w/2 & v/2 & c \end{bmatrix}$$

be a ternary. Then there is a top left heavy matrix G such that (1) the upper left determinant j of F equals that of GFG^T, such that (2) the absolute value of the top left entry of GFG^T is $\leq \sqrt{4|j|/3}$, and such that (3) the bottom right entry of $\overline{F}$ (namely, j) equals the bottom right entry of $\overline{GFG^T}$.

Proof: From Theorem 6.2.1 it follows that there is a top left heavy matrix G such that, in GFG^T, $|u| \leq |a| \leq |b|$ (see Theorem 6.1.6), and such that the upper left determinant is the same for GFG^T and for F (and hence the bottom right entry is the same for $\overline{F}$ and $\overline{GFG^T}$). (If, during the 'reduction', we have $a = 0$, then we can stop there, since $0 \leq \sqrt{4|j|/3}$.) Since $|u| \leq |a| \leq |b|$, it follows that

$$4a^2 \leq u^2 - u^2 + 4|ab| \leq a^2 - u^2 + 4|ab|$$

so that $3a^2 \leq -u^2 + 4|ab|$. If $ab \geq 0$ it follows that $a \leq \sqrt{4|j|/3}$. If $ab < 0$, we have

$$3a^2 \leq -4ab - u^2 \leq -4ab + u^2$$

and, again, the result follows.

Theorem 6.2.4 *Let F be a ternary. Then there is a bottom right heavy matrix H such that (1) the lower right determinant k of $\overline{F}$ equals that of $H\overline{F}H^T$, such that (2) the absolute value of the lower right entry of $H\overline{F}H^T$ is $\leq \sqrt{4|k|/3}$, and such that (3) the upper left entry of $\overline{H\overline{F}H^T}$ equals that of F. (This upper left entry is $k/\det F$.)*

Proof: Use Theorem 6.2.2 (as Theorem 6.2.1 was used in the proof of Theorem 6.2.3.) Note that if $\overline{F}$ is not a ternary, we must first prove the result for $4\overline{F}$ and then for $\overline{F}$. Note also that (3) follows since $\overline{H}$ is bottom right heavy if H is.

The F Sequence

Now let

$$F = \begin{bmatrix} a & u/2 & w/2 \\ u/2 & b & v/2 \\ w/2 & v/2 & c \end{bmatrix}$$

be a ternary. Starting with F, we shall generate a sequence of ternaries, equivalent to F, called F_1, F_2, F_3, ... The symbol a_n shall denote the top left entry of F_n, while j_n shall denote the top left determinant of F_n, so that j_n is also the bottom right entry C_n of $\overline{F_n}$. The symbol k_n shall denote the bottom right determinant of $\overline{F_n}$ — so that, in fact, $k_n = a_n D$, where D is the determinant of F_n (and of all the other F's).

If $|a| \leq \sqrt{4|j|/3}$ — where $j = ab - u^2/4$ — let $F_1 = F$. Otherwise, let G be as in Theorem 6.2.3, and let $F_1 = GFG^T$. Then, if a_1 is the top left entry of F_1, we have $|a_1| \leq \sqrt{4|j|/3}$. If C_1 is the lower right entry of $\overline{F_1}$, and C the lower right entry of $\overline{F}$, then $C_1 = C$. (Both equal j.)

Let k_1 be the lower right determinant of $\overline{F_1}$. If $|C_1| \leq \sqrt{4|k_1|/3}$, we halt this process. Otherwise, let H be as in Theorem 6.2.4, and let $F_2 = \overline{H}F_1\overline{H}^T$. Then, if C_2 is the lower right entry of $H\overline{F_1}H^T$,

$$|C_2| \leq \sqrt{4|k_1|/3} < |C_1|.$$

Also if a_2 is the top left entry of F_2, then $a_2 = a_1$. Note that $\overline{F_2} = H\overline{F_1}H^T$ since $\overline{\overline{H}} = H$ (because H, being bottom right heavy, has determinant 1, and, in general, $\overline{\overline{M}} = (\det M)M$).

Let j_2 be the upper left determinant of F_2. If $|a_1| \leq \sqrt{4|j_2|/3}$, we halt this process. Otherwise, let G be as in Theorem 6.2.3 (relative to F_2), and let $F_3 = GF_2G^T$. Then, if a_3 is the top left entry of F_3,

$$|a_3| \leq \sqrt{4|j_2|/3} < |a_2|.$$

Also the lower right entry of $\overline{F_3}$ equals that of $\overline{F_2}$, that is, $C_3 = C_2$.

Continuing in this way — loop back to the paragraph beginning 'let k_1 be ...' — we produce a sequence of equivalent ternaries, F, F_1, F_2, F_3, ... with

$$|a_1| = |a_2| > |a_3| = |a_4| > \dots$$

and

$$|C| = |C_1| > |C_2| = |C_3| > |C_4| = \ldots$$

— which must halt, since the a's are integers and the C's are quarter integers. Thus, for some n, $|a_n| \leq \sqrt{4|j_n|/3}$ and also $|C_n| \leq \sqrt{4|k_n|/3}$.

Thus every ternary F is equivalent to one in which

$$|a| \leq \sqrt{|u^2 - 4ab|/3}$$

and, where D is the determinant of F,

$$|C| = |ab - u^2/4| \leq \sqrt{\frac{4|(ac - w^2/4)(ab - u^2/4) - (uw/4 - av/2)^2|}{3}}$$

$$\leq \sqrt{4|aD|/3}$$

Hence

Theorem 6.2.5 *Every ternary F is equivalent to a ternary*

$$\begin{bmatrix} a & u/2 & w/2 \\ u/2 & b & v/2 \\ w/2 & v/2 & c \end{bmatrix}$$

in which

$$|a| \leq \sqrt{|u^2 - 4ab|/3} \text{ and } |u^2 - 4ab| \leq 8\sqrt{|aD|/3}$$

and hence $|a| \leq \frac{4}{3}\sqrt[3]{|D|}$ — where D is the determinant of F.

Proof: $3a^2 \leq 8\sqrt{|aD|/3}$ so $9a^4 \leq 64|aD|/3$.

EXAMPLE

Consider the ternary

$$F = \begin{bmatrix} 2 & 1 & -1 \\ 1 & 54 & -16 \\ -1 & -16 & 5 \end{bmatrix}$$

F has determinant 1, and

$$\overline{F} = \begin{bmatrix} 14 & 11 & 38 \\ 11 & 9 & 31 \\ 38 & 31 & 107 \end{bmatrix}$$

The top left determinant j of F is 107 and $2 = |a_1| \leq \sqrt{4|j|/3}$, so we let $F_1 = F$. $C_1 = 107$.

The lower right determinant k_1 of $\overline{F_1}$ is $9 \times 107 - 31^2 = 2$, so we have to 'reduce' this matrix using Theorem 6.2.4. We can use

$$H = \begin{bmatrix} 1 & 0 & 0 \\ 0 & 3 & -1 \\ 0 & 7 & -2 \end{bmatrix}$$

(To get this matrix H, we reduce $M = \begin{bmatrix} 9 & 31 \\ 31 & 107 \end{bmatrix}$ as in the previous section. Using $G_1 = \begin{bmatrix} 1 & 0 \\ -3 & 1 \end{bmatrix}$, we get $G_1 M G_1^T = \begin{bmatrix} 9 & 4 \\ 4 & 2 \end{bmatrix}$. Using $G_2 = \begin{bmatrix} 0 & 1 \\ -1 & 0 \end{bmatrix}$ we get $G_2 G_1 M G_1^T G_2^T = \begin{bmatrix} 2 & -4 \\ -4 & 9 \end{bmatrix}$. Finally, using $G_3 = \begin{bmatrix} 1 & 0 \\ 2 & 1 \end{bmatrix}$, we get

$$G_3 G_2 G_1 M G_1^T G_2^T G_3^T = \begin{bmatrix} 2 & 0 \\ 0 & 1 \end{bmatrix}$$

Now $G_3 G_2 G_1 = \begin{bmatrix} -3 & 1 \\ -7 & 2 \end{bmatrix}$, and we obtain H.) With H as above,

$$H\overline{F_1}H^T = \begin{bmatrix} 14 & -5 & 1 \\ -5 & 2 & 0 \\ 1 & 0 & 1 \end{bmatrix} \text{ and } F_2 = \overline{H}F\overline{H}^T = \begin{bmatrix} 2 & 5 & -2 \\ 5 & 13 & -5 \\ -2 & -5 & 3 \end{bmatrix}$$

We thus have $C_2 = 1$ and $a_2 = 2$. The upper left determinant j_2 of F_2 is 1, so we must find a matrix G to reduce F_2 (since we do not have $|a_1| \leq \sqrt{4|j_2|/3}$). We can take

$$G = \begin{bmatrix} -2 & 1 & 0 \\ -3 & 1 & 0 \\ 0 & 0 & 1 \end{bmatrix} \text{ so that } F_3 = GF_2G^T = \begin{bmatrix} 1 & 0 & -1 \\ 0 & 1 & 1 \\ -1 & 1 & 3 \end{bmatrix}$$

Then

$$\overline{F_3} = \overline{GF_2G^T} = \begin{bmatrix} 2 & -1 & 1 \\ -1 & 2 & -1 \\ 1 & -1 & 1 \end{bmatrix}$$

and here the bottom right entry, C_3, is less than $\sqrt{4|k|/3}$.

In the next two theorems, we derive a couple of results for ternaries with small determinants. These theorems are close to the heart of our proof of Fermat's conjecture.

Theorem 6.2.6 *Every ternary F with determinant $-1/4$ is equivalent to*

$$\begin{bmatrix} 0 & 0 & -1/2 \\ 0 & 1 & 0 \\ -1/2 & 0 & 0 \end{bmatrix}$$

Proof: By Theorem 6.2.5, F is equivalent to a ternary F' with $|a| \leq .9$ and hence $a = 0$, and thus also with $u^2 \leq 0$ and hence $u = 0$. If

$$F' = \begin{bmatrix} a & u/2 & w/2 \\ u/2 & b & v/2 \\ w/2 & v/2 & c \end{bmatrix}$$

then $\det F' = (w/2)(-bw/2) = -1/4$, so that $w^2b = 1$ and $w^2 = 1$, $b = 1$. If

$$G = \begin{bmatrix} 1 & 0 & 0 \\ -wv & 1 & 0 \\ -wc & 0 & 1 \end{bmatrix}$$

then

$$GF'G^T = \begin{bmatrix} 0 & 0 & w/2 \\ 0 & 1 & 0 \\ w/2 & 0 & 0 \end{bmatrix}$$

Call the latter matrix H. If $w = -1$, we are done. However, suppose $w = 1$. If

$$G' = \begin{bmatrix} 1 & 0 & 0 \\ 0 & 1 & 0 \\ 0 & 0 & -1 \end{bmatrix}$$

then

$$G'HG'^T = \begin{bmatrix} 0 & 0 & -w/2 \\ 0 & 1 & 0 \\ -w/2 & 0 & 0 \end{bmatrix}$$

as required.

Theorem 6.2.7 *Every ternary F with integer entries and determinant 1 is equivalent to*

$$\begin{bmatrix} 1 & 0 & 0 \\ 0 & 1 & 0 \\ 0 & 0 & 1 \end{bmatrix} \quad \text{or to} \quad \begin{bmatrix} 0 & 0 & 1 \\ 0 & -1 & 0 \\ 1 & 0 & 0 \end{bmatrix}$$

Proof: From the above, F is equivalent to a ternary F' with integer entries such that

$$\begin{aligned} |a| &\leq \sqrt{|u^2 - 4ab|/3} \\ |u^2 - 4ab| &\leq 8\sqrt{|a|/3} \\ |a| &\leq 4/3 \end{aligned}$$

Case 1. $a = \pm 1$.

By Theorem 6.2.1, we may take it that $u/2 = 0$. (Note that $u \neq 1$ since u is even — because the matrix has integer entries.) Since $3 \leq |u^2 - 4b| \leq 4$, this implies that $b = \pm 1$. Let

$$G = \begin{bmatrix} 1 & 0 & 0 \\ 0 & 1 & 0 \\ -\frac{1}{2}wa & 0 & 1 \end{bmatrix}$$

Since $\frac{1}{2}w$ is an integer, and $a = \pm 1$, it follows that $G \in GL_3(\mathbf{Z})$. If F' is the ternary to which F is equivalent,

$$GF'G^T = \begin{bmatrix} a & 0 & 0 \\ 0 & b & v/2 \\ 0 & v/2 & c - w^2/(4a) \end{bmatrix}$$

If

$$H = \begin{bmatrix} 1 & 0 & 0 \\ 0 & 1 & 0 \\ 0 & -\frac{1}{2}vb & 1 \end{bmatrix} \text{ then } HGF'(HG)^T = \begin{bmatrix} a & 0 & 0 \\ 0 & b & 0 \\ 0 & 0 & c' \end{bmatrix}$$

Since the determinant of the latter matrix is still 1, $c' = \pm 1$.

If a, b, and c' are all positive, we are done. Otherwise, exactly two of them equal -1. In that case, if $a = 1$, apply

$$G_1 = \begin{bmatrix} 1 & 1 & 0 \\ 1 & 1 & 1 \\ 1 & 0 & 0 \end{bmatrix}$$

If $b = 1$, apply

$$G_2 = \begin{bmatrix} -1 & 1 & 0 \\ -1 & 1 & 1 \\ 0 & 1 & 1 \end{bmatrix}$$

If $c = 1$, apply

$$G_3 = \begin{bmatrix} 0 & 1 & 1 \\ -1 & 1 & 1 \\ -1 & 0 & 1 \end{bmatrix}$$

Then, for the appropriate i, $G_i HGF'(G_i HG)^T$ has the second of the two forms given in the theorem.

Case 2. $a = 0$.

Then $u = 0$ and, since the determinant of F' is $-bw^2/4 = 1$, it follows that $b = -1$ and $w = \pm 2$. Since v is even, the following matrix is in $GL_3(\mathbf{Z})$:

$$G = \begin{bmatrix} 1 & 0 & 0 \\ -v/w & 1 & 0 \\ 0 & 0 & 1 \end{bmatrix}$$

Moreover,

$$GF'G^T = \begin{bmatrix} 0 & 0 & w/2 \\ 0 & -1 & 0 \\ w/2 & 0 & c \end{bmatrix}$$

If

$$H = \begin{bmatrix} 1 & 0 & 0 \\ w/2 & 1 & 0 \\ 0 & 1 & 1 \end{bmatrix} \text{ then } HGF'G^T H^T = \begin{bmatrix} 0 & 0 & w/2 \\ 0 & -1 & 0 \\ w/2 & 0 & c-1 \end{bmatrix}$$

so the given matrix F is equivalent to a matrix

$$F'' = \begin{bmatrix} 0 & 0 & w/2 \\ 0 & -1 & 0 \\ w/2 & 0 & c \end{bmatrix}$$

with c even.

Since c is even,

$$J = \begin{bmatrix} 1 & 0 & 0 \\ 0 & 1 & 0 \\ -c/w & 0 & 1 \end{bmatrix}$$

is in $GL_3(\mathbf{Z})$, and we have

$$JF''J^T = \begin{bmatrix} 0 & 0 & w/2 \\ 0 & -1 & 0 \\ w/2 & 0 & 0 \end{bmatrix}$$

If $w = 2$, we are done. Otherwise, apply

$$J' = \begin{bmatrix} 1 & 0 & 0 \\ 0 & 1 & 0 \\ 0 & 0 & -1 \end{bmatrix}$$

The matrix $J'JF''J^TJ'^T$ has the second of the two given forms.

Theorem 6.2.8 *Let T_1, T_2, T_3, and X be any integers (with $X \neq 0$). If $\gcd(T_1, T_2, T_3) = 1$ then there are relatively prime integers U and V such that*

$$\gcd(T_1V^2 - T_2UV + T_3U^2,\ 2X) = 1.$$

Proof: Let $u_1, \ldots, u_h$ be the distinct primes which divide $2X$ but not T_1. Let U be their product (or let $U = 1$ if there are no such primes). Let $v_1, \ldots, v_i$ be the distinct primes which divide $2X$ and T_1 but not T_3. Let V be their product (or let $V = 1$ if there are no such primes). Let $w_1, \ldots, w_k$ be the distinct primes which divide $2X$, T_1, and T_3. Then no w divides $Y = T_1V^2 - T_2UV + T_3U^2$ lest it divide T_2UV and hence T_2 — against the fact that $\gcd(T_1, T_2, T_3) = 1$.

From the definition of U and V, $\gcd(U, V) = 1$, and $\gcd(Y, 2X) = 1$ as well. (If a prime p divides $2X$ then it is a u, v, or w. If it is a u it does not divide Y, lest it divide T_1 — which it does not. So p is not a u. Similarly, it is not a v or w.)

The last theorem in this section is the key to the next.

Theorem 6.2.9 *Suppose*

$$A = \begin{bmatrix} a & b/2 & k/2 \\ b/2 & c & m/2 \\ k/2 & m/2 & n \end{bmatrix}$$

is a ternary with determinant $-1/4$. *Suppose* a, $c > 0$ *and* b *is odd. Suppose that* $b^2 - 4ac < 0$ *and* $\gcd(a, b, c) = 1$. *Then*

$$\begin{bmatrix} a & b/2 \\ b/2 & c \end{bmatrix}$$

is properly equivalent to a matrix

$$\begin{bmatrix} N^2 & b'/2 \\ b'/2 & c' \end{bmatrix}$$

where $\gcd(N, 2(b^2 - 4ac)) = 1$.

Proof: By Theorem 6.2.6, there is a matrix T in $GL_3(\mathbf{Z})$ such that, where

$$M = \begin{bmatrix} 0 & 0 & -\frac{1}{2} \\ 0 & 1 & 0 \\ -\frac{1}{2} & 0 & 0 \end{bmatrix}$$

we have $TMT^T = A$. Let $\begin{bmatrix} t_1 & t_2 & t_3 \end{bmatrix}$ be the bottom row of T, and let $\begin{bmatrix} T_1 & T_2 & T_3 \end{bmatrix}$ be the bottom row of $\overline{T}$. Then

$$\pm 1 = \det T = t_1 T_1 + t_2 T_2 + t_3 T_3$$

and hence $\gcd(T_1, T_2, T_3) = 1$.

By Theorem 6.2.8, there are relatively prime integers U and V such that

$$\gcd(T_1 V^2 - T_2 UV + T_3 U^2,\ 2(b^2 - 4ac)) = 1.$$

Let H and J be integers such that $UJ - VH = 1$. Let

$$S = \begin{bmatrix} U^2 & UH & H^2 \\ 2UV & UJ + VH & 2HJ \\ V^2 & VJ & J^2 \end{bmatrix}$$

By brute calculation, we find that $\det S = 1$ and $SMS^T = M$. Also the right column of $\overline{S}$ is

$$\begin{bmatrix} V^2 \\ -UV \\ U^2 \end{bmatrix}$$

so that the bottom right entry of $\overline{T}\,\overline{S}$ is $T_1 V^2 - T_2 UV + T_3 U^2$ — the (nonzero) integer relatively prime to $2(b^2 - 4ac)$.

Let

$$TS = \begin{bmatrix} r_1 & r_2 & r_3 \\ s_1 & s_2 & s_3 \\ * & * & * \end{bmatrix}$$

Since $TSM(TS)^T = A$, we obtain

$$\begin{aligned} a &= r_2{}^2 - r_1 r_3 \\ b/2 &= r_2 s_2 - r_1 s_3/2 - r_3 s_1/2 \\ c &= s_2{}^2 - s_1 s_3 \end{aligned}$$

Hence $as_1{}^2 - bs_1 r_1 + cr_1{}^2 = (r_1 s_2 - r_2 s_1)^2$.

Furthermore, $r_1 s_2 - r_2 s_1$ is the lower right entry of $\overline{TS}$, which we have seen equals $T_1 V^2 - T_2 UV + T_3 U^2$. Thus $r_1 s_2 - r_2 s_1$ is nonzero and relatively prime to $2(b^2 - 4ac)$.

Let

$$s'' = \frac{s_1}{\gcd(s_1, r_1)} \text{ and } r'' = \frac{r_1}{\gcd(s_1, r_1)}.$$

Let t'' and u'' be integers so that

$$-s''t'' - r''u'' = 1$$

and hence

$$G = \begin{bmatrix} -s'' & r'' \\ u'' & t'' \end{bmatrix}$$

is a member of $SL_2(\mathbf{Z})$. Then

$$G \begin{bmatrix} a & b/2 \\ b/2 & c \end{bmatrix} G^T = \begin{bmatrix} N^2 & b'/2 \\ b'/2 & c' \end{bmatrix}$$

where

$$N = \frac{r_1 s_2 - r_2 s_1}{\gcd(s_1, r_1)}$$

is relatively prime to $2(b^2 - 4ac)$.

Exercises 6.2

1. Show that if F is a ternary, and $G \in GL_3(\mathbf{Z})$ then GFG^T is a ternary.
2. Prove that 'equivalent' is an equivalence relation for ternaries.
3. Prove that if G is top left heavy, then so is $\overline{G}$.
4. Illustrate Theorem 6.2.5 in the case of the ternary

$$F = \begin{bmatrix} 2 & 3 & -\frac{5}{2} \\ 3 & 3 & 3 \\ -\frac{5}{2} & 3 & -4 \end{bmatrix}$$

5. Show that Theorem 6.2.9 applies in the case of the matrix

$$M = \begin{bmatrix} 1 & 1/2 & 1 \\ 1/2 & 2 & 7/2 \\ 1 & 7/2 & 6 \end{bmatrix}$$

6.3 Omega Kernel or Square Forms

Let D be an integer which is negative and congruent to 1 modulo 4. Let H be the set of residue classes x mod D such that $\gcd(x, D) = 1$ and $z^2 \equiv x \pmod{D}$ has a solution. Then H is a multiplicative group. Note also that if $x \in H$ then $x^2 \in H$.

Theorem 6.3.1 *H has $\phi(D)/2^r$ members, where r is the number of distinct prime factors of D.*

Proof: As we noted in connection with the Chinese Remainder Theorem, $z^2 \equiv x \pmod{D}$ (with $\gcd(x, D) = 1$) has either no solution or 2^r solutions. If x is in H, it has 2^r solutions. All these solutions are among the $\phi(D)$ residue classes relatively prime to D. If x_1 and x_2 are distinct members of H then no solution of $z^2 \equiv x_1 \pmod{D}$ is a solution of $z^2 \equiv x_2 \pmod{D}$. Hence, with each member of H we can associate a set of 2^r residues relatively prime to D, and these sets are pairwise disjoint. Moreover, if u is any residue relatively prime to D, u^2 is in H, and u is in the set of solutions associated with u^2. Thus the set of residues relatively prime to D is partitioned, via the members of H, into sets each containing 2^r members. Hence H has $\phi(D)/2^r$ members.

The gaussian form $F = \begin{bmatrix} a & b & c \end{bmatrix}$ *represents* an integer m just in case there are integers x and y such that

$$ax^2 + bxy + cy^2 = m$$

If F represents m, and F' is properly equivalent to F, then F' also represents m. For if $G \in SL_2(\mathbf{Z})$ then $(x\ y)M(x\ y)^T = m$ implies that

$$(x'\ y')GMG^T(x'\ y')^T = m$$

— if $(x'\ y') = (x\ y)G^{-1}$. Thus if $\begin{bmatrix} a & b & c \end{bmatrix}$ represents m, we say that the equivalence class $[\begin{bmatrix} a & b & c \end{bmatrix}]$ represents m.

We give the name C to the set of these equivalence classes. We shall see, in the next section, that C is a group.

Suppose x is in H. Then $z^2 \equiv x \pmod{D}$ has a solution z, and, with $D = b^2 - 4ac$,

$$z^2 + bz \times 0 + \frac{b^2 - D}{4} \times 0^2 = x + QD$$

for some integer Q. Thus if x is in H, the gaussian form $\begin{bmatrix} 1 & b & \frac{b^2-D}{4} \end{bmatrix}$ (with b odd) represents an integer congruent to x mod D. A gaussian form which represents an integer of the form $x + QD$, with x in H, is an *omega kernel form.* Note that if two gaussian forms are properly equivalent, and one is an omega kernel form, so is the other. Hence it makes sense to define an *omega kernel class* as an equivalence class in C which contains an omega kernel form (that is, one which represents an integer of the form $x+QD$ with x in H). Let K be the set of omega kernel classes.

In Section 5 of this chapter we shall define an ω function, with domain C and codomain U/H, where U is the set of residue classes relatively prime to D. This function will take an equivalence class $[f]$ of gaussian forms to the coset mH, where m is any residue in U such that a number of the form $m + QD$ is represented by $[f]$. We shall prove that the kernel of ω (the subset of C which ω maps to H) is, precisely, the set of omega kernel classes.

We now address the matter of 'square forms'.

Two gaussian forms $F_1 = \begin{bmatrix} a_1 & b_1 & c_1 \end{bmatrix}$ and $F_2 = \begin{bmatrix} a_2 & b_2 & c_2 \end{bmatrix}$ are *concordant* iff $b_1 = b_2$ and $a_2|c_1$. Note that, since $b_1^2-4a_1c_1 = b_2^2-4a_2c_2$, it follows that, for concordant forms, $a_1(c_1/a_2) = c_2$, and hence $a_1|c_2$.

The *composition* $F_1 \circ F_2$ of two concordant forms is the form

$$\begin{bmatrix} a_1a_2 & b_1 & c_1/a_2 \end{bmatrix}$$

Note that $b_1^2 - 4a_1a_2c_1/a_2 = D$. For example, the form

$$\begin{bmatrix} N & b & Nc \end{bmatrix}$$

is concordant with itself.

Form $\begin{bmatrix} a & b & c \end{bmatrix}$ is a *square form* iff there are two properly equivalent, concordant gaussian forms, F and F', such that $\begin{bmatrix} a & b & c \end{bmatrix}$ is properly equivalent to $F \circ F'$. For example,

$$\begin{bmatrix} N^2 & b & c \end{bmatrix} = \begin{bmatrix} N & b & Nc \end{bmatrix} \circ \begin{bmatrix} N & b & Nc \end{bmatrix}$$

is a square form.

Theorem 6.3.2 *If concordant forms F_1 and F_2 represent integers m_1 and m_2 respectively, then $F_1 \circ F_2$ represents m_1m_2.*

Proof: Suppose $c = c_2/a_1 = c_1/a_2$. If

$$\begin{aligned} X &= x_1x_2 - cy_1y_2 \\ Y &= a_1x_2y_2 + a_2y_1x_2 + by_1y_2 \end{aligned}$$

then, as was discovered by Gauss, brute calculation yields

$$(a_1x_1^2 + b_1x_1y_1 + c_1y_1^2)(a_2x_2^2 + b_2x_2y_2 + c_2y_2^2) = a_1a_2X^2 + bXY + cY^2$$

— where $b = b_1 = b_2$.

Theorem 6.3.3 *A gaussian form $\begin{bmatrix} a & b & c \end{bmatrix}$ is an omega kernel form iff it is a square form.*

Proof: If it is a square form then there are two properly equivalent, concordant forms F and F' such that $\begin{bmatrix} a & b & c \end{bmatrix}$ is properly equivalent to $F \circ F'$. If F represents x, then F' represents x, and $F \circ F'$ represents x^2 (Theorem 6.3.2), and $x^2 \in H$. Hence $\begin{bmatrix} a & b & c \end{bmatrix}$ is an omega kernel form.

Conversely, suppose $\begin{bmatrix} a & b & c \end{bmatrix}$ represents some integer $h + QD$, with h an element of H. Now h^{-1} is also in H, and, from the above, the gaussian form

$$\begin{bmatrix} 1 & b & (b^2 - D)/4 \end{bmatrix}$$

represents an integer j which is congruent to h^{-1}, mod D.

Now $\begin{bmatrix} a & b & c \end{bmatrix}$ and $\begin{bmatrix} 1 & b & (b^2 - D)/4 \end{bmatrix}$ are concordant, and, by Theorem 6.3.2, their composition $\begin{bmatrix} a & b & c \end{bmatrix}$ represents hj, which is congruent to 1, mod D. Thus there are integers m and k and Q such that

$$am^2 - bmk + ck^2 = (-Q)D + 1$$

Let

$$A = \begin{bmatrix} a & b/2 & k/2 \\ b/2 & c & m/2 \\ k/2 & m/2 & Q \end{bmatrix}$$

Then A has determinant

$$Q(-D/4) - \left(\frac{m}{2}\right)\left(\frac{am}{2} - \frac{bk}{4}\right) + \left(\frac{k}{2}\right)\left(\frac{bm}{4} - \frac{ck}{2}\right) = -\frac{1}{4}$$

Hence, by Theorem 6.2.9, $\begin{bmatrix} a & b & c \end{bmatrix}$ is properly equivalent to a form $\begin{bmatrix} N^2 & b' & c' \end{bmatrix}$ with $\gcd(N, 2D) = 1$.

Take $N > 0$. Since $\gcd(N, 2D) = 1$, it follows that $\gcd(N, b', Nc') = 1$. Hence $\begin{bmatrix} N & b' & Nc' \end{bmatrix}$ is a gaussian form (with discriminant D). It is self-concordant, and

$$\begin{bmatrix} N & b' & Nc' \end{bmatrix} \circ \begin{bmatrix} N & b' & Nc' \end{bmatrix} = \begin{bmatrix} N^2 & b' & c' \end{bmatrix}$$

Hence $\begin{bmatrix} a & b & c \end{bmatrix}$ is a square form.

Exercises 6.3

1. If $D = -55$, what are the members of H?
2. If $D = -55$, what are the reduced square forms?
3. Suppose $\gcd(m, D) = \gcd(n, D) = 1$, and m and n are both represented by $\begin{bmatrix} a & b & c \end{bmatrix}$. Then mn is in H.

6.4 Ambiguous or Self-Inverse Forms

In this section we first define 'ambiguous' forms. Then we define a group operation for the set C of equivalence classes of gaussian forms. Next we define 'self-inverse' forms, in terms of this group operation, and show that a form is ambiguous just in case it is a self-inverse. We end this section by using this fact to gain some information about the number of elements in C. (This is the *class number* for D.)

A gaussian form $\left[\begin{array}{ccc} a & b & c \end{array}\right]$ is *ambiguous* iff it is properly equivalent to a special ambiguous form $\left[\begin{array}{ccc} a' & a' & c' \end{array}\right]$.

Theorem 6.4.1 $\left[\begin{array}{ccc} a & b & c \end{array}\right]$ *is ambiguous iff* $\left[\begin{array}{ccc} a & b & c \end{array}\right]$ *is properly equivalent to* $\left[\begin{array}{ccc} c & b & a \end{array}\right]$

Proof: First suppose that $\left[\begin{array}{ccc} a & b & c \end{array}\right]$ is ambiguous, being properly equivalent to $\left[\begin{array}{ccc} a' & a' & c' \end{array}\right]$. Let G be in $SL_2(\mathbf{Z})$ such that, where

$$M = \begin{bmatrix} a & b/2 \\ b/2 & c \end{bmatrix} \quad \text{and} \quad M' = \begin{bmatrix} a' & a'/2 \\ a'/2 & c' \end{bmatrix}$$

(so that M' is the matrix associated with $\left[\begin{array}{ccc} a' & a' & c' \end{array}\right]$), we have $M = GM'G^T$. Let

$$H = \begin{bmatrix} 0 & 1 \\ 1 & 0 \end{bmatrix} \quad \text{and} \quad J = \begin{bmatrix} 1 & 0 \\ 1 & -1 \end{bmatrix}$$

Then HGJ is in $SL_2(\mathbf{Z})$. Also $JM'J^T = M'$. Thus M is properly equivalent to

$$HGJM'(HGJ)^T = HGM'G^TH^T = HMH^T$$

which is the matrix corresponding to $\left[\begin{array}{ccc} c & b & a \end{array}\right]$.

To prove the converse, suppose that $G * \left[\begin{array}{ccc} a & b & c \end{array}\right] = \left[\begin{array}{ccc} c & b & a \end{array}\right]$ with G in $SL_2(\mathbf{Z})$. Then

$$GMG^T = \begin{bmatrix} c & b/2 \\ b/2 & a \end{bmatrix}$$

Let

$$G' = \begin{bmatrix} r & s \\ t & u \end{bmatrix} = \begin{bmatrix} 0 & -1 \\ -1 & 0 \end{bmatrix} G$$

Then $G'MG'^T = M$, so that $G'M = M(G'^T)^{-1}$, and hence, comparing the top left entries of

$$G'M = \begin{bmatrix} ra + sb/2 & rb/2 + sc \\ ta + ub/2 & tb/2 + uc \end{bmatrix}$$

and

$$M(G'^T)^{-1} = \begin{bmatrix} -au + sb/2 & at - rb/2 \\ -ub/2 + cs & tb/2 - cr \end{bmatrix}$$

we have

$$ra + sb/2 = -au + sb/2$$

Thus $r = -u$, and so $ru - st = -1$ implies that $r^2 + st = 1$.

Case 1. $s \neq 0$.

Let $g = \gcd(r + 1, s)$. Since

$$(r-1)\left(\frac{r+1}{g}\right) = -\left(\frac{s}{g}\right)t$$

it follows that $(r + 1)/g$ divides t, and s/g divides $r - 1$. Let

$$\begin{aligned} x &= \frac{r+1}{g} \\ y &= \frac{s}{g} \\ w &= \frac{g+y}{2} \\ z &= \frac{xw-1}{y} \end{aligned}$$

We prove next that w and z are integers.

If s is odd, g is odd, and so is $y = s/g$, and hence w is an integer. If s is even then, since $r^2 + st = 1$, r is odd, and so g is even. Comparing the entries of $G'M$ and $M(G'^T)^{-1}$, we see that

$$\frac{rb}{2} + sc = at - \frac{rb}{2}$$

Since b and r are odd, and s is even, it follows that t is odd. Since $(r - 1)(r + 1) = -st$, it follows that s is divisible by a higher power of 2 than $r + 1$ is. Hence $y = s/g$ is even, and thus w is an integer.

From the definition of z,

$$z = \frac{x(g+y)/2 - 1}{y} = \frac{1}{2}(x + (xg-2)/y) = \frac{1}{2}\left(x + \frac{r-1}{s/g}\right)$$

To show that z is an integer, it suffices to show that the two summands have the same parity. (Recall that s/g divides $r-1$.) Suppose, for example, that $y = s/g$ is even. Then x is odd (since $\gcd(x, y) = 1$). Also r is odd (since $s = yg$ is even, and $r^2 + st = 1$). Hence, as above, t is odd. Since $(r-1)x = -yt$, it follows that $(r-1)/y$ is odd. Hence z is an integer.

Now let

$$T = \begin{bmatrix} x & y \\ z & w \end{bmatrix}$$

Then T is in $SL_2(\mathbf{Z})$, and, by brute calculation,

$$TG' = \begin{bmatrix} rx + yt & sx + yu \\ zr + wt & sz + wu \end{bmatrix}$$

$$= \begin{bmatrix} x & y \\ x - z & y - w \end{bmatrix} = JT$$

where J is as above. Since

$$TG'T^{-1}TMT^T(TG'T^{-1})^T = TG'MG'^TT^T = TMT^T$$

it follows that $JTMT^TJ^T = TMT^T$, and hence TMT^T has the form $\begin{bmatrix} a' & a' & c' \end{bmatrix}$. Thus M is properly equivalent to a special ambiguous form.

Case 2. $s = 0$.
Then $r = \pm 1$. If $r = 1$, let $x = 1$, $y = 0$, $w = 1$, and $z = (1-t)/2$. If $r = -1$, let $x = t$, $y = 2$, $w = 1$, and $z = (t-1)/2$. Then $TG' = JT$, and the result follows as above.

We are going to define a group operation on C. To show that it is 'well-defined', we need the following two theorems.

Theorem 6.4.2 *Let F_1 and F_2 be gaussian forms (with the same discriminant D), and N a nonzero integer. Then there are gaussian forms H_1 and H_2 such that H_1 is properly equivalent to F_1, H_2 is properly equivalent to F_2, H_1 and H_2 are concordant, and, where a_1 is the first coefficient of H_1, and a_2 is the first coefficient of H_2, we have* $\gcd(a_1, a_2) = \gcd(a_1a_2, N) = 1$.

Proof: Suppose $F_1 = \begin{bmatrix} T_1 & T_2 & T_3 \end{bmatrix}$. By Theorem 6.2.8 there are relatively prime integers U and V such that

$$\gcd(T_1U^2 + T_2UV + T_3V^2, 2N) = 1$$

Let P and Q be integers such that $UQ - VP = 1$, and let

$$G = \begin{bmatrix} U & V \\ P & Q \end{bmatrix}$$

so that G is in $SL_2(\mathbf{Z})$. Let

$$F_1' = G * F_1 = \begin{bmatrix} T_1' & T_2' & T_3' \end{bmatrix}$$

Then $T_1' = T_1U^2 + T_2UV + T_3V^2$ is nonzero and relatively prime to N.

Similarly, there is a gaussian form $F_2' = \begin{bmatrix} S_1' & S_2' & S_3' \end{bmatrix}$ which is properly equivalent to F_2, and such that S_1' is nonzero and relatively prime to $T_1'N$.

Let n_1 and n_2 be integers such that $T_1'n_1 - S_1'n_2 = (S_2' - T_2')/2$. Then

$$b = T_2' + 2T_1'n_1 = S_2' + 2S_1'n_2$$

Let

$$G_j = \begin{bmatrix} 1 & 0 \\ n_j & 1 \end{bmatrix}$$

Then

$$H_1 = G_1 * F_1' = \begin{bmatrix} T_1' & b & T_1'n_1^2 + T_2'n_1 + T_3' \end{bmatrix}$$

and

$$H_2 = G_2 * F_2' = \begin{bmatrix} S_1' & b & S_1'n_2^2 + S_2'n_2 + S_3' \end{bmatrix}$$

meet the requirements.

Theorem 6.4.3 *Suppose that gaussian forms f_1 and g_1 are properly equivalent, and that gaussian forms f_2 and g_2 are properly equivalent. Suppose that f_1 and f_2 are concordant, and that g_1 and g_2 are concordant. Then $f_1 \circ f_2$ is properly equivalent to $g_1 \circ g_2$.*

Proof: Let

$$\begin{aligned} f_1 &= \begin{bmatrix} a_1 & b & c_1 \end{bmatrix} \\ f_2 &= \begin{bmatrix} a_2 & b & c_2 \end{bmatrix} \\ g_1 &= \begin{bmatrix} a_1' & b' & c_1' \end{bmatrix} \\ g_2 &= \begin{bmatrix} a_2' & b' & c_2' \end{bmatrix} \end{aligned}$$

Case 1. $f_1 = g_1$ and $\gcd(a_1, a_2') = 1$.

Let

$$G = \begin{bmatrix} r & s \\ t & u \end{bmatrix}$$

be a matrix in $SL_2(\mathbf{Z})$ such that $G * f_2 = g_2$. Then, since $b' = b$, we have

$$G \begin{bmatrix} a_2 & b/2 \\ b/2 & c_2 \end{bmatrix} = \begin{bmatrix} a_2' & b/2 \\ b/2 & c_2' \end{bmatrix} (G^T)^{-1}$$

The top right entry is $rb/2 + sc_2 = a_2'(-t) + br/2$, so that $sc_2 = -ta_2'$. Since f_1 is concordant with f_2, $a_1|c_2$. Thus $a_1|ta_2'$, and hence $a_1|t$.

Let

$$G' = \begin{bmatrix} r & sa_1 \\ t/a_1 & u \end{bmatrix}$$

Then G' is in $SL_2(\mathbf{Z})$. Furthermore, by calculation, $G' * (f_1 \circ f_2) = g_1 \circ g_2$.

Case 2. $b = b'$ and $\gcd(a_1, a_2') = 1$.

Hence f_1 and g_2 are concordant. Since $\gcd(a_2', a_1) = 1$, an application of Case 1 shows that $g_2 \circ g_1$ is properly equivalent to $g_2 \circ f_1$. Since $f_1 \circ f_2$ is properly equivalent to $f_1 \circ g_2$ (because of Case 1), it follows that $f_1 \circ f_2$ is properly equivalent to $g_1 \circ g_2$. (Since $a_1|(b^2 - D)$, it follows that $a_1|4a_2'c_2'$ and hence $a_1|c_2'$.)

Case 3. $\gcd(a_1 a_2, a_1' a_2') = 1$.

Since b and b' are both odd, there are integers n and n' such that

$$a_1 a_2 n - a_1' a_2' n' = (b' - b)/2$$

and hence

$$B = b + 2a_1 a_2 n = b' + 2a_1' a_2' n'$$

We make the following definitions:

$$
\begin{aligned}
F_1 &= \begin{bmatrix} 1 & 0 \\ a_2 n & 1 \end{bmatrix} * f_1 = \begin{bmatrix} a_1 & B & * \end{bmatrix} \\
F_2 &= \begin{bmatrix} 1 & 0 \\ a_1 n & 1 \end{bmatrix} * f_2 = \begin{bmatrix} a_2 & B & * \end{bmatrix} \\
H_1 &= \begin{bmatrix} 1 & 0 \\ n & 1 \end{bmatrix} * (f_1 \circ f_2) = \begin{bmatrix} a_1 a_2 & B & * \end{bmatrix} \\
G_1 &= \begin{bmatrix} 1 & 0 \\ a_2' n' & 1 \end{bmatrix} * g_1 = \begin{bmatrix} a_1' & B & * \end{bmatrix} \\
G_2 &= \begin{bmatrix} 1 & 0 \\ a_1' n' & 1 \end{bmatrix} * g_2 = \begin{bmatrix} a_2' & B & * \end{bmatrix} \\
H_2 &= \begin{bmatrix} 1 & 0 \\ n' & 1 \end{bmatrix} * (g_1 \circ g_2) = \begin{bmatrix} a_1' a_2' & B & * \end{bmatrix}
\end{aligned}
$$

The discriminant equation ($b^2 - 4ac = D$) applied to H_1 shows that $a_1 a_2$ divides $(B^2 - D)/4$. From the discriminant equations for F_1 and F_2 it then follows that F_1 and F_2 are concordant. Similarly, G_1 and G_2 are concordant. By Case 2, $F_1 \circ F_2$ is properly equivalent to $G_1 \circ G_2$. (Since $\gcd(a_1 a_2, a_1' a_2') = 1$, we have $\gcd(a_1, a_2') = 1$.)

Now since the discriminant fixes the third coefficient given the first two, $H_1 = F_1 \circ F_2$ and $H_2 = G_1 \circ G_2$. Thus H_1 and H_2 are properly equivalent. But H_1 is properly equivalent to $f_1 \circ f_2$, while H_2 is properly equivalent to $g_1 \circ g_2$.

Case 4. No special restrictions.

By Theorem 6.4.2, there are gaussian forms

$$
\begin{aligned}
F_1 &= \begin{bmatrix} A_1 & B_1 & * \end{bmatrix} \\
F_2 &= \begin{bmatrix} A_2 & B_2 & * \end{bmatrix}
\end{aligned}
$$

such that F_1 is properly equivalent to f_1 and g_1, while F_2 is properly equivalent to f_2 and g_2, and also F_1 and F_2 are concordant, and $\gcd(A_1, A_2) = 1$, and

$$\gcd(A_1 A_2,\ a_1 a_2 a_1' a_2') = 1$$

Hence $\gcd(a_1a_2, A_1A_2) = 1$, so that, as in Case 3, $f_1 \circ f_2$ is properly equivalent to $F_1 \circ F_2$. Similarly, $g_1 \circ g_2$ is properly equivalent to $F_1 \circ F_2$.

This completes the proof.

Let $[F]$ be the equivalence class represented by the gaussian form F. Let F_1 and F_2 be any gaussian forms (with the same discriminant D). Let H_1 and H_2 be gaussian forms which are properly equivalent to F_1 and F_2 respectively, and concordant (such forms exist by Theorem 6.4.2). Define

$$[F_1][F_2] = [H_1 \circ H_2]$$

By Theorem 6.4.3, this binary operation is well defined. It is also commutative. We can prove, moreover, that it is associative:

Theorem 6.4.4 $([f_1][f_2])[f_3] = [f_1]([f_2][f_3])$

Proof: Suppose $f_3 = \begin{bmatrix} a_3 & b_3 & c_3 \end{bmatrix}$. By Theorem 6.4.2 there are gaussian forms H_1 and H_2 such that H_1 is properly equivalent to f_1, H_2 is properly equivalent to f_2, and, where a_1 is the first coefficient of H_1 and a_2 is the first coefficient of H_2, $\gcd(a_1, a_2) = \gcd(a_1a_2, a_3) = 1$.

Let b_1, b_2 be the second coefficients of H_1, H_2 respectively. Let n_1 and n_2 be integers such that

$$a_2n_2 - a_1n_1 = \frac{b_1 - b_2}{2}$$

Let n_3 and k be integers such that

$$\frac{b_1 - b_3}{2} + a_1n_1 = a_3n_3 - a_1a_2k$$

(Recall that all the b's are odd.) Let

$$\begin{aligned} n_1' &= n_1 + ka_2 \\ n_2' &= n_2 + ka_1 \\ n_3' &= n_3 \end{aligned}$$

We have

$$b_1 + 2a_1n_1' = b_2 + 2a_2n_2' = b_3 + 2a_3n_3'$$

Call this number B. For $i = 1, 2, 3$, let

$$G_i = \begin{bmatrix} 1 & 0 \\ n'_i & 1 \end{bmatrix}$$

Let $F_1 = G_1 * H_1$, $F_2 = G_2 * H_2$, and $F_3 = G_3 * f_3$. Then F_1, F_2, and F_3 all have the same second coefficient B. Since their first coefficients are pairwise relatively prime, they are pairwise concordant. Now

$$([f_1][f_2])[f_3] = [F_1 \circ F_2][F_3] = [\begin{bmatrix} a_1a_2 & B & * \end{bmatrix}][F_3] = [\begin{bmatrix} a_1a_2a_3 & B & * \end{bmatrix}]$$

and the same is true of $[f_1]([f_2][f_3])$.

Theorem 6.4.5 *The finite set of equivalence classes of gaussian forms, together with the above binary operation, forms a commutative group with identity* $[\begin{bmatrix} 1 & 1 & (1-D)/4 \end{bmatrix}]$.

The inverse of $[\begin{bmatrix} a & b & c \end{bmatrix}]$ *is* $[\begin{bmatrix} c & b & a \end{bmatrix}]$.

Proof: Theorem 6.4.4 shows that the binary operation is associative.

Using

$$G = \begin{bmatrix} 1 & 0 \\ (b-1)/2 & 1 \end{bmatrix}$$

it can be shown that $\begin{bmatrix} 1 & 1 & (1-D)/4 \end{bmatrix}$ and $\begin{bmatrix} 1 & b & (b^2-D)/4 \end{bmatrix}$ are properly equivalent. Thus

$$\begin{aligned} [\begin{bmatrix} a & b & c \end{bmatrix}][\begin{bmatrix} 1 & 1 & (1-D)/4 \end{bmatrix}] &= [\begin{bmatrix} a & b & c \end{bmatrix}][\begin{bmatrix} 1 & b & (b^2-D)/4 \end{bmatrix}] \\ &= [\begin{bmatrix} a & b & c \end{bmatrix}] \end{aligned}$$

Also

$$[\begin{bmatrix} a & b & c \end{bmatrix}][\begin{bmatrix} c & b & a \end{bmatrix}] = [\begin{bmatrix} ac & b & 1 \end{bmatrix}]$$

But, using

$$G = \begin{bmatrix} 0 & 1 \\ -1 & (b+1)/2 \end{bmatrix}$$

it can be shown that $\begin{bmatrix} ac & b & 1 \end{bmatrix}$ and $\begin{bmatrix} 1 & 1 & (1-D)/4 \end{bmatrix}$ are properly equivalent.

We give the name C to the group of equivalence classes defined above.

For example, when $D = -39$, the reduced forms are

$$\begin{aligned} f_0 &= \begin{bmatrix} 1 & 1 & 10 \end{bmatrix} \\ f_1 &= \begin{bmatrix} 2 & 1 & 5 \end{bmatrix} \\ f_2 &= \begin{bmatrix} 3 & 3 & 4 \end{bmatrix} \\ f_3 &= \begin{bmatrix} 2 & -1 & 5 \end{bmatrix} \end{aligned}$$

$[f_0]$ is the identity, and we have $[f_1][f_3] = [f_0]$ and $[f_1][f_1] = [f_2]$.

We now define 'self-inverses'.

A gaussian form f is a *self-inverse* iff $[f][f] = [\begin{bmatrix} 1 & 1 & (1-D)/4 \end{bmatrix}]$. Moreover, the equivalence class $[f]$ is a *self-inverse iff f is.*

Theorem 6.4.6 *A gaussian form is a self-inverse iff it is ambiguous.*

Proof: Suppose f is a self-inverse. Let $f = \begin{bmatrix} a & b & c \end{bmatrix}$. By Theorem 6.4.5, $[f] = [\begin{bmatrix} a & b & c \end{bmatrix}]$, so that f and $\begin{bmatrix} c & b & a \end{bmatrix}$ are properly equivalent. Hence, by Theorem 6.4.1, f is ambiguous.

Suppose now that f is ambiguous. Then, using Theorems 6.4.1 and 6.4.5, we may conclude that f is a self-inverse.

If C is the group of equivalence classes defined above, let $sq : C \to C$ such that $sq([f]) = [f][f]$. Then sq is a group homomorphism. The kernel of sq is precisely the set of self-inverse classes. If $\text{im}(sq)$ is the set of equivalence classes in C which can be written in the form $[f][f]$ — that is, the 'squares' — and if $\ker(sq)$ is the kernel of sq, then, by the First Isomorphism Theorem for groups,

$$|C| = |\ker(sq)||\text{im}(sq)|$$

— where, in general, $|G|$ is defined as the number of elements in the finite set G.

By Theorem 6.4.6, an equivalence class $[f]$ is a self-inverse iff f is ambiguous iff $[f]$ contains a special ambiguous form. We saw above that there are exactly 2^{r-1} equivalence classes containing special ambiguous forms — where r is the number of distinct prime divisors of D. Thus

$$|C| = 2^{r-1}|\mathrm{im}(sq)|$$

Linking this section with the previous one, we have

Theorem 6.4.7 *A gaussian form g is a square form iff $[g] \in \mathrm{im}(sq)$.*

Proof: Suppose g is a square form. Then there are two properly equivalent, concordant forms F and F' with g properly equivalent to $F \circ F'$. Thus

$$[g] = [F \circ F'] = [F][F'] = [F][F]$$

Thus $[g]$ is in $\mathrm{im}(sq)$.

Conversely, if $[g] \in \mathrm{im}(sq)$ then $[g] = [f][f] = [f' \circ f'']$ for some concordant forms f' and f'', with f' properly equivalent to f, and f'' properly equivalent to f. Thus g is properly equivalent to $f' \circ f''$, where f' and f'' are properly equivalent. Hence g is a square form.

Recall that an *omega kernel class* is an equivalence class in C which contains an omega kernel form (that is, one which represents a member of H). Recall also that we used K to denote the set of omega kernel classes. By Theorem 6.3.3, $[f] \in \mathrm{im}(sq)$ iff f is a square form iff f is an omega kernel form iff $[f] \in K$. Thus $|C|/|K| = 2^{r-1}$.

Theorem 6.4.8 *The number of equivalence classes in C is 2^{r-1} times the number of classes representing some member or other of H — where r is the number of distinct prime factors in D.*

Exercises 6.4

1. If $D = -55$, which reduced gaussian forms are ambiguous?

6.5 Sums of Triangular Numbers

Let U be the set of residue classes relatively prime to D. If $m \in D$, let $J(m) = \left(\frac{m}{-D}\right)$. Since $-D \equiv 3 \pmod 4$, it follows that $J(-1) = -1$ (Theorem 3.10.2). Now if x is in U, so is $-x$, and $J(-x) = J(-1)J(x) = -J(x)$. Thus the members of U can be partitioned into 2 equal sets — those for which the Jacobi symbol is 1, and those for which the Jacobi symbol is -1. Each set has $\phi(D)/2$ members.

Let $\ker(J)$ be the set with Jacobi symbol 1. From the properties of the Jacobi symbol, it follows that $\ker(J)$ is a subgroup of the multiplicative group U. Moreover, H is a subgroup of $\ker(J)$. In Theorem 6.3.1 we saw that H has $\phi(D)/2^r$ members, where r is the number of distinct prime factors of D. Since $\ker(J)$ has $\phi(D)/2$ members, the quotient group $\ker(J)/H$ has 2^{r-1} members.

Theorem 6.5.1 *If an integer m is relatively prime to D, and represented by a gaussian form, then $J(m) = 1$.*

Proof: Let $f = \begin{bmatrix} a & b & c \end{bmatrix}$ be a gaussian form representing m, and let $m = ar^2 + brs + cs^2$. Let $k = \gcd(r, s)$, and let t and u be integers such that $(r/k)u - (s/k)t = 1$. Let

$$G = \begin{bmatrix} r/k & s/k \\ t & u \end{bmatrix}$$

Then $G \in SL_2(\mathbf{Z})$, and $G * f$ has the form $\begin{bmatrix} m/k^2 & p & q \end{bmatrix}$. Hence $p^2 - 4(m/k^2)q = D$, and $D \equiv p^2 \pmod{m/k^2}$.

Let $m = 2^e m'$, where e is a nonnegative integer, and m' is odd. If e is odd then m/k^2 is even, and $D \equiv 1 \pmod 8$, and hence $J(2) = 1$ (Theorem 3.10.2).

Since $D \equiv 1 \pmod 4$, $(-D-1)/2$ is odd. Thus, by Jacobi Reciprocity,

$$J(m') = \left(\frac{-D}{m'}\right)(-1)^{(m'-1)/2} = \left(\frac{-p^2}{m'}\right)(-1)^{(m'-1)/2} = 1$$

— using Theorems 3.10.1 and 3.10.2 and the fact that

$$\left(\frac{-D}{m'}\right) = \left(\frac{-Dk^2}{m'}\right) = \left(\frac{-p^2k^2}{m'}\right) = \left(\frac{-p^2}{m'}\right)$$

since $Dk^2 \equiv p^2 \pmod{m}$. Thus $J(m) = J(2^e)J(m') = 1$.

Theorem 6.5.2 *Suppose a gaussian form f represents integers m and n with $\gcd(m, D) = \gcd(n, D) = 1$. Then, if n^{-1} is an inverse of n mod D, $mn^{-1} \in H$.*

Proof: Let

$$\begin{aligned} f &= \begin{bmatrix} a & b & c \end{bmatrix} \\ m &= ar^2 + brs + cs^2 \\ n &= at^2 + btu + cu^2 \end{aligned}$$

Let

$$G = \begin{bmatrix} r & s \\ t & u \end{bmatrix}$$

Then

$$G \begin{bmatrix} a & b/2 \\ b/2 & c \end{bmatrix} G^T = \begin{bmatrix} m & k/2 \\ k/2 & n \end{bmatrix}$$

for some integer k. Equating determinants, we obtain

$$(\det G)^2(-D/4) = mn - k^2/4$$

and hence $4mn = k^2 + QD$. Since every square is in H, $4mn$ is in H. And so is the inverse B of $(2n)^2$. Since H is a group, it also contains $4mnB = mn^{-1}$.

Since H is a subgroup of U (the set of residue classes relatively prime to D), we can form the quotient group U/H. By Theorem 6.5.2, if $f = \begin{bmatrix} a & b & c \end{bmatrix}$ represents m and n, both relatively prime to D, then $mH = nH$. Since forms in the same equivalence class $[\begin{bmatrix} a & b & c \end{bmatrix}]$ represent the same integers, we can, thanks also to Theorem 6.5.2, define a function $\omega : C \to U/H$ as $\omega[f] = mH$, where m is any integer relatively prime to D and represented by f. For example,

$$\omega[\begin{bmatrix} 1 & 1 & (1-D)/4 \end{bmatrix}] = H$$

Theorem 6.5.3 *ω is a group homomorphism with kernel K (the set of omega kernel classes).*

Proof: $(\omega[f])(\omega[g]) = pqH$, where p is any integer in U represented by f, and q is any integer in U represented by g. Note that pq is in U.

Let f' and g' be concordant forms with f' properly equivalent to f, and g' properly equivalent to g (this possible by Theorem 6.4.2). Then $f' \circ g'$ represents pq (Theorem 6.3.2). Thus $\omega([f][g]) = \omega([f' \circ g']) = pqH$, as before.

Furthermore, $\omega[f] = H$ just in case f represents a number of the form $x + QD$ with $x \in H$ — that is, f is an omega kernel form.

From Theorem 6.5.3 and elementary group theory, it follows that

$$|\text{im}(\omega)| = |C|/|K|$$

which, we saw above, equals 2^{r-1} where r is the number of distinct prime factors in D (Theorem 6.4.8).

By Theorem 6.5.1, $\text{im}(\omega)$ is a subset of $\ker(J)/H$ which, as we noted at the beginning of this section, has 2^{r-1} members. Hence

Theorem 6.5.4 $\text{im}(\omega) = \ker(J)/H$.

Finally we have,

Theorem 6.5.5 *Every positive integer Z is a sum of three triangular numbers.*

Proof: Let $u = 8Z + 3$, and let $D = -u$. Then $J(-2) = 1$ and thus, by Theorem 6.5.4, there is a gaussian form f such that $\omega[f] = -2H$. By Theorem 6.4.2 there is a gaussian form $\begin{bmatrix} a' & b & c' \end{bmatrix}$ in $[f]$ such that $\gcd(a', D) = 1$. By Theorem 6.5.2, $a' \equiv -2h \pmod{D}$ for some h in H (since $a'(-2 + QD)^{-1} \in H$).

Let $a = 2a'$, and $c = 2c'$. Then $ac - b^2 = -D = u$.

Suppose $z^2 \equiv h \pmod{D}$ has solution z. Then $z \in U$ and z has some inverse z^{-1} mod D. Moreover,

$$-a = -2a' \equiv 4h \equiv (2z)^2 \pmod{u}$$

Let $N = 2z$. Since N is in U, there is an integer M (which is congruent to $N^{-1}b$ mod u) such that $b \equiv MN \pmod{u}$. Moreover,

$$-c \equiv -(N^{-1})^2 N^2 c \equiv (N^{-1})^2 ac \equiv (N^{-1})^2 b^2 \equiv M^2 \pmod{u}$$

We define 6 integers:

$$\begin{aligned}
C &= \frac{a+N^2}{u} \\
B &= \frac{MN-b}{u} \\
A &= \frac{c+M^2}{u} \\
m &= BN - CM = \frac{-aM-bN}{u} \\
n &= BM - AN = \frac{-bM-cN}{u} \\
s &= AC - B^2 = \frac{1-mM-nN}{u}
\end{aligned}$$

Then $bn - cm = M$, $an - bm = -N$, and $1 - mM - nN = su$, so that

$$R = \begin{bmatrix} a & b & m \\ b & c & n \\ m & n & s \end{bmatrix}$$

has determinant 1. (To see this, expand starting from the bottom.)

Moreover,

$$su = 1 - mM - nN = 1 - bmn - cm^2 + an^2 - bmn$$

$$= 1 - 2bmn + an^2 + \left(\frac{u+b^2}{a}\right) m^2 = \frac{m^2 u}{a} + \frac{(an-bm)^2}{a} + 1$$

Thus the coefficient of z^2 in

$$F(x,y,z) = \frac{(ax+by+mz)^2}{a} + \frac{(uy+(an-bm)z)^2}{au} + \frac{z^2}{u}$$

is s. Indeed, by straightforward calculation, we have

$$\begin{bmatrix} x & y & z \end{bmatrix} R \begin{bmatrix} x & y & z \end{bmatrix}^T = F(x,y,z)$$

Since $a = 2a'$ and $a' > 0$ (since a' is the first coefficient in a gaussian form), and since $u > 0$, it follows that $F(x, y, z)$ is always nonnegative for any integers x, y, and z.

By Theorem 6.2.7, R is equivalent to

$$\begin{bmatrix} 1 & 0 & 0 \\ 0 & 1 & 0 \\ 0 & 0 & 1 \end{bmatrix} \quad \text{or to} \quad \begin{bmatrix} 0 & 0 & 1 \\ 0 & -1 & 0 \\ 1 & 0 & 0 \end{bmatrix}$$

Call the latter matrix Q. If R is equivalent to Q then, for some matrix G in $GL_3(\mathbf{Z})$, we have $GQG^T = R$. Let

$$\begin{bmatrix} x & y & z \end{bmatrix} = \begin{bmatrix} 0 & 1 & 0 \end{bmatrix} G^{-1}$$

Then

$$\begin{bmatrix} x & y & z \end{bmatrix} R \begin{bmatrix} x & y & z \end{bmatrix}^T$$
$$= \begin{bmatrix} 0 & 1 & 0 \end{bmatrix} G^{-1}GQG^T(G^{-1})^T \begin{bmatrix} 0 & 1 & 0 \end{bmatrix}^T = -1$$

Since $F(x, y, z)$ never represents negative integers, this is impossible. Hence R is equivalent to the identity matrix.

Thus there is a matrix H in $GL_3(\mathbf{Z})$ such that HRH^T is the identity matrix, so that $R = H^{-1}(H^T)^{-1}$, and hence $\overline{R} = H^T H$. Now the bottom right entry of $\overline{R}$ is $ac - b^2$, while the bottom right entry of $H^T H$ is the sum of three squares: $x_1^2 + x_2^2 + x_3^2$. Thus u is the sum of these three squares, and so

$$8Z + 3 = x_1^2 + x_2^2 + x_3^2$$

By considerations mod 8, all the x's are odd. Thus we have

$$8Z + 3 = (2y_1 + 1)^2 + (2y_2 + 1)^2 + (2y_3 + 1)^2$$

and hence

$$Z = \frac{y_1(y_1 + 1)}{2} + \frac{y_2(y_2 + 1)}{2} + \frac{y_3(y_3 + 1)}{2}$$

— the sum of three triangular numbers.

For example, suppose we want to write 13 as the sum of three triangular numbers, following the procedure of the above proof. Let $u = 8 \times 13 + 3 = 107$, and $D = -107$.

Taking $a = 2$, $b = 1$, and $c = 54$, we can use 31 for N. Then we have $M = 38$, $C = 9$, $B = 11$, $A = 14$, $m = -1$, $n = -16$, and $s = 5$. If

$$R = \begin{bmatrix} 2 & 1 & -1 \\ 1 & 54 & -16 \\ -1 & -16 & 5 \end{bmatrix} \quad \text{and} \quad G = \begin{bmatrix} -2 & -2 & -7 \\ -3 & -2 & -7 \\ 1 & 1 & 3 \end{bmatrix}$$

then G is in $GL_3(\mathbf{Z})$, and GRG^T is the identity matrix. The bottom right entry of G^TG is

$$(-7)^2 + (-7)^2 + 3^2 = 107$$

Thus

$$8 \times 13 + 3 = (2 \times 3 + 1)^2 + (2 \times 3 + 1)^2 + (2 \times 1 + 1)^2$$

and hence

$$13 = \frac{3 \times 4}{2} + \frac{3 \times 4}{2} + \frac{1 \times 2}{2}$$

— the sum of three triangular numbers.

ALL THE WAYS OF EXPRESSING THE GIVEN INTEGER AS A NONDECREASING SUM OF THREE TRIANGULAR NUMBERS

1	$0+0+1$		
2	$0+1+1$		
3	$0+0+3$	$1+1+1$	
4	$0+1+3$		
5	$1+1+3$		
6	$0+0+6$	$0+3+3$	
7	$0+1+6$	$1+3+3$	
8	$1+1+6$		
9	$0+3+6$	$3+3+3$	
10	$0+0+10$	$1+3+6$	
11	$0+1+10$		
12	$0+6+6$	$1+1+10$	$3+3+6$
13	$0+3+10$	$1+6+6$	
14	$1+3+10$		
15	$0+0+15$	$3+6+6$	

Exercises 6.5

1. Use the above theory to write 1000 as a sum of 3 triangular numbers.

6.6 Cauchy's Proof

Cauchy's proof of Fermat's polygonal number conjecture is found in volume 6 of the second series of his *Oeuvres complètes.* In this section we give a shortened and simplified version of it. Recall that a polygonal number is a nonnegative integer of the form $m(t^2-t)/2+t$, where m is a positive integer, and t is a nonnegative integer. Fermat's conjecture is the statement that, for any positive integer m, every positive integer is a sum of $m+2$ of these numbers. We begin with a theorem based on Gauss's result for the triangular numbers.

Theorem 6.6.1 *Let k and s be odd positive integers such that*

$$\sqrt{3k-2}-1 \leq s \leq \sqrt{4k}$$

Then there are nonnegative integers t, u, v, and w such that

$$\begin{aligned} k &= t^2+u^2+v^2+w^2 \\ s &= t+u+v+w \end{aligned}$$

Proof: Since every positive integer is a sum of three triangular numbers, every positive integer of the form $8n+3$ is a sum of three squares. Thus

$$4k-s^2 = x^2+y^2+z^2$$

for some odd positive integers x, y, and z. Now

$$(x+y+z)^2 \leq (x+y+z)^2+(x-y)^2+(x-z)^2+(y-z)^2 = 3(4k-s^2)$$

Since $\sqrt{3k-2}-1 \leq s$, it follows that $3(4k-s^2) < (s+4)^2$. Hence

$$x+y+z < s+4$$

and

$$\frac{s \pm x \pm y \pm z}{4} > -1$$

Let $c = s-x-y-z$ and $d = s+x+y+z$. Since s, x, y, and z are all odd, c and d are even. Moreover, $c+d = 2s$ being twice an odd number, one of c and d is divisible by 4.

Case 1. $4|c$.

Let

$$\begin{aligned} t &= \frac{c}{4} \\ u &= t+\frac{y+z}{2} \\ v &= t+\frac{x+z}{2} \\ w &= t+\frac{x+y}{2} \end{aligned}$$

These are all nonnegative integers, their sum is s, and the sum of their squares is

$$\frac{s^2+x^2+y^2+z^2}{4}$$

which equals k.

Case 2. $4|d$.

Let

$$\begin{aligned} t &= \frac{d}{4} \\ u &= t-\frac{y+z}{2} \\ v &= t-\frac{x+z}{2} \\ w &= t-\frac{x+y}{2} \end{aligned}$$

Since $(s \pm x \pm y \pm z)/4 > -1$, all these are nonnegative integers. Their sum is s, and the sum of their squares is k.

The key part of Cauchy's proof is the following.

Theorem 6.6.2 *Let k and s be odd positive integers such that*

$$\sqrt{3k-2}-1 \leq s \leq \sqrt{4k}$$

Let m be an integer > 2. Let r be a nonnegative integer with $r \leq m-2$. Then

$$\frac{m(k-s)}{2}+s+r$$

is a sum of $m+2$ $(m+2)$-gonal numbers.

Proof: By Theorem 6.6.1, there are nonnegative integers t, u, v, w such that

$$\begin{aligned} k &= t^2+u^2+v^2+w^2 \\ s &= t+u+v+w \end{aligned}$$

Now

$$\frac{m(k-s)}{2} + s + r$$

$$= \frac{m(t^2-t)}{2} + t + \frac{(u^2-u)}{2} + u + \frac{m(v^2-v)}{2} + v + \frac{m(w^2-w)}{2} + w$$

$$+1+1+\cdots+1$$

where there are r 1's. Since $r \leq m-2$, the sum on the right of the equal sign has fewer than $m+2$ $(m+2)$-gonal number terms.

In all that follows, k is an odd positive integer. For a given k, s is an odd integer between $\sqrt{3k-2}-1$ and $\sqrt{4k}$ inclusive. (There is always at least one odd integer between these two numbers.) We define $s_1(k)$ as the least odd positive integer between these two numbers, and $s_2(k)$ as the largest odd positive integer between these two numbers. (For certain k, $s_1(k) = s_2(k)$.)

Where m is an integer > 2, we define

$$\begin{aligned} g(k) &= \frac{m(k-s_2(k))}{2} + s_2(k) \\ &= \frac{mk}{2} - (\frac{m}{2}-1)s_2(k) \\ h(k) &= \frac{m(k-s_1(k))}{2} + s_1(k) + m - 2 \\ &= \frac{mk}{2} - (\frac{m}{2}-1)s_1(k) + m - 2 \end{aligned}$$

Theorem 6.6.3 *Let m be an integer > 2. Let N be an integer $\geq$ $44m+19$. Then N is a sum of $m+2$ $(m+2)$-gonal numbers.*

Proof: As s runs down the sequence of odd numbers

$$s_2(k),\ \ s_2(k)-2,\ \ \ldots,\ \ s_1(k)$$

and as, for each s, r varies from 0 to $m-2$, the form

$$\frac{m(k-s)}{2} + s + r$$

takes all the integer values between $g(k)$ and $h(k)$ inclusive. For we have

$$\begin{array}{lll}\frac{mk}{2}-(\frac{m}{2}-1)s_2(k) & \ldots & \frac{mk}{2}-(\frac{m}{2}-1)s_2(k)+m-2 \\ \frac{mk}{2}-(\frac{m}{2}-1)(s_2(k)-2) & \ldots & \frac{mk}{2}-(\frac{m}{2}-1)(s_2(k)-2)+m-2 \\ \vdots & & \vdots \\ \frac{mk}{2}-(\frac{m}{2}-1)s_1(k) & \ldots & \frac{mk}{2}-(\frac{m}{2}-1)s_1(k)+m-2\end{array}$$

— with the last entry in each row equal to the first entry in the next row.

Suppose $k \geq 107$. Then

$$\sqrt{4(k+2)}-2>\sqrt{3k-2}-1+2$$

and thus $s_2(k+2) > s_1(k)$. (It is possible, for small k, to have $s_2(k+2) = s_1(k)$.) Hence $h(k) > g(k+2)-2$, or $h(k) \geq g(k+2)-1$.

Consider the intervals

$$[g(107),h(107)],\quad [g(109),h(109)],\quad [g(111),h(111)],\quad \ldots$$

The sequence

$$g(107),\quad g(109),\quad g(111),\quad \ldots$$

tends to infinity. Since $h(k) \geq g(k+2)-1$, the union of these intervals includes all the integers $\geq g(107) = 44m+19$.

The theorem now follows by Theorem 6.6.2.

Since $g(105) = 43m+19$ and $h(105) = 45m+15$, and since $h(105) \geq g(107)-1$, we can lower the bound of Theorem 6.6.3 from $44m+19$ to $43m+19$. Indeed, by calculations of this kind, we can lower the bound to $g(89) = 36m+17$. However, $h(87) = 36m+15$, so that our gh intervals do not cover $36m+16$. This is not a problem, though, since

$$36m+16=(28m+8)+(6m+4)+(m+2)+(m+2)$$

and these 4 summands are $(m+2)$-gonal numbers (having the form $m(t^2-t)/2+t$ with $t = 8$, 4, 2, and 2, respectively).

The gh intervals

$$[g(71), h(71)], \quad \ldots, \quad [g(87), h(87)]$$

include all the integers from $28m + 15$ to $36m + 15$ inclusive, and so we can lower the bound on N to $28m + 15$. Indeed, since

$$28m + 14 = (21m + 7) + (6m + 4) + (m + 2) + 1$$

we can lower it to $28m + 14$.

Indeed, continuing in this way, we can lower the bound right down to 1. For the only integers not covered by the gh intervals are the following — and they can each be expressed directly as a sum of $m+2$ $(m + 2)$-gonal numbers.

$m + 2$	$8m + 8$	$19m + 12$
$2m + 4$	$9m + 8$	$20m + 12$
$3m + 4$	$10m + 8$	$21m + 12$
$4m + 6$	$13m + 10$	$27m + 14$
$5m + 6$	$14m + 10$	$28m + 14$
$6m + 6$	$15m + 10$	$36m + 16$

Fermat was right.

Exercises 6.6

1. Derive Lagrange's Four Square Theorem as a corollary to Cauchy's Theorem 6.6.1.
2. Complete Cauchy's proof by showing just what numbers the gh intervals do cover, and by handling all the cases not taken care of by the gh intervals.
3. Prove that all hexagonal numbers are triangular.
4. What is the smallest number that has 5 distinct expressions as a sum of 5 pentagonal numbers?
5. Write 100 in every possible way as a sum of $m + 2$ $(m + 2)$-gonal numbers.
6. Find a formula for triangular pentagonal numbers.
7. Show that every integer > 169 is a sum of 5 positive squares.

Chapter 7

Analytic Number Theory

In this chapter we draw on real and complex analysis to present four beautiful theorems. The first is P. Dirichlet's theorem that there are infinitely many primes in any arithmetic progression

$$a,\quad a+b,\quad a+2b,\quad a+3b,\quad \ldots$$

(assuming a and b are relatively prime). The second, due to J. Lambek, L. Moser, and R. Wild, gives the order of the number of primitive Pythagorean triangles with area less than n. The third is the Prime Number Theorem, first proved, independently, by J. Hadamard and C. J. de la Vallée Poussin. This states that if $\pi(n)$ is the number of primes less than or equal to the positive number n, then

$$\lim_{n\to\infty} \frac{\pi(n)}{n/\ln n} = 1$$

The fourth beautiful theorem is H. Rademacher's theorem establishing an exact formula for the number $p(n)$ of partitions of a natural number n — a partition being a way of writing n as a sum of nonincreasing positive integer summands.

It would be nice if we could establish these theorems without using the heavy machinery of analysis, but we have not yet found a way of doing so.

7.1 Characters

To prove Dirichlet's theorem that there are infinitely many primes in an arithmetic progression P, we show that there is a function $\Lambda(a)$ such that

$$\lim_{s\downarrow 1} \sum_{primes\ a\ in\ P} \frac{\Lambda(a)}{a^s} = \infty$$

This would not, of course, be true if there were only finitely many primes in the arithmetic progression P.

In order to establish this fact about Λ we make use of the 'Dirichlet L-series'

$$\sum_{a=1}^{\infty} \frac{X(a)}{a^s}$$

where s is any real ≥ 1, and X is any function, often a 'k-character'. In this section, we explore the basic properties of these k-characters. In the next section, we give some results concerning Dirichlet L-series, and in the section after that, we define the Λ function, and demonstrate some of its properties. All this will put us in position to derive the key lemma that, if X is any k-character, then $\sum_{a=1}^{\infty} X(a)/a \neq 0$. With this lemma, it will not take long to prove that

$$\lim_{s\downarrow 1} \sum_{primes\ a\ in\ P} \frac{\Lambda(a)}{a^s} = \infty$$

Consider the sequence

$$l, \ l+k, \ l+2k, \ l+3k, \ \ldots$$

where l and k are positive integers, and $\gcd(l,k) = 1$. In Chapter 5 we defined p-characters, where p was a prime, and used them to prove that if p has the form $2^n + 1$ then the regular p-gon is constructible using only straightedge and compass. In order to prove Dirichlet's theorem about the infinitude of primes in an arithmetic progression we extend this definition.

If k is any positive integer, a *k-character* is a function $X : \mathbf{Z} \to \mathbf{C}$ such that

(1) $a \equiv b \pmod{k}$ implies $X(a) = X(b)$

(2) $X(ab) = X(a)X(b)$

(3) $X(a) = 0$ iff $\gcd(a, k) \neq 1$.

Note that every p-character is a k-character, with $k = p$.

As an example, suppose p is an odd prime, and $k = 4p$. Suppose that if a is even or a multiple of p then $X(a) = 0$. However, if a is odd and not a multiple of p, then $X(a) = (-1)^{(a-1)/2}\left(\frac{a}{p}\right)$. Then $X(a)$ is a k-character. We shall call this character H.

In Chapter 5 we proved various properties of p-characters. In a similar way we can establish analogous properties of the more general k-characters. In particular, if X is a k-character and $\gcd(a, k) = 1$ then

$$X(1) = 1$$

$$(X(a))^{\phi(k)} = 1$$

$$(X(a))^{-1} = X(a^{-1}) \text{ if } a^{-1} \equiv a \pmod{k}$$

$$(X(a))^{-1} = \overline{X(a)} \text{ (the complex conjugate).}$$

We define X_0 as the k-character that maps a to 1 (unless $\gcd(a, k) \neq 1$, in which case $X_0(a) = 0$). X_0 is the *principal character.*

We define $X^{-1}(a)$ as $X(a^{-1})$ (or 0 if $\gcd(a, k) \neq 1$). Then X^{-1} is also a k-character. For example, the $4p$-character H, defined above, is its own inverse.

Since the product of two k-characters is a k-character, it follows that the k-characters form a group with identity X_0. Since each character maps a domain of k distinct elements into a codomain of $1 + \phi(k)$ elements (0 and roots of unity), this is a finite group. We shall prove it contains $\phi(k)$ characters.

As in Chapter 5, if the k-character $X \neq X_0$ then

$$\sum_{a=0}^{k-1} X(a) = 0$$

Also as in Chapter 5, if $\gcd(a,k) = 1$ and a is *not* congruent to 1 mod k, then

$$\sum_{characters} X(a) = 0$$

The proof of this relies on the fact that there is some k-character X_1 such that $X_1(a) \neq 1$. But why should this be? Suppose k has prime factorisation $k = p_1^{m_1} \dots p_n^{m_n}$. Since a is not congruent to 1 mod k, there is some i such that a is not congruent to 1 mod $p_i^{m_i}$. There are three cases.

Case 1. p_i is odd.

In this case, let g be a primitive root of $p_i^{m_i}$. Define X_1 such that $X_1(b) = 0$ if $\gcd(b,k) \neq 1$, but otherwise

$$X_1(b) = e^{2\pi i h/\phi(p_i^{m_i})}$$

where h is the exponent such that $g^h \equiv b \pmod{p_i^{m_i}}$. Then X_1 is a k-character. Moreover, $X_1(a) \neq 1$ (since a is not congruent to 1 mod $p_i^{m_i}$).

Case 2. $p_i = 2$ and $a \equiv 1 \pmod 4$.

Then $m_i > 2$ (since a is not congruent to 1 mod 2^{m_i}). As the reader will be asked to show in the exercises, (1) $5^x \equiv 1 \pmod{2^{m_i}}$ iff $2^{m_i-2} \mid x$ and (2) for any odd number b there is a unique nonnegative integer $h(b) < 2^{m_i-2}$ such that $5^{h(b)} \equiv (-1)^{(b-1)/2} b \pmod{2^{m_i}}$. Define X_1 such that $X_1(b) = 0$ if $\gcd(b,k) \neq 1$, but otherwise

$$X_1(b) = e^{2\pi i h/2^{m_i-2}}$$

where $h = h(b)$. Then X_1 is a k-character and $X_1(a) \neq 1$.

Case 3. $p_i = 2$ and $a \equiv 3 \pmod 4$.

Define X_1 such that $X_1(b) = 0$ if $\gcd(b,k) \neq 1$, but otherwise

$$X_1(b) = (-1)^{(b-1)/2}$$

Then X_1 is a k-character (since $m > 1$) and $X_1(a) \neq 1$.

This completes the proof of the fact that if $\gcd(a,k) \neq 1$ and a is not congruent to 1 mod k then, for some k-character X_1, $X_1(a) \neq 1$.

Adding up all the values of all the k-characters, in two different ways, we have

$$\sum_{a=0}^{k-1} \sum_{characters} X(a) = \sum_{characters} X(1) = \sum_{characters} 1$$

$$\sum_{characters} \sum_{a=0}^{k-1} X(a) = \sum_{a=0}^{k-1} X_0(a) = \phi(k)$$

Hence there are exactly $\phi(k)$ k-characters.

We close this section with a theorem we shall need later.

Theorem 7.1.1 *If* $\gcd(t,k) = 1$ *(so that t has an inverse t^{-1} mod k), and a is not congruent to t mod k, then*

$$\sum_{characters} \frac{X(a)}{X(t)} = \sum_{characters} X(a)X(t^{-1}) = \sum_{characters} X(at^{-1}) = 0$$

Exercises 7.1

1. Prove $H^{-1} = H$.
2. If $m > 2$ then $5^x \equiv 1 \pmod{2^m}$ iff $2^{m-2} \mid x$. (Hint: by MI on $t > 2$,

$$5^{2^{t-3}} = 1 + 2^{t-1} \times \text{ odd number}$$

Hence $5^{2^{m-3}}$ is not congruent to 1 mod 2^m. But $5^{2^{m-2}} \equiv 1 \pmod{2^m}$.)
3. If $m > 2$ then to every odd integer b there is a unique integer $h(b)$ such that $0 \leq h(b) < 2^{m-2}$ and $5^{h(b)} \equiv (-1)^{(b-1)/2} b \pmod{2^m}$. (Hint: from the previous exercise, the numbers $1, 5, 5^2, \ldots 5^{2^{m-2}}$ are distinct mod 2^m. They are all congruent to 1 mod 4. Any complete set of residues mod 2^m contains exactly 2^{m-2} integers congruent to 1 mod 4. Hence one and only one of the 2^{m-2} powers of 5 is congruent to b mod 2^m if $b \equiv 1 \pmod 4$. And one and only one is congruent to $-b$ if $b \equiv 3 \pmod 4$.)

7.2 Dirichlet Series

We define the 'Dirichlet L-series' as follows:

$$L(s,X) = \sum_{a=1}^{\infty} \frac{X(a)}{a^s}$$

where s is any real ≥ 1, and X is any function (e.g. a k-character). In this section we derive some useful properties of these L-series. We begin with a lemma about k-characters.

Theorem 7.2.1 *If X is a k-character, $X \neq X_0$, and u, v are any positive integers,*

$$\left|\sum_{a=u}^{v} X(a)\right| \leq \frac{\phi(k)}{2}$$

Proof: Since $\sum X(a) = 0$ when a ranges over a complete set of residues, we may assume that the number $v-u+1$ of terms of the sum is $\leq k-1$ (so that $v+1 \leq u+k-1$). If the sum contains at most $\phi(k)/2$ nonzero terms, we are done (since nonzero terms have modulus 1). Suppose, then, that it contains more than $\phi(k)/2$ nonzero terms. Its terms, together with those of

$$A = \sum_{v+1}^{u+k-1} X(a)$$

contain exactly $\phi(k)$ nonzero terms, and hence A contains fewer than $\phi(k)/2$ nonzero terms, so that $|A| < \phi(k)/2$. But

$$\left|\sum_{a=u}^{v} X(a)\right| = \left|\sum_{a=u}^{u+k-1} X(a) - A\right| = |0 - A| \leq \phi(k)/2$$

Using Theorem 7.2.1, we can obtain our first result about L-series:

Theorem 7.2.2 *If X is a k-character, and $X \neq X_0$ then $L(s,X)$ converges uniformly for $s \geq 1$.*

Proof: Let $R(w) = \sum_{b=u}^{w} X(b)$ (with $R(u-1) = 0$). Then, from Theorem 7.2.1, $|R(w)| \le \phi(k)/2$, and we obtain

$$\left|\sum_{a=u}^{v} \frac{X(a)}{a^s}\right| = \left|\frac{1}{v^s}R(v) + \sum_{a=u}^{v-1}\left(\frac{1}{a^s} - \frac{1}{(a+1)^s}\right) R(a)\right|$$

$$\le \phi(k)/2u^s \le \phi(k)/2u$$

and the result follows by the Cauchy criterion for uniform convergence.

Corollary. If X is a nonprincipal k-character, and $s \ge 1$ then

$$\left|\sum_{a=u}^{v} \frac{X(a)}{a^s}\right| \le \frac{\phi(k)}{2u}$$

Similarly, we have the following.

Theorem 7.2.3 *If X is a k-character and $X \ne X_0$ then $L(s, X\ln)$ converges uniformly for $s \ge 1$. Moreover, on this interval, $L'(s,X) = -L(s, X\ln)$ and $|L(s, X\ln)| < \phi(k)$.*

Proof: If $x \ge 3 > e$ then the function $(\ln x)/x^s$ decreases. Hence if $a \ge 3$,

$$\frac{\ln a}{a^s} - \frac{\ln(a+1)}{(a+1)^s}$$

is nonnegative. Hence, if $u \ge 3$, we obtain, as in the previous proof,

$$\left|\sum_{a=u}^{v} \frac{X(a)\ln a}{a^s}\right| \le \frac{\phi(k)\ln u}{2u}$$

and so we have uniform convergence as s varies on the interval $[1, \infty]$. Hence we can differentiate term by term, obtaining

$$\frac{d\, L(s,X)}{ds} = -L(s, X\ln)$$

Letting $v \to \infty$ in

$$\left|\frac{X(1)\ln 1}{1^s} + \frac{X(2)\ln 2}{2^s} + \sum_{a=3}^{v} \frac{X(a)\ln a}{a^s}\right| \le \frac{\ln 2}{2} + \frac{\phi(k)\ln 3}{6} \le \phi(k)$$

we obtain the final result.

Corollary. If $X \neq X_0$ then $\frac{d}{ds}L(s,X)$ is continuous on $[1,\infty]$.

If we include the principal character, our result is almost as sharp.

Theorem 7.2.4 *If ϵ is any positive real, and X is any k-character then $L(s, X\ln)$ converges uniformly for $s > 1+\epsilon$. Moreover, on this interval, $L'(s,X) = -L(s, X\ln)$.*

Proof: This follows from the Weierstrass M-test since

$$\left|\frac{X(a)\ln a}{a^s}\right| < \frac{\ln a}{a^{1+\epsilon}}$$

The Möbius function proves to be important at this point:

Theorem 7.2.5 *If X is any k-character, $L(s, X\mu)$ converges absolutely when $s > 1$, and $L(s,X)L(s,X\mu) = 1$.*

Proof: Because the two L series converge absolutely (since $s > 1$), we can rearrange the terms in their product.

$$L(s,X)L(s,X\mu) = \sum_{l=1}^{\infty}\sum_{ab=l}\frac{C(b)C(a)\mu(a)}{b^s a^s} = \sum_{l=1}^{\infty}\frac{X(l)}{l^s}\sum_{a|l}\mu(a)$$

By Theorem 1.11.1, $\sum_{a|l}\mu(a) = 0$ unless $l = 1$. Hence the above product is just $C(1)\mu(1) = 1$.

Corollary: If $s > 1$ then $L(s,X) \neq 0$.

The Möbius function also plays a role in our next theorem.

Theorem 7.2.6 *If $s > 1$, and X is any k-character,*

$$L(s,X) = \frac{1}{\prod_{primes}\left(1 - \frac{X(p)}{p^s}\right)}$$

Proof: First note that the product can be understood as

$$e^{\sum \ln(1-X(p)/p^s)}$$

the exponent being absolutely convergent.

Let $N > 2$. Then, if p, p', p'', ... denote primes

$$\prod_{primes \leq N} \left(1 - \frac{X(p)}{p^s}\right)$$

$$= 1 - \sum_{p \leq N} \frac{X(p)}{p^s} + \sum_{p,\, p' \leq N} \frac{X(pp')}{(pp')^s} - \sum_{p,\, p',\, p'' \leq N} \frac{X(pp'p'')}{(pp'p'')^s} + \cdots$$

$$= \sum_{p|a \Rightarrow p \leq N} \frac{X(a)\mu(a)}{a^s}$$

$$= \sum_{a=1}^{N} \frac{X(a)\mu(a)}{a^s} + \sum_{a>N \text{ and } p|a \Rightarrow p \leq N} \frac{X(a)\mu(a)}{a^s}$$

As $N \to \infty$, the first of the last two sums tends to $1/L(s, X)$ (Theorem 7.2.5), while the second of the last two sums tends to 0.

Corollary. The 1-character is simply the function $X(1) = 1$. For this character, $L(s, X) = \sum_{a=1}^{\infty} 1/a^s$. This is the *Riemann zeta function* ζ. What the preceding theorem tells us about the Riemann zeta function is that

$$\zeta(s) = \prod_{primes} (1 - 1/p^s)^{-1}$$

— a result due to Euler.

Exercises 7.2

1. Let p be an odd prime of the form $4m + 3$, and let H be the $4p$-character defined by $H(a) = (-1)^{(a-1)/2} \left(\frac{a}{p}\right)$ if $\gcd(a, 4p) = 1$. Consider

$$L(1, H) = \sum_{m=0}^{\infty} \frac{(-1)^m \left(\frac{2m+1}{p}\right)}{2m+1}$$

Show that as m goes from 1 to $p-1$, the numerators of the terms are

$$\left(\frac{1}{p}\right), \left(\frac{p-3}{p}\right), \left(\frac{5}{p}\right), \left(\frac{p-7}{p}\right), \left(\frac{9}{p}\right), \ldots, \left(\frac{4}{p}\right), \left(\frac{p-2}{p}\right), 0,$$

$$\left(\frac{2}{p}\right), \left(\frac{p-4}{p}\right), \left(\frac{6}{p}\right), \left(\frac{p-8}{p}\right), \ldots, \left(\frac{3}{p}\right), \left(\frac{p-1}{p}\right)$$

covering all the $\left(\frac{4t}{p}\right)$ and $\left(\frac{4t+1}{p}\right)$ before the 0, and all the $\left(\frac{4t+2}{p}\right)$ and $\left(\frac{4t+3}{p}\right)$ after the 0.

2. For $L(1, H)$, show that the sum of the terms with $m \geq p$ is bounded by $\pi^2/24$.

3.* Pick any single hearted prime p of the form $4m+3$. Let $x = a$, $y = b$ be the smallest positive integer solution of $x^2 - py^2 = 1$. Show that $\sqrt{p}L(1, H) = \ln(a + b\sqrt{p})$.

4. Using the fact that $\zeta(2) = \pi^2/6$, show that

$$\sum_{a=1}^{\infty} \frac{\mu(a)}{a^2} = \frac{6}{\pi^2}$$

7.3 Mangoldt Function

If n is a positive integer, we define $\Lambda(n) = \ln p$ if n is a power of prime p. Otherwise, $\Lambda(n) = 0$. This is the *Mangoldt function*. For example, $\Lambda(8) = \ln 2$ and $\Lambda(10) = 0$. Note that $\ln n = \sum_{d|n} \Lambda(d)$.

The Mangoldt function is related to the L-series in the following theorem.

Theorem 7.3.1 *If X is any k-character, and $s > 1$ then*

$$L(s, X\Lambda)L(s, X) = L(s, X \ln)$$

(with $L(s, X) \neq 0$).

Proof: First note that $L(s, X\Lambda)$ converges absolutely. Now

$$L(s, X)L(s, X\Lambda) = \sum_{b=1}^{\infty} \frac{X(b)}{b^s} \sum_{a=1}^{\infty} \frac{X(a)\Lambda(a)}{a^s}$$

$$= \sum_{l=1}^{\infty} \frac{X(l)}{l^s} \sum_{a|l} \Lambda(a) = \sum_{l=1}^{\infty} \frac{X(l)\ln l}{l^s} = L(s, X\ln)$$

(The rearrangement of terms is justified by the fact that the series converge absolutely.)

We use the Mangoldt function to obtain the infinity we shall later use to show that

$$\sum_{primes\ a\ in\ P} \frac{\Lambda(a)}{a^s}$$

is infinite:

Theorem 7.3.2 *If X_0 is the principal character,*

$$\lim_{s\downarrow 1} \frac{L(s, X_0\ln)}{L(s, X_0)} = \infty$$

Proof: By Theorem 7.3.1,

$$\frac{L(s, X_0\ln)}{L(s, X_0)} = L(s, X_0\Lambda) = \sum_{(a,k)=1} \frac{\Lambda(a)}{a^s}$$

$$= \sum_{a=1}^{\infty} \frac{\Lambda(a)}{a^s} - \sum_{p|k} \ln p \left(\frac{1}{p^s} + \frac{1}{p^{2s}} + \frac{1}{p^{3s}} + \cdots\right)$$

The last sum, taken over all primes p dividing k, equals

$$\sum_{p|k} \frac{\ln p}{p^s - 1}$$

which is finite, and remains finite as $s \to 1$. Hence it suffices to prove that $\sum_{a=3}^{\infty} \Lambda(a)/a^s$ diverges. But this sum is greater than $\sum_{primes} 1/p^s$ which tends to ∞ as $s \downarrow 1$ (see Exercises 7.3).

Exercises 7.3

1. Prove that if n is a positive integer, $\sum_{d|n} \Lambda(d) = \ln n$. Hence, $\Lambda(n) = -\sum_{d|n} \mu(d) \ln d$.
2. Prove that if $0 \leq x \leq 1/2$ then $1/(1-x) \leq 1 + 2x \leq e^{2x}$.
3. If N is a large positive integer, s is a real ≥ 1, and p varies over primes, then, using the previous exercise,

$$\sum_{a=1}^{N} \frac{1}{a^s} \leq \prod_{p \leq N} \left(1 + \frac{1}{p^s} + \frac{1}{p^{2s}} + \frac{1}{p^{3s}} + \cdots\right)$$

$$= \prod_{p \leq N} \frac{1}{1 - 1/p^s} \leq \exp\left(\sum_{p \leq N} \frac{2}{p^s}\right)$$

4. With the above notation, and $s > 1$,

$$\frac{1}{s-1}\left(1 - \frac{1}{N^{s-1}}\right) = \int_1^N \frac{1}{a^s} da \leq \sum_{a=1}^{N} \frac{1}{a^s}$$

When $s = 1$, $\ln N \leq \sum_{a=1}^{\infty} \frac{1}{a}$.
5. With the above notation, and $s > 1$,

$$-\ln(s-1) \leq 2 \sum \frac{1}{p^s}$$

and hence

$$\lim_{s \downarrow 1} \sum_{primes} \frac{1}{p^s} = \infty$$

And $\sum 1/p$ diverges.

7.4 $L(1, X) \neq 0$

In this section we prove that if X is a k-character then $L(1, X) \neq 0$. Note that $L(1, X_0) = \infty$, so we can restrict our attention to the nonprincipal characters. We begin with three lemmas.

Theorem 7.4.1 *If a, b, $c > 0$ then $3abc \leq a^3 + b^3 + c^3$.*

Proof:

$$\begin{aligned} a^2 + b^2 &\geq 2ab \\ b^2 + c^2 &\geq 2bc \\ c^2 + a^2 &\geq 2ac \end{aligned}$$

and hence, adding, $a^2 + b^2 + c^2 - ab - bc - ca \geq 0$. Multiplying this by the nonnegative number $a + b + c$, we obtain the result.

Theorem 7.4.2 *If x is real and $0 < y < 1$,*

$$(1-y)^3|1-ye^{xi}|^4|1-ye^{2xi}|^2 < 1$$

Proof: Since $1 - ye^{xi} = (1 - y\cos x) - iy\sin x$, we have

$$|1-ye^{xi}|^2 = (1-y\cos x)^2 + y^2\sin^2 x = 1 - 2y\cos x + y^2$$

(this being some positive real), and

$$(|1-ye^{xi}|^2)^2|1-ye^{2xi}|^2 = (1-2y\cos x + y^2)^2(1-2y\cos 2x + y^2)$$

With $a = b = (1-2y\cos x + y^2)^{1/3}$ and $c = (1-2y\cos 2x + y^2)^{1/3}$, the previous theorem implies that the preceding product is bounded above by

$$\begin{aligned} &\frac{(3-4y\cos x - 2y\cos 2x + 3y^2)^3}{27} \\ &\leq \frac{(3-4y\cos x - 4y\cos^2 x + 2y + 3y^2)^3}{27} \\ &\leq \frac{(3-4y(\cos x + 1/2)^2 + 3y + 3y^2)^3}{27} \\ &\leq \frac{(3+3y+3y^2)^3}{27} \leq (1+y+y^2)^3 < (1-y)^{-3} \end{aligned}$$

Theorem 7.4.3 *If $s > 1$ and X is any k-character,*

$$(L(s,X_0))^3|L(s,X)|^4|L(s,X^2)| \geq 1$$

Proof: Let p be a prime not dividing k. Suppose $X(p) = e^{xi}$. Then $X^2(p) = e^{2xi}$. Also $X_0(p) = 1$. By the previous theorem,

$$\left(1 - \frac{X_0(p)}{p^s}\right)^3 \left|1 - \frac{X(p)}{p^s}\right|^4 \left|1 - \frac{X^2(p)}{p^s}\right|^2 \leq 1$$

The same is also true if p is a prime dividing k (when the inequality reduces to $1 \leq 1$). Taking the product for all primes p, and using Theorem 7.2.6, we obtain the result.

Theorem 7.4.4 *If X is a k-character at least one of whose values is not real, then $L(1,X) \neq 0$.*

Proof: For this proof, let

$$s = 1 + \frac{1}{16(\phi(k))^6}$$

Since X has a non-real value, $X^2 \neq X_0$, and the corollary to Theorem 7.2.2 implies that $|L(s,C^2)| < \phi(k)$.

Also

$$L(s,X_0) \leq \sum_{a=1}^{\infty} \frac{1}{a^s} < 1 + \int_1^{\infty} \frac{1}{a^s}\, da = 1 + \frac{1}{s-1} < \frac{2}{s-1}$$

Hence, by Theorem 7.4.3,

$$|L(s,X)| \geq (L(s,X_0))^{-3/4}|L(s,X^2)|^{-1/2} > \frac{(s-1)^{3/4}}{2\sqrt{\phi(k)}} = \frac{1}{8(\phi(k))^5}$$

Now

$$\int_1^s \frac{d}{dt} L(t,X)\, dt = L(s,X) - L(1,X)$$

and hence

$$|L(s,X) - L(1,X)| \leq (s-1)\phi(k) = \frac{1}{16(\phi(k))^5}$$

(Theorem 7.2.3).

If $L(1,X) = 0$ then

$$\frac{1}{16(\phi(k))^5} \geq |L(s,X)| > \frac{1}{8(\phi(k))^5}$$

which is impossible. Hence $L(1,X) \neq 0$.

Theorem 7.4.5 *If X is a nonprincipal k-character all of whose values are real, then* $L(1,X) > 0$.

Proof:

The f function

Let

$$f(a) = \sum_{d|a} X(d)$$

this being real-valued, since X is real-valued. If p is a prime,

$$f(p^l) = 1 + X(p) + (X(p))^2 + \cdots + (X(p))^l$$

Regardless of whether $X(p) = 0$, 1, or -1, this is nonnegative. If l is even, $f(p^l) \geq 1$.

f is 'multiplicative' in the sense that if $\gcd(m,n) = 1$ then $f(mn) = f(m)f(n)$. For if $\gcd(m,n) = 1$,

$$f(mn) = \sum_{d|mn} X(d) = \sum_{d_1|m,\ d_2|n} X(d_1d_2) = \sum_{d_1|m,\ d_2|n} X(d_1)X(d_2)$$

$$= \sum_{d_1|m} X(d_1) \times \sum_{d_2|n} X(d_2) = f(m)f(n)$$

Hence for any integer a, $f(a) \geq 0$ and if a is a square, $f(a) \geq 1$.

A Lower Bound on $z = z_1 + z_2$

Abbreviate $(4\phi(k))^6$ as m.

$$z = \sum_{n=1}^{m} 2(m-n)f(n) \geq \sum_{b=1}^{\sqrt{m}} 2(m-b^2) \geq \sum_{b=1}^{\sqrt{m}/2} 2(m-b^2)$$

$$\geq \sum_{b=1}^{\sqrt{m}/2} 2(m-m/4) = (\sqrt{m}/2)(3m/2) = \frac{3}{4}(4\phi(k))^9$$

An Upper Bound on z_1

Also

$$z = \sum_{n=1}^{m}\sum_{b|n} 2(m-n)X(b) = \sum_{ab\leq m} 2(m-ab)X(b)$$

$$= \sum_{a=1}^{m^{1/3}-1} \sum_{m^{2/3}<b\leq m/a} 2(m-ab)X(b) + \sum_{b=1}^{m^{2/3}} \sum_{0<a\leq m/b} 2(m-ab)X(b)$$

Call the last two sums z_1 and z_2. The limits of summation in these two sums are explained by the fact that the region

$$\{(a,b) \mid 1 \leq a,\ b \text{ and } ab \leq m\}$$

is the disjoint union of the regions

$$\{(a,b) \mid 1 \leq a < m^{1/3}, \text{ and } m^{2/3} < b \leq m/a\}$$

and

$$\{(a,b) \mid 1 \leq b \leq m^{2/3} \text{ and } 1 \leq a \leq m/b\}$$

— as can be seen by graphing $b = m/a$. If w is an integer $> m^{2/3}$, let

$$R(w) = \sum_{b=m^{2/3}+1}^{w} X(b)$$

and let $R(m^{2/3}) = 0$. Then

$$\sum_{m^{2/3}<b\leq m/a} 2(m-ab)X(b) = \sum_{b=m^{2/3}+1}^{[m/a]} 2(m-ab)(R(b)-R(b-1))$$

$$= \sum_{b=m^{2/3}+1}^{[m/a]-1} 2aR(b) + 2(m - a[m/a])R([m/a])$$

and hence, by Theorem 7.2.1, the absolute value of this is bounded by

$$\sum_{b=m^{2/3}+1}^{[m/a]-1} 2a\frac{\phi(k)}{2} + 2(m - a[m/a])\frac{\phi(k)}{2}$$

$$= \phi(k)(a[m/a] - a - am^{2/3} - a + a + m - a[m/a])$$

$$= \phi(k)(m - am^{2/3} - a) < \phi(k)m$$

Thus

$$|z_1| \leq \sum_{a=1}^{m^{1/3}-1} \phi(k)m < \phi(k)mm^{1/3} = m^{4/3}\phi(k)$$

An Upper Bound on z_2

Let $d = m/b - [m/b]$. We have

$$\sum_{a=1}^{[m/b]} (2m - 2ab) = 2m \sum_{a=1}^{[m/b]} 1 - 2b \sum_{a=1}^{[m/b]} a$$

$$= 2m[m/b] - b[m/b]([m/b] + 1)$$

$$= 2m(m/b - d) - b(m/b - d)(m/b - d + 1)$$

$$= m^2/b - m + bd(1 - d)$$

Hence

$$z_2 = \sum_{b=1}^{m^{2/3}} (m^2/b - m + bd(1 - d))X(b)$$

$$= m^2 \sum_{b=1}^{m^{2/3}} \frac{X(b)}{b} - m \sum_{b=1}^{m^{2/3}} X(b) + \sum_{b=1}^{m^{2/3}} bd(1 - d)X(b)$$

$$\leq m^2 \left(L(1, X) - \sum_{b=m^{2/3}+1}^{\infty} \frac{X(b)}{b} \right) + m\phi(k)/2 + \sum_{b=1}^{m^{2/3}} b$$

by Theorem 7.2.1 and the fact that $|d(1-d)X(b)| \leq 1$. By the Corollary to Theorem 7.2.2,

$$\begin{aligned} z_2 &\leq m^2 L(1,X) + m^2 \frac{\phi(k)}{2(m^{2/3}+1)} + \frac{m\phi(k)}{2} + \frac{m^{2/3}(m^{2/3}+1)}{2} \\ &\leq m^2 L(1,X) + m^{4/3}\phi(k)/2 + m\phi(k)/2 + m^{4/3} \\ &\leq m^2 L(1,X) + m^{4/3}\phi(k)/2 + m^{4/3}\phi(k)/2 + m^{4/3}\phi(k) \\ &= m^2 L(1,X) + 2m^{4/3}\phi(k) \end{aligned}$$

A Lower Bound on $L(1,X)$

From the previous results, we have

$$\begin{aligned} \frac{3}{4}(4\phi(k))^9 \leq z = z_1 + z_2 &< m^{4/3}\phi(k) + m^2 L(1,X) + 2m^{4/3}\phi(k) \\ &= m^2 L(1,X) + \frac{3}{4}(4\phi(k))^9 \end{aligned}$$

Hence $0 < m^2 L(1,X)$ and the result follows.

Theorem 7.4.6 *If $X \neq X_0$ then $L'(s,X)/L(s,X)$ is continuous on* $[1,\infty]$.

Proof: By Theorems 7.2.5, 7.4.4, and 7.4.5, $L(s,X)$ is nonzero on $[1,\infty]$. By Theorems 7.2.2 and 7.2.3, both numerator and denominator are continuous on $[1,\infty]$.

Exercises 7.4

1. Prove $L(1,X_0)$ diverges.
2. Let $p = 7$. Graph $L(s,H)$ with s on $[1,\infty]$.
3. Let X be the 5-character $\left(\frac{a}{5}\right)$. With $f(a) = \sum_{d|a} X(d)$, show that $f(15) = 0$ and $f(15 \times 33) = 2$.

7.5 Dirichlet's Theorem on Primes in AP

We can now prove the first of our four beautiful theorems. Suppose $\gcd(l, k) = 1$ with $l > 0$. If $s > 1$ then, by Theorems 7.2.3 and 7.3.1,

$$-\sum_{characters} \frac{1}{X(l)} \frac{L'(s,X)}{L(s,X)}$$

$$= \sum_{characters} \frac{1}{X(l)} L(s, X\Lambda) = \sum_{characters} \sum_{a=1}^{\infty} \frac{X(a)}{X(l)} \frac{\Lambda(a)}{a^s}$$

The above double sum converges absolutely, so we can switch the order of the summation signs. Using Theorem 7.1.1, we obtain

$$\sum_{a=1}^{\infty} \frac{\Lambda(a)}{a^s} \sum_{characters} \frac{X(a)}{X(l)} = \sum_{a \equiv l \pmod{k}} \frac{\Lambda(a)\phi(k)}{a^s}$$

since there are $\phi(k)$ k-characters.

If $X \neq X_0$ then

$$\lim_{s \downarrow 1} \frac{1}{X(l)} \frac{L'(s,X)}{L(s,X)}$$

is some finite number (by Theorem 7.4.6 — the fruit of those long proofs about $L(1, X) \neq 0$). However, by Theorems 7.2.3 and 7.3.2,

$$\lim_{s \downarrow 1} \frac{1}{X(l)} \frac{-L'(s,X)}{L(s,X)} = \infty$$

Hence

$$\lim_{s \downarrow 1} \sum_{a \equiv l \pmod{k}} \frac{\Lambda(a)}{a^s} = \infty$$

Now for any $s > 1$, we have the following relation:

$$\sum_{a \equiv l \pmod{k}} \frac{\Lambda(a)}{a^s} = \sum_{a \equiv l \pmod{k},\, prime} \frac{\Lambda(a)}{a^s} + \sum_{a \equiv l \pmod{k},\, prime\ power} \frac{\Lambda(a)}{a^s}$$

(since all the series converge absolutely, it is possible to have rearrangements). Call the latter two sums $J(s)$ and $K(s)$.

For K we have (with p varying over primes, and $m > 1$)

$$K(s) = \sum_{a \equiv l \pmod{k},\, a=p^m} \frac{\Lambda(a)}{a^s} < \sum_{a=p^m} \frac{\Lambda(a)}{a^s}$$

$$= \sum_{p\ prime,\, m>1} \frac{\ln p}{p^{sm}} < \sum_{p\ prime,\, m>1} \frac{\ln p}{p^m}$$

$$< \sum_{p\ prime,\, m>1} \frac{1}{p^{m-1/2}} = B$$

Hence for any $s > 1$,

$$\sum_{a \equiv l \pmod{k}} \frac{\Lambda(a)}{a^s} < J(s) + B$$

Taking the limit as $s \downarrow 1$, we obtain

$$\infty \le \lim_{s \downarrow 1} J(s) + B$$

and thus

$$\lim_{s \downarrow 1} \sum_{a \equiv l \pmod{k},\, a\ prime} \frac{\Lambda(a)}{a^s} = \infty$$

This sum cannot have merely finitely many terms.

Exercises 7.5

1. Find the smallest prime of the form 13m + 9.

7.6 How Many Pythagorean Triangles?

In this section we estimate the number $P(n)$ of primitive Pythagorean triangles with area less than n.[1] A primitive Pythagorean triangle, recall, is, in effect, a triple of positive integers (a, b, c) such that $c^2 =$

[1] The results are due to J. Lambek, L. Moser, and R. Wild. See *Pacific Journal of Mathematics*, 5 (1955), 73-91.

$a^2 + b^2$ and $\gcd(a, b, c) = 1$. In order to count these triangles, we make use of the notion of 'quasi-primitive' Pythagorean triangles. A *quasi-primitive Pythagorean triangle* is a triple of positive integers (a, b, c) such that $c^2 = a^2 + b^2$ and $\gcd(a, b, c) \leq 2$. The quasi-primitive Pythagorean triangles can all be obtained from the primitive ones, simply by multiplying the sides of the latter by 2. Since this has the effect of multiplying the area by 4, the number $Q(n)$ of quasi-primitive Pythagorean triangles with area less than n is $P(n) + P(n/4)$. We can express $P(n)$ in terms of the following finite sum:

$$P(n) = P(n) + P\left(\frac{n}{4}\right) - \left(P\left(\frac{n}{4}\right) + P\left(\frac{n}{4^2}\right)\right) + \left(P\left(\frac{n}{4^2}\right) + P\left(\frac{n}{4^3}\right)\right) - \cdots$$

$$= Q(n) - Q\left(\frac{n}{4}\right) + Q\left(\frac{n}{4^2}\right) - Q\left(\frac{n}{4^3}\right) + \cdots$$

Thus the problem reduces to that of estimating $Q(n)$.

In any primitive Pythagorean triangle (a, b, c) with c the hypotenuse, exactly one of a and b is even. In any quasi-primitive Pythagorean triangle with $\gcd(a, b, c) = 2$, and with c the hypotenuse, exactly one of a and b is a multiple of 4. The following theorem thus gives us a one-to-one correspondence between quasi-primitive Pythagorean triangles and primitive lattice points (x, y) with $x > y > 0$.

Theorem 7.6.1 (a, b, c) *is either (1) a primitive Pythagorean triangle with hypotenuse c, and a even or (2) a Pythagorean triangle with* $\gcd(a, b, c) = 2$, *hypotenuse c, and b a multiple of 4*
iff there are relatively prime positive integers x and y, with $x > y$, such that

$$\begin{aligned} a &= 2xy \\ b &= x^2 - y^2 \\ c &= x^2 + y^2 \end{aligned}$$

Proof: Suppose the right hand side of the equivalence is true. If x and y have different parity we get a primitive Pythagorean triangle. If x and y are both odd we get a Pythagorean triangle with $\gcd(a, b, c) = 2$.

Suppose the left hand side of the equivalence is true. If the triangle is primitive the right hand side follows with x and y having different parity. If the triangle has gcd 2, the right hand side follows with x and y both odd. For let $a = 2a'$, $b = 2b'$ (with b' even) and $c = 2c'$. Then there are relatively prime integers x' and y' with different parity and $x' > y' > 0$ such that $a' = x'^2 - y'^2$, $b' = 2x'y'$ and $c' = x'^2 + y'^2$. If $x = x' + y'$ and $y = x' - y'$ then x and y are relatively prime integers, with $x > y$, such that $a = 2xy$, $b = x^2 - y^2$, and $c = x^2 + y^2$.

As a result of the preceding theorem, $Q(t)$ is the number $L_1(t)$ of primitive lattice points in the region

$$R(t) = \{(x, y) \mid \frac{1}{2} 2xy(x^2 - y^2) < t,\ x > y > 0\}$$

In general, let $L_i(t)$ be the number of lattice points (x, y) in $R(t)$ with $\gcd(x, y) = i$. The number $L_i(t)$ of lattice points (x, y) in $R(t)$ with $\gcd(x, y) = i$ equals the number $L_1(t/i^4)$ of primitive lattice points in $R(t/i^4)$. Thus, where $L(t)$ is the total number of lattice points in $R(t)$,

$$L(t) = \sum_{i=1}^{\infty} L_i(t) = \sum_{i=1}^{\infty} L_1(t/i^4)$$

Note that if $t/i^4 < 4$ there are no lattice points in $R(t/i^4)$, so that this sum is finite.

Let us abbreviate $L_1(t/i^4)$ as $F(i)$. Consider

$$\sum_{j=1}^{\infty} \mu(j) L(t/j^4) = \sum_{j=1}^{\infty} \sum_{i=1}^{\infty} \mu(j) F(ij) = \sum_{h=1}^{\infty} \sum_{j|h} \mu(j) F(h)$$

$$= \sum_{h=1}^{\infty} F(h) \sum_{j|h} \mu(j) = F(1)$$

by Theorem 1.11.1. Thus

$$Q(t) = L_1(t) = F(1) = \sum_{j=1}^{\infty} \mu(j) L(t/j^4)$$

— the sum being finite — and the problem reduces to one of estimating the number of lattice points in $R(t)$. Note that if we draw the boundary of $R(t)$ on an x versus y graph, it consists of the positive x-axis (vertical), the line $x = y$, and a curve that has these two straight lines as its asymptotes. Since this curve is a relatively 'nice' curve, the number of lattice points in $R(t)$ is approximately equal to its area.

Translating $xy(x^2 - y^2) = t$ into polar coordinates, we get

$$r^4 \sin u \cos u(\sin^2 u - \cos^2 u) = t$$

with $\pi/4 \leq u \leq \pi/2$. This is equivalent to

$$r = \frac{(4t)^{1/4}}{(-\sin 4u)^{1/4}}$$

and using the area formula $A = \int \frac{1}{2} r^2 \, du$, we obtain

$$A(t) = (4t)^{1/2} \frac{1}{2} \int_0^{\pi/4} (\sin 4u)^{-1/2} \, du$$

Let $v = \sqrt{\sin 4u}$. Then $v' = 4(1/2v)\sqrt{1 - v^4}$ and

$$A(t) = t^{1/2} 2 \int_0^1 \frac{1}{2} (1 - v^4)^{-1/2} \, dv$$

$$= t^{1/2} \frac{1}{4} \int_0^1 w^{1/4-1} (1 - w)^{1/2-1} \, dw$$

(with $w = u^4$). The integral is the beta function

$$B(1/4, 1/2) = \frac{\Gamma(1/4)\Gamma(1/2)}{\Gamma(3/4)}$$

A standard gamma function formula gives

$$\Gamma\left(\frac{3}{4}\right) \Gamma\left(\frac{1}{4}\right) = \pi\sqrt{2}$$

and hence the area of $R(t)$ is

$$A(t) = 2^{-5/2} \pi^{-1/2} (\Gamma(1/4))^2 \sqrt{t} \approx 1.31103\sqrt{t}$$

Graphed on an x versus y coordinate system, $R(t)$ looks like a bug with two 'antennae'. The antennae are the parts of $R(t)$ with $y < 1$ and

$y > t^{1/3}$. As we shall show, the antennae do not contain lattice points and have a combined area proportionate to $t^{1/3}$. The body of the bug has a perimeter also proportionate to $t^{1/3}$ and hence the number $L(t)$ of lattice points in $R(t)$ equals the area $A(t)$ of $R(t)$ plus or minus an error proportionate to $t^{1/3}$. We argue for these assertions as follows.

The left antenna contains no lattice point since $y < 1$. Moreover, the vertical line $y = 1$ meets the curve $xy(x^2 - y^2) = t$ in a point with x-coordinate between $t^{1/3}$ and $t^{1/3} + 1$. The left side of the body of the bug thus has length $< t^{1/3} + 1$. When $y < 1$, x on the boundary curve is such that $xy(x-1)x < t$ and hence $(x-1)^3 y < t$, so that $x < (t/y)^{1/3} + 1$. This implies that the area of the left antenna is bounded by

$$\int_0^1 (t/y)^{1/3} + 1 \, dy = \frac{3}{2}t^{1/3} + 1 < \frac{5}{2}t^{1/3}$$

The right antenna contains no lattice points since $x > y > t^{1/3}$ and the boundary curve is $xy(x-y)(x+y) = t$. (If (x, y) is a lattice point, and $x > y$ then $x - y \geq 1$.) The area of the right antenna is

$$\int_{t^{1/3}}^{\infty} x - y \, dy = \int_{t^{1/3}}^{\infty} \frac{t}{xy(x+y)} \, dy < \int_{t^{1/3}}^{\infty} \frac{t}{2y^3} \, dy = \frac{1}{4}t^{1/3}$$

Thus the sum of the areas of the two antennae is bounded by $3t^{1/3}$.

The length of the part of the upper curve $xy(x^2 - y^2) = t$ which bounds the body of the bug is less than the length of the V which is the lower boundary of the body of the bug. Thus the perimeter of the body of the bug is bounded by

$$2 \times (t^{1/3} + 1 + \sqrt{2}t^{1/3}) < 7t^{1/3}$$

From the above, then, it follows that there is a constant K (e.g. 100) such that

$$|L(t) - A(t)| < Kt^{1/3}$$

(Lambek and Moser prove this as a consequence of a more general theorem — for which they give a fully rigorous proof.)

We can now estimate

$$Q(t) = L_1(t) = \sum_{j=1}^{\infty} \mu(j) L(t/j^4)$$

Switching the L for the A, our error is bounded by

$$\sum_{j=1}^{\infty} K(t/j^4)^{1/3} = Kt^{1/3}\sum_{j=1}^{\infty} j^{-4/3}$$

$$< Kt^{1/3}\left(1 + \int_1^{\infty} j^{-4/3}\,dj\right) = Kt^{1/3}(1+3) = 4Kt^{1/3}$$

Thus, with the above error possible,

$$Q(t) = L_1(t) \approx \sum_{j=1}^{\infty} \mu(j)A(t/j^4)$$

$$= \sum_{j=1}^{\infty} \mu(j)A(1)(t/j^4)^{1/2} = A(1)t^{1/2}\sum_{j=1}^{\infty} \mu(j)/j^2 = A(1)t^{1/2}6/\pi^2$$

using Theorem 7.2.5 and Exercise 7.2 # 4.

Now, as shown above,

$$P(n) = \sum_{i=0}^{\infty} (-1)^i Q(n/4^i)$$

The total error caused by replacing $Q(t)$ with $A(1)t^{1/2}6/\pi^2$ is bounded by

$$\sum_{i=0}^{\infty} 4K(n/4^i)^{1/3} = 4Kn^{1/3}\sum_{i=0}^{\infty} (1/4^{1/3})^i = \frac{4Kn^{1/3}4^{1/3}}{4^{1/3}-1}$$

Thus, with the above error possible,

$$P(n) \approx \sum_{i=0}^{\infty} (-1)^i A(1)(n/4^i)^{1/2}6/\pi^2$$

$$= A(1)n^{1/2}\frac{6}{\pi^2} \times \frac{2}{3} = \frac{\Gamma(\frac{1}{4})^2}{\sqrt{2\pi^5}}\sqrt{n} \approx 0.531\sqrt{n}$$

R. Wild sharpened this result to

$$P(n) \approx \frac{\Gamma(\frac{1}{4})^2}{\sqrt{2\pi^5}}\sqrt{n} - \frac{\zeta(1/3)(1+2^{-1/3})}{\zeta(4/3)(1+4^{-1/3})}\sqrt[3]{n} \approx 0.531\sqrt{n} - 0.297\sqrt[3]{n}$$

with an error at worst proportionate to $n^{1/4}\ln n$. To do this, Wild used Cardano's solution to the cubic equation. The reader is encouraged to consult Wild's paper for the details.

Exercises 7.6

1. By actually counting the lattice points in the relevant regions, compute the exact number of primitive Pythagorean triangles with area < 100.
2. How does this exact number compare with Lambek and Moser's original estimate?
3. How does it compare with Wild's sharper estimate?
4. What is the lowest point on the bug's head?

7.7 Prime Preliminaries

The Prime Number Theorem is the fact that

$$\lim_{x\to\infty} \frac{\pi(x)}{x/\ln x} = 1$$

(where $\pi(x)$ is the number of primes $\leq x$). In this section we prove a series of theorems about various functions related to $\pi(x)$, and their approximate magnitudes. These theorems will allow us, in the next section, to prove the prime number theorem itself.

The letter p shall range over primes only, the letter n shall range over positive integers, and x shall denote a real not less than 1. We define

$$\psi(x) = \sum_{p^k \leq x} \ln p$$

For example,

$$\psi(14.3) = \ln(2 \times 4 \times 8 \times 3 \times 9 \times 5 \times 7 \times 11 \times 13)$$

Note that, since $\Lambda(d) = \ln d$ just in case d is a power of a prime,

$$\psi(x) = \sum_{d \leq x} \Lambda(d)$$

We define $R(x) = \psi(x) - x$ and we shall prove the Prime Number Theorem by first proving that

$$\lim_{x\to\infty} R(x)/x = 0$$

Most elementary proofs of the prime number theorem use the 'big O' notation. We have chosen to be more concrete and more exact, actually giving error bounds. By $f(x) = g(x) \pm h(x)$ we mean $|f(x) - g(x)| \le h(x)$. Readers who prefer the big O notation have only to replace, say, $\pm 5 \ln x$ by $O(\ln x)$. The arguments are the same.

We begin our preliminaries with a theorem about logarithms.

Theorem 7.7.1 *If $x \ge 2$,*

$$\sum_{n \le x} \ln n = x \ln x - x + 1 \pm \ln x$$

Proof: The antiderivative of $\ln x$ is $x \ln x - x$, so by geometry,

$$\begin{aligned} -\ln x &< x \ln x - x + 1 - \sum_{n \le x} \ln n \\ &< \ln 2 + (\ln 3 - \ln 2) + (\ln 4 - \ln 3) + \\ &\quad \cdots + (\ln[x] - \ln([x] - 1)) + (\ln x - \ln[x]) \\ &= \ln x \end{aligned}$$

Theorem 7.7.2

$$\sum_{n \le x} \ln^2 n = x \ln^2 x - 2x \ln x + 2x - 2 \pm \ln^2 x$$

Proof: Note that $\pm \ln^2 x$ is to be understood as an error bound.

Consider the graph of $y = \ln^2 x$. The antiderivative of this function is $g(x) = x \ln^2 x - 2x \ln x + 2x$ and hence the area bounded by the x-axis and the curve $y = \ln^2 x$ and the vertical line $y = x$ is $g(x) - g(1)$. The difference between this area and the given sum is bounded above by

$$\begin{gathered} \ln^2 2 + (\ln^2 3 - \ln^2 2) + (\ln^2 4 - \ln^2 3) + \\ \cdots + (\ln^2[x] - \ln^2([x] - 1)) + (\ln^2 x - \ln^2[x]) \\ = \ln^2 x \end{gathered}$$

Theorem 7.7.3 *There is a positive constant γ (called Euler's constant)* such that if $x \ge 1$,

$$\sum_{n \le x} \frac{1}{n} = \ln x + \gamma \pm \frac{1}{x}$$

Proof: Note that the $\pm\frac{1}{x}$ is an error bound.

If t is a positive integer, let

$$\gamma(t) = \frac{1}{t} - \int_t^{t+1} \frac{1}{u}\, du = 1/t - \ln(1 + 1/t)$$

This is always positive and less than $1/t - 1/(t+1)$. Hence $\sum \gamma(t)$ converges to a positive number γ less than 1. Now

$$\gamma(1) + \gamma(2) + \cdots + \gamma([x]) + (\ln([x] + 1) - \ln x) = \sum_{n\leq x} \frac{1}{n} - \ln x$$

Hence

$$\sum_{n\leq x} \frac{1}{n} - \ln x - \gamma = -(\gamma([x]+1) + \gamma([x]+2) + \cdots) + (\ln([x]+1) - \ln x)$$

The sum of the γs is a positive number less than $1/x$ while the difference of the logs is also a positive number less than $1/x$. The result follows.

Theorem 7.7.4 *If $x \geq 2$, there is a constant δ such that*

$$\sum_{n\leq x} \frac{\ln n}{n} = \frac{\ln^2 x}{2} + \delta \pm \frac{\ln x}{x}$$

Proof: The theorem is true when $x < 3$. Suppose $x \geq 3$.

The curve $y = (\ln x)/x$ rises from $(1, 0)$ to a maximum of $(e, 1/e)$ and then descends towards the positive x-axis as asymptote. It becomes concave up at $x = e^{3/2} \approx 4.5$. The antiderivative of $(\ln x)/x$ is $(\ln^2 x)/2$ so that $(\ln^2 x)/2$ is the area of the region bounded by $y = (\ln x)/x$, the x-axis and the vertical line through x.

If t is a positive integer, let

$$\delta(t) = \frac{\ln t}{t} - \int_t^{t+1} \frac{\ln u}{u}\, du$$

If $t \geq 3$, this is always positive and less than $(\ln t)/t - (\ln(t+1))/(t+1)$. Hence $\sum \delta(t)$ converges to a number δ (which is about -0.07).

By geometry,

$$\delta(1) + \delta(2) + \cdots + \delta([x]) + (\frac{1}{2}\ln^2([x]+1) - \frac{1}{2}\ln^2 x) = \sum_{n \le x} \frac{\ln n}{n} - \frac{\ln^2 x}{2}$$

Hence

$$\sum_{n \le x} \frac{\ln n}{n} - \frac{\ln^2 x}{2} - \delta = -(\delta([x]+1)+\delta([x]+2)+\cdots) + \frac{1}{2}\ln^2([x]+1) - \frac{1}{2}\ln^2 x$$

Since $x \ge 3$, the sum of the δs is a positive number bounded above by $(\ln([x]+1))/([x]+1)$ while the difference of the logs is an area less than the area of a 1 by $(\ln x)/x$ rectangle. Hence, because $(\ln x)/x$ decreases when $x \ge 3$, the result follows.

Corollary. If $x_m > e^2$ and $x/x_m \ge 2$ then

$$\sum_{x/x_m < n \le x} \frac{\ln n}{n} \le \frac{\ln^2 x - \ln^2(x/x_m)}{2} + 2 < \ln x_m \ln x$$

Theorem 7.7.5

$$\left| \sum_{n \le x} \frac{\mu(n)}{n} \ln \frac{x}{n} \right| < 2 + \gamma$$

Proof: By Theorem 7.7.3,

$$f(x) = \sum_{n \le x} \frac{x}{n} = x \ln x + \gamma x \pm 1$$

and hence by Theorem 1.11.5,

$$x = \sum_{n \le x} \mu(n) \left(\frac{x}{n} \ln \frac{x}{n} + \gamma \frac{x}{n} \pm 1 \right)$$

$$= x \sum_{n \le x} \frac{\mu(n)}{n} \ln \frac{x}{n} + x\gamma \sum_{n \le x} \frac{\mu(n)}{n} \pm x$$

Dividing by x and applying Theorem 1.11.4, we obtain the result.

Theorem 7.7.6

$$\sum_{n\le x} \frac{\mu(n)}{n} \ln^2 \frac{x}{n} = 2\ln x \pm 8$$

Proof: If $x < 3$ the theorem is true. Suppose $x \geq 3$. By Theorem 7.7.3 and Theorem 7.7.4,

$$\sum_{n\le x} \frac{x}{n} \ln \frac{x}{n} = x\ln x \sum_{n\le x} \frac{1}{n} - x\sum_{n\le x} \frac{\ln n}{n}$$

$$= x\ln x(\ln x + \gamma \pm 1/x) - x\left(\frac{\ln^2 x}{2} + \delta \pm \frac{\ln x}{x}\right)$$

$$= \frac{x\ln^2 x}{2} + \gamma x\ln x - \delta x \pm 2\ln x$$

By Theorem 1.11.5,

$$x\ln x = \sum_{n\le x} \mu(n)\left(\frac{x\ln^2(x/n)}{2n} + \gamma(x/n)\ln(x/n) - \delta(x/n) \pm 2\ln(x/n)\right)$$

$$= \frac{x}{2}\sum_{n\le x} \frac{\mu(n)}{n}\ln^2\frac{x}{n} + \gamma x\sum_{n\le x}\frac{\mu(n)}{n}\ln\frac{x}{n} - \delta x\sum_{n\le x}\frac{\mu(n)}{n} \pm 2\sum_{n\le x}\mu(n)\ln\frac{x}{n}$$

Dividing through by x and using Theorem 7.7.5 and Theorem 1.11.4, we obtain

$$\left|\ln x - \frac{1}{2}\sum_{n\le x}\frac{\mu(n)}{n}\ln^2\frac{x}{n}\right| \le \gamma(\gamma+2) + |\delta| + (2/x)\sum_{n\le x}\ln\frac{x}{n}$$

$$\le \gamma(\gamma+2) + |\delta| + (2/x)(x\ln x - (x\ln x - x + 1 - \ln x))$$

$$\le \gamma(\gamma+2) + |\delta| + 2 + (2/x)(\ln x - 1) \le 4$$

using Theorem 7.7.1.

Theorem 7.7.7

$$\sum_{n\le x} \ln^2 \frac{x}{n} < 2.5x$$

Proof: The sum equals

$$\sum_{n\le x}(\ln^2 x - 2\ln x\ln n + \ln^2 n)$$

Using Theorem 7.7.1 and Theorem 7.7.2, we see this is bounded above by $2x + 3\ln^2 x - 2\ln x - 2$. For $x > 200$ this, in turn, is bounded by $2.5x$, and for smaller x the result holds by straight computation.

The next theorem is a version of the *Selberg Symmetry Formula.*

Theorem 7.7.8

$$\sum_{n\le x}\Lambda(n)\ln n + \sum_{rs\le x}\Lambda(r)\Lambda(s) = 2x\ln x \pm 20x$$

Proof: The theorem is true when $x < 2$. Suppose $x \ge 2$.

If $w(n) = 1$, then by Theorem 1.11.2, $(w * \Lambda) * \Lambda = w * (\Lambda * \Lambda)$. Hence, using Exercise 7.3 # 1,

$$\sum_{d|n}\Lambda(n/d)\ln d = \sum_{d|n}\left(\sum_{r|d}\Lambda(r)\right)\Lambda(n/d) = \sum_{d|n}\sum_{r|d}\Lambda(r)\Lambda(d/r)$$

Using that same exercise,

$$\begin{aligned}\ln^2 n &= \sum_{d|n}\Lambda(d)\ln n\\ &= \sum_{d|n}\Lambda(d)(\ln d + \ln n/d)\\ &= \sum_{d|n}\Lambda(d)\ln d + \sum_{d|n}\Lambda(n/d)\ln d\\ &= \sum_{d|n}\left(\Lambda(d)\ln d + \sum_{r|d}\Lambda(r)\Lambda(d/r)\right)\end{aligned}$$

by the previous set of equations. By Theorem 1.11.3 (the Möbius Inversion Formula),

$$\Lambda(n)\ln n + \sum_{d|n}\Lambda(d)\Lambda(n/d) = \sum_{d|n}\mu(d)\ln^2(n/d)$$

Summing over $n \le x$, we obtain

$$\sum_{n\le x} \Lambda(n)\ln n + \sum_{d\le x} \Lambda(d)(\Lambda(1)+\Lambda(2)+\cdots+\Lambda([x/d]))$$

$$= \sum_{n\le x}\sum_{d|n} \mu(d)\ln^2(n/d)$$

Hence

$$\sum_{n\le x} \Lambda(n)\ln n + \sum_{rs\le x} \Lambda(r)\Lambda(s)$$

$$= \sum_{d\le x} \mu(d)\left(\ln^2 1 + \ln^2 2 + \cdots + \ln^2[x/d]\right)$$

$$= \sum_{d\le x} \mu(d)\left(\frac{x}{d}\ln^2\frac{x}{d} - \frac{2x}{d}\ln\frac{x}{d} + \frac{2x}{d} - 2 \pm \ln^2\frac{x}{d}\right)$$

$$= x\sum_{d\le x} \frac{\mu(d)}{d}\ln^2\frac{x}{d} - 2x\sum_{d\le x}\frac{\mu(d)}{d}\ln\frac{x}{d}$$

$$+ 2x\sum_{d\le x}\frac{\mu(d)}{d} - 2\sum_{d\le x}\mu(d) \pm \sum_{d\le x}\mu(d)\ln^2\frac{x}{d}$$

using Theorem 7.7.2. By Theorems 7.7.6, 7.7.5, 1.11.4, and 7.7.7, the absolute value of the difference of this number (which equals the LHS of the given equation) and $2x\ln x$ is bounded by

$$8x + 2x(2+\gamma) + 2x + 2x + 2.5x < 20x$$

Theorem 7.7.9 $\psi(x) \le 2x$.

Proof: Let $N(y) = [y] + [y/6] - [y/2] - 2[y/3]$. Then $N(y)$ is always positive, and, with $1 \le y \le 3$, $N(y)$ is a constant 1. Hence

$$\sum_{d\le x} \Lambda(d)N(x/d) \ge \sum_{x/3<d\le x} \Lambda(d)N(x/d)$$

$$= \sum_{x/3<d\le x} \Lambda(d) = \psi(x) - \psi(x/3)$$

Furthermore,

$$\sum_{n\le x} \ln n - \sum_{n\le x/2} \ln n - 2\sum_{n\le x/3} \ln n + \sum_{n\le x/6} \ln n$$

$$= \sum_{n\le x}\sum_{d|n} \Lambda(d) - \sum_{n\le x/2}\sum_{d|n} \Lambda(d) - 2\sum_{n\le x/3}\sum_{d|n} \Lambda(d) + \sum_{n\le x/6}\sum_{d|n} \Lambda(d)$$

$$= \sum_{d\le x} \Lambda(d)[x/d] - \sum_{d\le x} \Lambda(d)[x/2d] - 2\sum_{d\le x} \Lambda(d)[x/3d] + \sum_{d\le x} \Lambda(d)[x/6d]$$

$$= \sum_{d\le x} \Lambda(d) N(x/d)$$

Thus, by Theorem 7.7.1, $\psi(x) - \psi(x/3)$ is bounded above by

$$x\ln x - x + 1 + \ln x - \left(\frac{x}{2}\ln\frac{x}{2} - \frac{x}{2} + 1 - \ln\frac{x}{2}\right)$$

$$-2\left(\frac{x}{3}\ln\frac{x}{3} - \frac{x}{3} + 1 - \ln\frac{x}{3}\right) + \frac{x}{6}\ln\frac{x}{6} - \frac{x}{6} + 1 + \ln\frac{x}{6}$$

$$= \left(\frac{\ln 2}{3} + \frac{\ln 3}{2}\right)x + 5\ln x - 2\ln 2 - 3\ln 3 - 1$$

Let a denote the coefficient of x, and let $b = 2\ln 2 + 3\ln 3 + 1 = 5.68\ldots$.

Now if $x \ge 3$,

$$\psi(x) = (\psi(x) - \psi(x/3)) + (\psi(x/3) - \psi(x/9)) + \cdots$$

where there are at most $[(\ln x)/\ln 3]$ terms. Hence

$$\psi(x) \le ax + 5\ln x - b + a(x/3) + 5\ln(x/3) - b + a(x/9) + 5\ln(x/9) - b + \cdots$$

$$\le ax(3/2) + 5(\ln^2 x)/\ln 3 - b[(\ln x)/\ln 3]$$

If $x > 100$ this is less than $2x$. And, as can be checked by the computer, the theorem is also true for $x \le 100$.

Theorem 7.7.9 enables us to deduce the following version of the Selberg Symmetry Formula.

Theorem 7.7.10

$$\psi(x)\ln x + \sum_{rs\leq x} \Lambda(r)\Lambda(s) = 2x\ln x \pm 22x$$

Proof:

$$\int_1^2 \frac{1}{t}\sum_{n\leq 1}\Lambda(n)\,dt + \int_2^3 \frac{1}{t}\sum_{n\leq 2}\Lambda(n)\,dt +$$

$$\cdots + \int_{[x]-1}^{[x]} \frac{1}{t}\sum_{n\leq [x]-1}\Lambda(n)\,dt + \int_{[x]}^{x} \frac{1}{t}\sum_{n\leq [x]}\Lambda(n)\,dt$$

$$= \sum_{n\leq 1}\Lambda(n)\ln 2 + \sum_{n\leq 2}\Lambda(n)(\ln 3 - \ln 2) +$$

$$\cdots + \sum_{n\leq [x]-1}\Lambda(n)(\ln[x] - \ln([x]-1)) + \sum_{n\leq [x]}\Lambda(n)(\ln x - \ln[x])$$

$$= (\ln 2)(-\Lambda(2)) + (\ln 3)(-\Lambda(3)) + \cdots + (\ln[x])(-\Lambda([x])) + \sum_{n\leq x}\Lambda(n)\ln x$$

Hence, using the fact that $\sum_{n\leq t}\Lambda(t) = \psi(t)$,

$$\sum_{n\leq x}\Lambda(n)\ln n = \psi(x)\ln x - \int_1^x \frac{\psi(t)}{t}\,dt = \psi(x)\ln x \pm 2x$$

using also Theorem 7.7.9. By Theorem 7.7.8,

$$\psi(x)\ln x + \sum_{rs\leq x}\Lambda(r)\Lambda(s) = \psi(x)\ln x - \sum_{n\leq x}\Lambda(n)\ln n + 2x\ln x \pm 20x$$

$$= \psi(x)\ln x - \psi(x)\ln x \pm 2x + 2x\ln x \pm 20x = 2x\ln x \pm 22x$$

Corollary.

$$\psi(x)\ln x - 2x + 2 \leq \sum_{n\leq x}\Lambda(n)\ln n \leq \psi(x)\ln x$$

Theorem 7.7.11

$$\sum_{n\leq x}\frac{\Lambda(n)}{n} = \ln x \pm 2$$

Proof:

$$\sum_{n\leq x}\ln n=\sum_{n\leq x}\sum_{d|n}\Lambda(d)=\sum_{d\leq x}\Lambda(d)[x/d]<x\sum_{d\leq x}\Lambda(d)/d$$

Hence by Theorems 7.7.1 and 7.7.9,

$$x\ln x-x+1-\ln x\leq\sum_{n\leq x}\ln n<x\sum_{d\leq x}\frac{\Lambda(d)}{d}$$

$$<\sum_{n\leq x}\ln n+\sum_{d\leq x}\Lambda(d)\leq x\ln x-x+1+\ln x+2x$$

(since $\sum\Lambda(d)=\psi(x)$).

Corollary. $\sum_{n\leq x}\frac{\Lambda(n)}{n}<2\ln x$.

Theorem 7.7.12

$$\sum_{n\leq x}\frac{\Lambda(n)}{n}\ln\frac{x}{n}=\frac{\ln^2 x}{2}\pm 2\ln x$$

Proof: As in the proof of Theorem 7.7.10,

$$\int_1^x\frac{1}{y}\sum_{n\leq y}\frac{\Lambda(n)}{n}\,dy=\sum_{n\leq x}\frac{\Lambda(n)}{n}\ln x-\sum_{n\leq x}\frac{\Lambda(n)}{n}\ln n$$

Thus, using Theorem 7.7.11, the difference of the sums (which is the left hand side of the theorem) is bounded above by

$$\int_1^x\frac{1}{y}(\ln y+2)\,dy=\frac{1}{2}\ln^2 x+2\ln x$$

Similarly, it is bounded below by $\frac{1}{2}\ln^2 x-2\ln x$.

Corollary.

$$\sum_{n\leq x}\frac{\Lambda(n)}{n}\ln n=\frac{1}{2}\ln^2 x\pm 4\ln x$$

Theorem 7.7.13

$$\sum_{rs\le x} \Lambda(r)\Lambda(s)\ln s = \frac{1}{2}\ln x \sum_{rs\le x} \Lambda(r)\Lambda(s) \pm 2x\ln x$$

Proof: Using the corollary to Theorem 7.7.10,

$$\sum_{rs\le x} \Lambda(r)\Lambda(s)\ln s = \sum_{r\le x} \Lambda(r) \sum_{s\le x/r} \Lambda(s)\ln s$$

$$= \sum_{r\le x} \Lambda(r)(\psi(x/r)\ln(x/r) \pm 2(x/r))$$

$$= \ln x \sum_{r\le x} \Lambda(r)\psi(x/r) - \sum_{r\le x} \Lambda(r)\psi(x/r)\ln r \pm 2\sum_{r\le x} \Lambda(r)(x/r)$$

By the Corollary to Theorem 7.7.11, this equals

$$\ln x \sum_{r\le x} \Lambda(r) \sum_{s|x/r} \Lambda(s) - \sum_{r\le x} \Lambda(r)\ln r \sum_{s|x/r} \Lambda(s) \ \pm 2x(2\ln x)$$

$$= \ln x \sum_{rs\le x} \Lambda(r)\Lambda(s) - \sum_{rs\le x} \Lambda(r)\Lambda(s)\ln r \ \pm 2x(2\ln x)$$

The middle term is just the left hand side, so, bringing it to the left and dividing by 2, we obtain the result.

Theorem 7.7.14

$$\psi(x)\ln^2 x = 2\sum_{rs\le x} \Lambda(r)\Lambda(s)\psi(x/rs) \pm 114x\ln x$$

Proof: By Theorem 7.7.8 (Selberg's Symmetry Formula) with $x = x/m$ we have

$$\Lambda(m) \sum_{n\le x/m} \Lambda(n)\ln n + \Lambda(m) \sum_{rs\le x/m} \Lambda(r)\Lambda(s)$$

$$= 2\Lambda(m)\frac{x}{m}\ln\frac{x}{m} \pm 20\frac{x}{m}\Lambda(m)$$

Summing over positive integers $m \leq x$ and using the Corollary to Theorem 7.7.11,

$$\sum_{mn \leq x} \Lambda(m)\Lambda(n)\ln n + \sum_{rst \leq x} \Lambda(r)\Lambda(s)\Lambda(t)$$

$$= 2x \sum_{m \leq x} \frac{\Lambda(m)}{m} \ln \frac{x}{m} \pm 20x(2\ln x)$$

$$= x\ln^2 x \pm 4x\ln x \pm 40x\ln x$$

(by Theorem 7.7.12). Thus, using Theorem 7.7.13,

$$\frac{1}{2}\ln x \sum_{rs \leq x} \Lambda(r)\Lambda(s) + \sum_{rst \leq x} \Lambda(r)\Lambda(s)\Lambda(t) = x\ln^2 x \pm 46x\ln x$$

Multiplying the identity of Theorem 7.7.10 by $(\ln x)/2$, we obtain

$$\frac{1}{2}\psi(x)\ln^2 x + \frac{1}{2}\ln x \sum_{rs \leq x} \Lambda(r)\Lambda(s) = x\ln^2 x \pm 11x\ln x$$

Taking the difference of the previous two equations, we obtain

$$\frac{1}{2}\psi(x)\ln^2 x - \sum_{rst \leq x} \Lambda(r)\Lambda(s)\Lambda(t) = \pm 57x\ln x$$

But the immediately preceding sum equals

$$\sum_{rs \leq x} \Lambda(r)\Lambda(s) \sum_{t \leq x/rs} \Lambda(t) = \sum_{rs \leq x} \Lambda(r)\Lambda(s)\psi(x/rs)$$

Theorem 7.7.15 *If* $1.1y \geq x \geq y > 1$ *then*

$$\psi(x) - \psi(y) \leq 2(x-y) + 50y/\ln y$$

Proof: By Theorem 7.7.10,

$$\psi(x)\ln x - \psi(y)\ln y + \sum_{rs \leq x} \Lambda(r)\Lambda(s) - \sum_{rs \leq y} \Lambda(r)\Lambda(s)$$

$$= 2x\ln x \pm 22x - 2y\ln y \pm 22y$$

Thus

$$\psi(x)\ln x - \psi(y)\ln y \le 2x\ln x - 2y\ln y + 22x + 22y$$

$$(\psi(x) - \psi(y))\ln x \le 2(x-y)\ln x + (2y - \psi(y))\ln(x/y) + 22x + 22y$$

$$\le 2(x-y)\ln x + 2x\ln 1.1 + 44x$$

and hence

$$\begin{aligned}\psi(x) - \psi(y) &\le 2(x-y) + 44.2x/\ln x \\ &\le 2(x-y) + 50y/\ln y\end{aligned}$$

For example, if m and n are real numbers such that $1.1n \ge m \ge n > 1$ then

$$\frac{\psi(m) - \psi(n)}{n} \le 2\,\frac{m-n}{n} + \frac{50}{\ln n}$$

Hence we have the following

Corollary. If $1.1n \ge m \ge n > 1$ and $n > e^{50/\epsilon}$ then

$$\frac{\psi(m) - \psi(n)}{n} \le 2\,\frac{m-n}{n} + \epsilon$$

Theorem 7.7.16

$$\left|\sum_{n\le x} \frac{\psi(n) - n}{n(n+1)}\right| < 4.5$$

Proof: By Theorem 7.7.11,

$$\sum_{n\le x} \frac{\Lambda(n)}{n} = \ln x \pm 2$$

But

$$\begin{aligned}\sum_{n\le x} \frac{\Lambda(n)}{n} &= \sum_{n=2}^{[x]} \frac{\psi(n) - \psi(n-1)}{n} \\ &= \sum_{n=2}^{[x]} \left(\frac{1}{n} + \frac{R(n) - R(n-1)}{n}\right)\end{aligned}$$

where $R(n) = \psi(n) - n$. By Theorem 7.7.3, this equals

$$-1 + \ln x + \gamma \pm 1/x + \sum_{n \leq x} \frac{R(n)}{n(n+1)} + \frac{R([x])}{[x]+1}$$

Hence, by Theorem 7.7.9,

$$\pm 2 = -1 + \gamma \pm 1/x + \sum_{n \leq x} \frac{R(n)}{n(n+1)} \pm 1$$

so that

$$\left| \sum_{n \leq x} \frac{R(n)}{n(n+1)} \right| < 4.5$$

Corollary. Let I be any interval of positive reals > 1. Then

$$\left| \sum_{n \in I} \frac{R(n)}{n(n+1)} \right| < 9$$

Exercises 7.7

1. Show that the LCM of the first n positive integers is $e^{\psi(n)}$.
2. Show that the LCM of the first n positive integers is less than 8^n.
3. Graph

$$\sum_{n \leq x} \frac{\mu(n)}{n} \ln^2 \frac{x}{n} \; - \; 2 \ln x$$

with x ranging over the positive integers less than 200.
4. Do a numerical study of Selberg's Symmetry Formula, graphing

$$\sum_{n \leq x} \Lambda(n) \ln n + \sum_{rs \leq x} \Lambda(r)\Lambda(s) - 2x \ln x$$

for the interval $[1, 100]$.

7.8 Prime Number Theorem Proof

We now move into the proof of the Prime Number Theorem itself. We define $R(x) = \psi(x) - x$. We shall use mathematical induction to show that $R(x)/x$ tends to 0 as $x \to \infty$. We begin by getting a bound on $|R(x)| \ln^2 x$.

Theorem 7.8.1

$$|R(x)| \ln^2 x \leq 2 \sum_{n \leq x} |R(x/n)| \ln n + 150x \ln x$$

Proof: By Theorem 7.7.9, $|R(x)| \leq x$ and hence $|R(x)| \ln^2 x \leq x \ln^2 x$. This is less than $150x \ln x$ if $x < e^{150}$, so, without loss of generality, assume $x \geq e^{150}$.

Since $\psi(x/n) = \sum_{t \leq x/n} \Lambda(t)$, Theorem 7.7.13 can be written

$$2 \sum_{n \leq x} \Lambda(n)\psi(x/n) \ln n = \ln x \sum_{rs \leq x} \Lambda(r)\Lambda(s) \pm 4x \ln x$$

Combining this with Theorem 7.7.10 (multiplied by $\ln x$), we obtain

$$\psi(x) \ln^2 x + 2 \sum_{n \leq x} \Lambda(n)\psi(x/n) \ln n = 2x \ln^2 x \pm 26x \ln x \qquad (*)$$

By the Corollary to Theorem 7.7.12,

$$\sum_{n \leq x} \frac{\Lambda(n)}{n} \ln n = \frac{1}{2} \ln^2 x \pm 4 \ln x$$

and hence

$$\begin{aligned} &2 \sum_{n \leq x} \Lambda(n)\psi(x/n) \ln n \\ &= 2 \sum_{n \leq x} \Lambda(n)R(x/n) \ln n + 2 \sum_{n \leq x} \Lambda(n)(x/n) \ln n \\ &= 2 \sum_{n \leq x} \Lambda(n)R(x/n) \ln n + x \ln^2 x \pm 8x \ln x \end{aligned}$$

and thus, from $(*)$,

$$R(x) \ln^2 x + 2 \sum_{n \leq x} \Lambda(n)R(x/n) \ln n = \pm 34x \ln x$$

so that

$$|R(x)|\ln^2 x \le 2\sum_{n\le x}\Lambda(n)|R(x/n)|\ln n + 34x\ln x \qquad (**)$$

By Theorems 7.7.11 and 7.7.12,

$$\sum_{rs\le x}\frac{\Lambda(r)}{r}\frac{\Lambda(s)}{s} = \sum_{r\le x}\frac{\Lambda(r)}{r}\sum_{x\le x/r}\frac{\Lambda(s)}{s}$$

$$= \sum_{r\le x}\frac{\Lambda(r)}{r}(\ln(x/r)\pm 2) = \frac{\ln^2 x}{2}\pm 2\ln x \pm 2(2\ln x)$$

$$= \frac{\ln^2 x}{2}\pm 6\ln x$$

By Theorem 7.7.14 (writing $R(x)+x$ for $\psi(x)$), we obtain

$$R(x)\ln^2 x + x\ln^2 x$$

$$= 2\sum_{rs\le x}\Lambda(r)\Lambda(s)R(x/rx) + x\ln^2 x \pm 12x\ln x \pm 114x\ln x$$

and hence

$$|R(x)|\ln^2 x \le 2\sum_{rs\le x}\Lambda(r)\Lambda(s)|R(x/rs)| + 126x\ln x$$

Thus, averaging the immediately preceding inequality and using $(**)$,

$$|R(x)|\ln^2 x \le \sum_{n\le x}\Lambda(n)|R(x/n)|\ln n + \sum_{rs\le x}\Lambda(r)\Lambda(s)|R(x/rs)| + 80x\ln x$$

$$= \sum_{n\le x}\left(\Lambda(n)\ln n + \sum_{rs=n}\Lambda(r)\Lambda(s)\right)|R(x/n)| + 80x\ln x$$

Now, by Theorem 7.7.8, if $x\ge 1$,

$$T(x) =_{df} \sum_{n\le x}\left(\Lambda(n)\ln n + \sum_{rs=n}\Lambda(r)\Lambda(s)\right) = 2x\ln x \pm 20x$$

Let $T(0) =_{df} 0$. Since

$$T(n) - T(n-1) = \Lambda(n)\ln n + \sum_{rs=n} \Lambda(r)\Lambda(s)$$

we have

$$\begin{aligned}
&|R(x)|\ln^2 x \\
\leq\ & \sum_{n\leq x}(T(n)-T(n-1))|R(x/n)| + 80x\ln x \\
=\ & T([x])|R(x/[x])| + \sum_{n\leq x} T(n)\,(|R(x/n)| - |R(x/(n+1))|) + 80x\ln x \\
\leq\ & 2[x]R(x/[x])\ln[x] + 20x \\
& + \sum_{n\leq x-1} 2n\ln n\,(|R(x/n)| - |R(x/(n+1))|) \\
& + \sum_{n\leq x-1} 20n\,||R(x/n)| - |R(x/(n+1))|| + 80x\ln x \\
\leq\ & 2[x](x/[x])\ln[x] + 20x + 2\sum_{n=2}^{[x]-1} |R(x/n)|(n\ln n - (n-1)\ln(n-1)) \\
& + \sum_{n\leq x-1} 20n|R(x/n) - R(x/(n+1))| + 80x\ln x \\
\leq\ & 2x\ln x + 20x + 2\sum_{n=2}^{[x]-1} |R(x/n)|(\ln n + 1) \\
& + \sum_{n\leq x} 20n|\psi(x/n) - x/n - \psi(x/(n+1)) + x/(n+1)| + 80x\ln x \\
\leq\ & 20x + 2\sum_{n\leq x} |R(x/n)|(\ln n + 1) + \sum_{n\leq x} 20n|\psi(x/n) - \psi(x/(n+1))| \\
& + \sum_{n\leq x} 20n(x/n - x/(n+1)) + 82x\ln x \\
\leq\ & 20x + 2\sum_{n\leq x} |R(x/n)|\ln n + 2\sum_{n\leq x} |R(x/n)| + \sum_{n\leq x} 20\psi(x/n) \\
& +20x(\ln x + 1/x) + 82x\ln x \\
\leq\ & 20x + 2\sum_{n\leq x} |R(x/n)|\ln n + 2\sum_{n\leq x} x/n + \sum_{n\leq x} 40x/n + 20x(\ln x) \\
& +20 + 82x\ln x
\end{aligned}$$

$$\le 2\sum_{n\le x}|R(x/n)|\ln n + 42x\sum_{n\le x}\frac{1}{n} + 20x + 20x\ln x + 20 + 82x\ln x$$
$$\le 2\sum_{n\le x}|R(x/n)|\ln n + 42x(\ln x + \gamma + 1/x) + 20x + 20 + 102x\ln x$$
$$\le 2\sum_{n\le x}|R(x/n)|\ln n + 144x\ln x + 45x + 62$$
$$\le 2\sum_{n\le x}|R(x/n)|\ln n + 150x\ln x$$

since $x \ge e^{150}$.

Theorem 7.8.2 *Given any ϵ between 0 and 1, if $x > e^{2/\epsilon}$ then the interval $I = (x, e^{10/\epsilon}x]$ contains an integer n with $|R(n)| < \epsilon n$.*

Proof: Since $\epsilon < 1$ the interval I does contain integers, and they are greater than 8.

Case 1. R takes different signs at different integers in the interval. Then for some n in the interval, the distance between $R(n+1)$ and $R(n)$ is greater than $|R(n)|$ and, using the definition of ψ, we have

$$|R(n)| \le |R(n+1) - R(n)| = |\psi(n+1) - \psi(n) - 1| < |\ln(n+1) - 1| < \ln n$$

since $n \ge 8$. Hence, since $n > x > e^2$,

$$|R(n)|/n < (\ln n)/n < (\ln x)/x < \epsilon$$

the latter inequality holding since

$$\ln(e^{2/\epsilon})/e^{2/\epsilon} < \epsilon$$

Case 2. R takes only positive values at the integers in the interval I. By the Corollary to Theorem 7.7.16,

$$\left(\min_I R(n)/n\right)\sum_{n\in I} 1/(n+1) \le \sum_{n\in I}\frac{R(n)}{n(n+1)} < 9$$

Now if $b = e^{10/\epsilon}x$ the sum

$$\sum_{n\in I}\frac{1}{n+1}$$

is bounded below by

$$\int_x^b \frac{1}{t}\,dt - 1/x = \ln(e^{10/\epsilon}) - 1/x > 10/\epsilon - 1/e^{2/\epsilon}$$

Hence for the integer n which minimises $R(n)/n$ we have

$$\frac{|R(n)|}{n} < \frac{9}{10/\epsilon - 1/e^{2/\epsilon}}$$

and hence

$$\frac{|R(n)|}{n} < \frac{9\epsilon e^{2/\epsilon}}{10e^{2/\epsilon} - \epsilon} < \epsilon$$

since $\epsilon < 1$.

Case 3. R takes only negative values at the integers in I. Then we consider $\min(-R(n)/n)$ and the result follows as before.

Theorem 7.8.3 *Given any ϵ between 0 and 1, and any $x > e^{200/\epsilon}$ there is an integer n such that $x < n \leq e^{40/\epsilon}x$ and such that if real number m is in the interval*

$$[n,\ (1+\epsilon/10)n]$$

then $|R(m)| < \epsilon m$.

Moreover, $(1+\epsilon/10)n < e^{41/\epsilon}x$.

Proof: Let $\epsilon' = \epsilon/4$.

By Theorem 7.8.2, then $(x,\ e^{10/\epsilon'}x]$ contains an integer n with $|R(n)| < \epsilon' n$.

Now suppose m is such that $n \leq m \leq (1+\epsilon/10)n$. Then

$$\left|\frac{R(m)}{m} - \frac{R(n)}{n}\right| \leq \left|\frac{R(m)}{m}\right|\left|1 - \frac{m}{n}\right| + \left|\frac{R(m)-R(n)}{n}\right|$$

$$\leq (\epsilon/10)|R(m)/m| + |\psi(m) - \psi(n) - (m-n)|/n$$
$$\leq (\epsilon/10)(|R(m)/m| + 1) + |\psi(m) - \psi(n)|/n$$
$$\leq (\epsilon/10)(|R(m)/m| + 1) + 2(m-n)/n + \epsilon'$$

by the Corollary to Theorem 7.7.15. Hence

$$\left|\frac{R(m)}{m} - \frac{R(n)}{n}\right| \leq (\epsilon/10)(|R(m)/m| + 3) + \epsilon'$$

By Theorem 7.7.9, $|R(x)/x| < 1$ for any x. Hence

$$\begin{aligned} \left|\frac{R(m)}{m}\right| &< \left|\frac{R(n)}{n}\right| + (\epsilon/10)(1+3) + \epsilon' \\ &\leq \epsilon' + (16/10)\epsilon' + \epsilon' < \epsilon \end{aligned}$$

This completes the proof.

Theorem 7.8.4

$$\lim_{x\to\infty} \frac{\psi(x)}{x} = 1$$

Proof: This is equivalent to showing that $|R(x)/x|$ tends to 0 as $x \to \infty$.

Consider the sequences

$$\begin{aligned} c_0 &= 1 \\ c_{m+1} &= c_m(1 - c_m^2/50000) \end{aligned}$$

and

$$\begin{aligned} x_0 &= e^{201} \\ x_{m+1} &= x_m^{(100/c_m)^3} \end{aligned}$$

The c_m's are all positive and decrease to some limit c. Indeed,

$$c = \lim_{m\to\infty} c_{m+1} = \lim_{m\to\infty} c_m(1 - c_m^2/50000) = c(1 - c^2/50000)$$

so that $c = 0$. The x_ms are all positive and increase without bound. Note that, for any nonnegative integer m, $\ln x_m > 200$.

Consider the statement

$$x > x_m \implies |R(x)/x| < c_m$$

By Theorem 7.7.9, this is true when $m = 0$. Suppose it true for m. In what follows we shall show that it is then true for $m+1$. Hence, by mathematical induction, it is true for all nonnegative integers m. Since c_m approaches 0, as m approaches ∞, it follows that, given any $\epsilon > 0$, there is an m with $c_m < \epsilon$ such that $x > x_m$ entails $|R(x)/x| < \epsilon$. In other words, $|R(x)/x|$ tends to 0 as $x \to \infty$.

Hence to prove the theorem, it suffices now to show that the above statement is true for $m+1$ (assuming it is true for m).

Let x be an arbitrary real number $> x_{m+1}$. Let $k = e^{82/c_m}$. Let t range over all the positive integers such that $k^t > x_m^{2/c_m}$ and $k^{t+1} < \sqrt{x}$. That is, let t range over the positive integers between

$$u = \left[\frac{2\ln x_m}{c_m \ln k}\right] + 1$$

and

$$v = \left[\frac{\ln x}{2\ln k}\right] - 1$$

inclusive. The gap between these two limits is bounded below as follows. Since $\ln x > (100/c_m)^3 \ln x_m$, we have

$$v - u \geq \frac{\ln x}{2\ln k} - \frac{2\ln x_m}{c_m \ln k} - 4 \geq \frac{\ln x}{3\ln k} > 1000$$

By Theorem 7.8.3, with $\epsilon = c_m/2$ and $x = k^t$, for each t between u and v inclusive, there is an integer s_t such that

$$k^t < s_t \leq e^{80/c_m} k^t < k^{t+1} < \sqrt{x}$$

and such that if m is a real number in the interval

$$I_t = [s_t,\ (1 + c_m/20)s_t]$$

then $|R(m)/m| < c_m/2$. (Note that $k^t > e^{200/\epsilon}$ since $x_m^{2/c_m} > e^{200(2/c_m)}$.)

Let n vary over the positive integers. Note that $x/n \in I_t$ iff

$$s_t \leq x/n \leq (1 + c_m/20)s_t$$

iff

$$\frac{x}{(1 + c_m/20)s_t} \leq n \leq \frac{x}{s_t}$$

Hence the number of numbers of the form x/n in I_t is greater than or equal to

$$\frac{21}{40}\frac{x}{s_t}\left(1-\frac{1}{1+c_m/20}\right)=\frac{21x}{40s_t}\frac{c_m/20}{1+c_m/20}>\frac{c_m x}{40s_t}$$

(Since $s_t < k^{t+1} < \sqrt{x}$, it follows that $x/s_t > \sqrt{x}$. Thus

$$\frac{c_m x}{40s_t} > \frac{2}{\ln k}\sqrt{x} > \frac{2k^{t+1}}{\ln k} > 2k^t > 1000\ )$$

Note that if $x/n \in I_t$ then $n < x/x_m$. For

$$n \le \frac{x}{s_t} < \frac{x}{x_m}$$

since $x_m < (k^t)^{c_m/2} < k^t < s_t$.

Let A be the set of all positive integers n such that x/n is in one of the intervals I_t. Then A is a subset of the set of positive integers less than or equal to x/x_m. Let B be the complement of A with respect to this larger set.

Since $(\ln n)/n$ decreases when $n \ge 3$,

$$\sum_A \frac{\ln n}{n} = \sum_{t=u}^{v}\sum_{x/n \in I_t}\frac{\ln n}{n}$$

$$> \sum_{t=u}^{v}\frac{c_m x}{40s_t}\frac{\ln(x/s_t)}{x/s_t} = \frac{c_m}{40}\sum_{t=u}^{v}\ln(x/s_t) \ge \frac{c_m}{40}(v-u)\ln\sqrt{x}$$

Since

$$v-u \ge \frac{\ln x}{3\ln k} = \frac{c_m \ln x}{246}$$

it follows that

$$\sum_A \frac{\ln n}{n} \ge \frac{c_m}{40}\frac{c_m \ln x}{246}\frac{\ln x}{2} \ge \frac{c_m^2 \ln^2 x}{20000}$$

By Theorem 7.8.1, and Theorem 7.7.9,

$$|R(x)|\ln^2 x \le 2\sum_{n\le x}|R(x/n)|\ln n + 150x\ln x$$

$$\le 2 \sum_{n\le x/x_m} |R(x/n)| \ln n + 2 \sum_{x/x_m<n\le x} (x/n) \ln n + 150x \ln x$$

$$\le 2\sum_A |R(x/n)| \ln n + 2\sum_B |R(x/n)| \ln n + (2\ln x_m + 150)\, x \ln x$$

by the Corollary to Theorem 7.7.4. Thus, using the definition of I_t and the induction hypothesis, $|R(x)| \ln^2 x$ is bounded above by

$$2\sum_A (x/n)(c_m/2) \ln n + 2\sum_B (x/n) c_m \ln n + (2\ln x_m + 150) x \ln x$$

$$= 2\sum_A (x/n) c_m \ln n + 2\sum_B (x/n) c_m \ln n$$

$$- \sum_A (x/n) c_m \ln n + (2\ln x_m + 150) x \ln x$$

$$= 2 \sum_{n\le x/x_m} (x/n) c_m \ln n - x c_m \sum_A \frac{\ln n}{n} + (2\ln x_m + 150) x \ln x$$

$$\le 2x c_m \sum_{n\le x/x_m} \frac{\ln n}{n} - x c_m \frac{c_m^2}{20000} \ln^2 x + (2\ln x_m + 150) x \ln x$$

Now by Theorem 7.7.4, the first term is bounded above by

$$2c_m x \left(\frac{\ln^2(x/x_m)}{2} + 1 \right) \le c_m x \ln^2 x$$

Thus

$$|R(x)| \ln^2 x \le c_m x \ln^2 x - \frac{c_m^2}{20000} c_m x \ln^2 x + (2\ln x_m + 150) x \ln x$$

Hence

$$|R(x)| < x(c_m - c_m^3/20000) + x(2\ln x_m + 150)/\ln x$$

$$< x(c_m - c_m^3/50000)$$

since $\ln x = (\ln x_m)(100/c_m)^3$. Thus $|R(x)/x| < c_{m+1}$.
This completes the proof.

Now let $\theta(x) = \sum_{p\leq x} \ln p$, with p varying over primes. For example, $\theta(4) = \ln 6$ and $\theta(1) = 0$.

If $a \leq 1/2$, then, by Theorem 7.7.4,

$$\lim_{x\to\infty} \frac{\theta(x^a)}{x} = 0$$

Now

$$\psi(x) = \theta(x) + \theta(x^{1/2}) + \theta(x^{1/3}) + \cdots + \theta(x^{1/k})$$

where $k = [\ln x/\ln 2]$. Dividing by x and taking the limit as $x \to \infty$, we obtain

$$1 = \lim_{x\to\infty} \frac{\theta(x)}{x} + 0$$

Note that if n is a positive integer,

$$\int_n^{n+1} \frac{\theta(x)}{x\ln^2 x}\, dx = \theta(n)\left(\frac{1}{\ln n} - \frac{1}{\ln(n+1)}\right)$$

Note also that if $\pi(x)$ is the number of primes $\leq x$,

$$\pi(x) = \sum_{2\leq n\leq x} \frac{\theta(n) - \theta(n-1)}{\ln n}$$

Hence

$$\pi(x) = \int_2^x \frac{\theta(t)}{t\ln^2 t}\, dt + \frac{\theta(x)}{\ln x}$$

Since $0 < \theta(t) < \psi(t) < 2t$ (Theorem 7.7.9), the above integral lies between 0 and

$$2\int_2^{\sqrt{x}} \frac{1}{\ln^2 t}\, dt + 2\int_{\sqrt{x}}^x \frac{1}{\ln^2 t}\, dt$$

$$\leq 2(\sqrt{x} - 2)\frac{1}{\ln^2 2} + 2(x - \sqrt{x})\frac{1}{\ln^2 \sqrt{x}}$$

$$\leq 2\left(\frac{4x}{\ln^2 x} + 3\sqrt{x}\right)$$

Hence if we multiply the integral by $(\ln x)/x$ and take the limit as $x \to \infty$, we shall get limit 0. Thus

$$\begin{aligned}\lim_{x\to\infty} \pi(x)(\ln x)/x &= \lim_{x\to\infty} \frac{\theta(x)}{\ln x}\frac{\ln x}{x} \\ &= 1\end{aligned}$$

This is the Prime Number Theorem.

Exercises 7.8

1. Show that

$$\lim_{n\to\infty} \frac{p_n}{n\ln n} = 1$$

where p_n is the nth prime.

7.9 Partitions

How many ways can you factor a whole number? For example, 12 has 4 distinct factorisations as 1×12, 2×6, 3×4 and $2\times 2\times 3$. In the case of pure powers, such as 2^n the answer is simply the number of *partitions* of n. For example, 5 can be partitioned as

$$5,\ \ 1+4,\ \ 2+3,\ \ 1+1+3,\ \ 1+2+2,\ \ 1+1+1+2,\ \ 1+1+1+1+1$$

— in 7 ways — and hence 32 factors in 7 ways, namely,

$$\begin{gathered}1\times 32,\ \ 2\times 16,\ \ 4\times 8,\ \ 2\times 2\times 8 \\ 2\times 4\times 4,\ \ 2\times 2\times 2\times 4,\quad \text{and}\quad 2^5\end{gathered}$$

If $p(n)$ is the number of partitions of a whole number n, that is, the number of ways of writing it in the form of a nondecreasing sequence of summands, then $p(1) = 1$, $p(2) = 2$, $p(3) = 3$, $p(4) = 5$ and $p(5) = 7$. We take it that $p(0) = 1$. The object of the next sections is to establish a formula for $p(n)$. This is not an easy task. We shall have to draw on Euler's power series for partitions, on Farey fractions, on Ford circles, on Möbius transformations, on Dedekind sums, and on the η function. The result we shall establish is due to Hans Rademacher, and our proof is a slight simplification of his.

7.10 Euler's Power Series

Let

$$F(x) = \prod_{m=1}^{\infty} \frac{1}{1-x^m}$$

What we show in this section is that

$$F(x) = \sum_{n=0}^{\infty} p(n)x^n$$

where p is the partition function. This was first done by Euler.

Theorem 7.10.1 *If* $|z| \leq R < 1$ *then*

$$\prod_{n=1}^{\infty} \frac{1}{1-z^n} = \frac{1}{\prod_{n=1}^{\infty}(1-z^n)}$$

converges absolutely and uniformly, and hence to an analytic function.

Proof. It suffices to show that

$$\sum_{n=1}^{\infty} \ln(1-z^n)$$

converges absolutely and uniformly. And this follows by the Weierstrass M-test, since

$$|\ln(1-z^n)|^2 = |\ln|1-z^n| + i\arg(1-z^n)|^2$$

$$= (\ln|1-z^n|)^2 + (\arg(1-z^n))^2 \leq (\ln(1-|z|^n))^2 + (\arctan(|z|^n/1))^2$$

$$\leq (|z|^n/(1-|z|))^2 + |z|^{2n} \leq R^{2n}\left(1+\frac{1}{1-R}\right)$$

— using the fact that if $0 < \alpha < 1$,

$$|\ln(1-\alpha)| = \alpha + \alpha^2/2 + \alpha^3/3 + \cdots \leq \frac{\alpha}{1-\alpha}$$

This completes the proof.

Now let $p_m(n)$ be the number of partitions of n into summands no larger than m (with $p_m(0) = 1$). Then

Theorem 7.10.2 $p_m(n) \leq (n+1)^m$.

Proof. A partition of n can be represented by columns of pebbles, each column with no more pebbles than the one to its left. If the summands are no larger than m, then the first column on the left has no more than m pebbles in it. Such a column diagram can also be read as a row diagram, giving a partition with no more than m summands. Hence $p_m(n)$ equals the number of partitions of n with no more than m summands. We can think of putting n pebbles into m boxes. For each box there is a prima facie choice of 0, 1, 2, ..., or n pebbles, and hence $p_m(n)$ is bounded by $(n+1)^m$.

Theorem 7.10.3 *Let $|z| \leq R < 1$. Then the following converges to an analytic function of z.*

$$\sum_{n=0}^{\infty} p_m(n) z^n$$

Proof. Use the Weierstrass M-test, Theorem 7.10.1, and the ratio test.

Theorem 7.10.4 *Let $0 \leq x < 1$. Then*

$$\sum_{n=0}^{\infty} p_m(n) x^n = \frac{1}{\prod_{n=1}^{m}(1-x^n)}$$

(Note that when $x = 0$ we have to take $0^0 = 1$ on the left side.)

Proof.

$$\begin{aligned}
\frac{(1-x^{m!k})^m}{\prod_{n=1}^{m}(1-x^n)} &= \prod_{n=1}^{m} \frac{1-(x^n)^{(m!/n)k}}{1-x^n} \\
&= \prod_{n=1}^{m}(1+x^n+x^{2n}+\cdots+x^{((m!/n)k-1)n}) \\
&= (1+x+x^2+x^3+\cdots+x^{m!k-1})
\end{aligned}$$

$$\begin{aligned}
&\times(1+x^2+x^4+x^6+\cdots+x^{2((m!/2)k-1)})\\
&\times(1+x^3+x^6+x^9+\cdots+x^{3((m!/3)k-1)})\\
&\times\cdots\\
&\times(1+x^m+x^{2m}+x^{3m}+\cdots+x^{m((m!/m)k-1)})\\
=&\sum_h c_h x^h
\end{aligned}$$

where the last sum is finite, and $0 \leq c_h \leq p_m(h)$ (because there are m factors in the product). The x^h term is found by adding up a number of products, each product being the result of 'threading' our way down the above product in such a way that the exponents add up to h. Such a 'threading' tells us how many 1's go into the sum h, and how many 2's, and so on — up to how many m's. If $h < m!k$ it tells us about a typical partition of h into summands none of which exceeds m. Hence, if $h < m!k$, then $c_h = p_m(h)$. Thus

$$\sum_{h=0}^{m!k-1} p_m(h)x^h \leq \frac{(1-x^{m!k})^m}{\prod_{n=1}^m(1-x^n)} \leq \sum_{h=0}^{\infty} p_m(h)x^h$$

As $k \to \infty$, $(1-x^{m!k})^m \to 1$ (since $0 \leq x < 1$) and hence

$$\frac{1}{\prod_{n=1}^m(1-x^n)} = \sum_{n=0}^{\infty} p_m(n)x^n$$

Theorem 7.10.5 *Let $|z| \leq R < 1$. Then the following converges to an analytic function of z:*

$$\sum_{n=0}^{\infty} p(n)z^n$$

Proof. This follows since the convergence is uniform (by the Weierstrass M-test). For

$$\sum_{n=0}^{\infty} p(n)R^n$$

converges. This is because, using Theorem 7.10.4, if $0 \leq x < 1$,

$$\sum_{n=0}^{m} p(n)x^n = \sum_{n=0}^{m} p_m(n)x^n \leq \sum_{n=0}^{\infty} p_m(n)x^n$$

$$= \frac{1}{\prod_{n=1}^{m}(1-x^n)} \leq \frac{1}{\prod_{n=1}^{\infty}(1-x^n)}$$

The left hand sum increases as $m \to \infty$, so

$$\sum_{n=0}^{\infty} p(n)x^n$$

exists and is $\leq$ the right hand reciprocal. Taking $x = R$, we get the result.

We now have two analytic functions on the disc $|z| \leq R < 1$. They are

$$\prod_{n=1}^{\infty} \frac{1}{1-z^n} = \frac{1}{\prod_{n=1}^{\infty}(1-z^n)}$$

and

$$\sum_{n=0}^{\infty} p(n)z^n$$

If these two functions agree when $0 \leq z < 1$, then, by analytic continuation, they are the same analytic function when $|z| < 1$ (Euler's result).

Theorem 7.10.6 *For* $0 \leq x < 1$,

$$\prod_{n=1}^{\infty} \frac{1}{1-x^n} = \frac{1}{\prod_{n=1}^{\infty}(1-x^n)} = \sum_{n=0}^{\infty} p(n)x^n$$

Proof. As above,

$$\sum_{n=0}^{\infty} p(n)x^n$$

exists and is

$$\leq \frac{1}{\prod_{n=1}^{\infty}(1-x^n)}$$

But now, by Theorem 7.10.4,

$$\sum_{n=0}^{\infty} p(n)x^n \geq \sum_{n=0}^{\infty} p_m(n)x^n = \frac{1}{\prod_{n=1}^{m}(1-x^n)}$$

Letting $m \to \infty$, we obtain

$$\sum_{n=0}^{\infty} p(n)x^n \geq \frac{1}{\prod_{n=1}^{\infty}(1-x^n)}$$

and so the result follows.

7.11 A Fractal Path of Ford Circles

Recall that the *Farey fractions* of order n, denoted by F_n, is the ascending sequence of reduced proper fractions with denominators $\leq n$. $0/1$ is included as the first fraction, and $1/1$ as the last. These fractions were treated in Section 2.4.

Given any proper fraction h/k in lowest terms, there is the associated *Ford circle* $C(h,k)$ in the complex plane with centre $h/k + i/2k^2$ and radius $1/2k^2$. (L. R. Ford first studied these circles in 1938.) Note that $C(h,k)$ is tangent to the real axis at h/k.

Theorem 7.11.1 *Two Ford circles $C(a,b)$ and $C(c,d)$ are either tangent to each other or they do not intersect. They are tangent iff $bc - ad = \pm 1$. In particular, Ford circles of consecutive Farey fractions are tangent to each other.*

Proof. The square of the distance D between centres is

$$D^2 = (a/b - c/d)^2 + (1/2b^2 - 1/2d^2)^2$$

while the square of the sum of their radii is

$$(1/2b^2 + 1/2d^2)^2$$

and the difference between these two squares is

$$\frac{(ad-bc)^2 - 1}{b^2d^2}$$

Since a, d, b, and c are integers, and $ad \neq bc$, this is nonnegative, equalling 0 if and only if $ad - bc = \pm 1$.

Theorem 7.11.2 *Let $h_1/k_1 < h/k < h_2/k_2$ be three consecutive Farey fractions (of some order n). The points of tangency of $C(h,k)$ with $C(h_1,k_1)$ and $C(h_2,k_2)$ are the points*

$$\alpha_1(h,k) = \frac{h}{k} - \frac{k_1}{k(k^2+k_1^2)} + \frac{i}{k^2+k_1^2}$$

$$\alpha_2(h,k) = \frac{h}{k} + \frac{k_2}{k(k^2+k_2^2)} + \frac{i}{k^2+k_2^2}$$

Moreover, the point of contact α_1 lies on the semicircle whose diameter is the interval $[h_1/k_1, h/k]$.

Proof. Let b be the rise and a the run as we go from α_1 to the centre of the Ford circle $C(h,k)$. Then, by similar triangles,

$$\frac{a}{h/k - h_1/k_1} = \frac{1/2k^2}{1/2k^2 + 1/2k_1^2}$$

and hence, since $hk_1 - h_1k = 1$,

$$a = \frac{k_1}{k(k^2+k_1^2)}$$

Similarly,

$$b = \frac{k_1^2 - k^2}{2k^2(k^2+k_1^2)}$$

and this leads straightforwardly to the result for α_1. The result for α_2 is similar.

Finally, the angle formed by going from h_1/k_1 to α_1 to h/k is right, this following from the fact that the imaginary part, $1/(k^2+k_1^2)$, of α_1 is the geometric mean of a and $h/k - h_1/k_1 - a$. Again, this follows by straight calculation.

For each positive integer N we construct a path $P(N)$ joining i and $i+1$: consider the Ford circles for the Farey series of order N; if $h_1/k_1 < h/k < h_2/k_2$ are consecutive in F_N, the points of tangency of $C(h_1,k_1)$, $C(h,k)$, and $C(h_2,k_2)$ divide $C(h,k)$ into an upper arc and a lower arc; $P(N)$ is the union of the upper arcs so obtained. (For the fractions $0/1$ and $1/1$ we use only the part of the upper arcs lying above the interval $[0,1]$.) From Theorem 7.11.1, it follows that $P(N)$ lies above the row of semicircles connecting adjacent Farey fractions in F_N. Now this continuous path $P(N)$ is used by Rademacher as a path of integration. The limit of $P(N)$ as $N \to \infty$ is a fractal of infinite length. We can prove that it is infinite as follows. As we fill in more Ford circles, we get at least half of each circumference, so the pathlength exceeds the sum of the radii of the Ford circles. Now for every prime p we have $p-1$ Ford circles with radius $1/2p^2$, so the pathlength exceeds

$$\sum_{primes} \frac{p-1}{2p^2} > \frac{1}{4} \sum_{primes} \frac{1}{p}$$

which diverges.

Theorem 7.11.3 *The transformation*

$$z = -ik^2(\tau - h/k)$$

maps the Ford circle $C(h,k)$ in the τ-plane onto a circle K in the z-plane of radius $\frac{1}{2}$ about the point $z = \frac{1}{2}$ as centre. The points of contact α_1 and α_2 of the previous theorem are mapped onto the points

$$z_1(h,k) = \frac{k^2}{k^2+k_1^2} + i\frac{kk_1}{k^2+k_1^2}$$

and (now for α_2)

$$z_2(h,k) = \frac{k^2}{k^2+k_2^2} - i\frac{kk_2}{k^2+k_2^2}$$

The upper arc joining α_1 and α_2 maps onto that arc of K which does not touch the imaginary z-axis.

Proof. The translation $\tau - h/k$ moves $C(h,k)$ to the left a distance h/k and thereby places its centre at $i/2k^2$. Multiplication by k^2

expands the radius to $1/2$, with the centre now at $i/2$. Multiplication by $-i$ rotates the circle $-\pi/2$ radians (a quarter turn clockwise). The expressions for z_1 and z_2 follow by straight calculation.

Theorem 7.11.4 *Suppose $h_1/k_1 < h/k < h_2/k_2$ are three consecutive fractions in F_N. Then*

$$|z_1(h,k)| = \frac{k}{\sqrt{k^2+k_1^2}}$$

$$|z_2(h,k)| = \frac{k}{\sqrt{k^2+k_2^2}}$$

Moreover, if z is on the chord joining z_1 and z_2 we have $|z| < \sqrt{2}k/N$. The length of this chord does not exceed $2\sqrt{2}k/N$.

Proof. The modulus equations are straightforward. If z is on the chord, then $|z| \le \max(|z_1|, |z_2|)$, so it suffices to show $|z_1| < \sqrt{2}k/N$ and $|z_2| < \sqrt{2}k/N$. Now

$$\frac{k+k_1}{2} \le \sqrt{\frac{k^2+k_1^2}{2}}$$

so that

$$\sqrt{k^2+k_1^2} \ge (k+k_1)/\sqrt{2} \ge (N+1)/\sqrt{2} > N/\sqrt{2}$$

since the sum of the denominators of two consecutive Farey fractions of order N is not less than $N+1$. (From Theorem 2.4.3, if x/y, x'/y' and x''/y'' are three successive terms in F_N then $y'' = [(N+y)/y']y' - y$ and hence $y''+y' > N$ if $((N+y)/y'-1)y' - y + y' \ge N$, which it is.) Thus

$$\sqrt{2}k/N > k/\sqrt{k^2+k_1^2} = |z_1|$$

and, similarly,

$$\sqrt{2}k/N > k/\sqrt{k^2+k_2^2} = |z_2|$$

Finally, the length of the chord is $|z_1 - z_2| \le |z_1| + |z_2| < 2\sqrt{2}k/N$.

7.12 Möbius Transformations

Let a, b, c, and d be integers such that $ad - bc = 1$. Then a complex function of the form $(az+b)(cz+d)$ is a *Möbius transformation.* These are named after A. F. Möbius (1790–1868), who gave us the famous strip.

These transformations map circles into circles (including the straight line as a circle here). For (1) $1/|z|^2$ gives an inversion in the unit circle about the origin; (2) $1/z = \overline{z}/|z|^2$ gives a reflection in the real axis followed by an inversion in the unit circle; (3) $1/(cz+d)$ gives a contraction or expansion, followed by a translation, followed by a reflection in the real axis and an inversion in the unit circle; and hence (4)

$$\frac{az+b}{cz+d} = \frac{a}{c} + \frac{bc-ad}{c}\frac{1}{cz+d}$$

gives a contraction or expansion (by a factor of c), followed by a translation (by d), followed by a reflection in the real axis and an inversion in the unit circle, followed by a contraction or expansion by $-1/c$, followed by a translation (by a/c). None of these transformations changes a circle or line into anything other than a circle or line.

Since

$$\frac{a' \frac{a\tau+b}{c\tau+d} + b'}{c' \frac{a\tau+b}{c\tau+d} + d'} = \frac{(aa'+b'c)\tau + a'b + b'd}{(c'a+d'c)\tau + c'b + dd'}$$

we can associate a Möbius transformation $(a\tau+b)/(c\tau+d)$ with a matrix

$$\begin{bmatrix} a & b \\ c & d \end{bmatrix}$$

and the composition of two such transformations is associated, in the same way, with the product of the matrices associated with each of the transformations individually. It is not hard to show that the set of Möbius transformations forms a group isomorphic with the multiplicative group of 2 by 2 matrices with integer coefficients and determinant 1 (identifying the matrices M and $-M$). The group of Möbius transformations is the *modular group* Γ.

Theorem 7.12.1 *The modular group* Γ *is generated by the two matrices*

$$T = \begin{bmatrix} 1 & 1 \\ 0 & 1 \end{bmatrix} \text{ and } S = \begin{bmatrix} 0 & -1 \\ 1 & 0 \end{bmatrix}$$

Proof. This proof is based on the idea of the 'reduction of binary quadratic forms'. Since we can identify M and $-M$, we can take it without loss of generality that the lower left entry c is nonnegative.

If $c = 0$ then $ad = 1$ and

$$\begin{bmatrix} \pm 1 & b \\ 0 & \pm 1 \end{bmatrix} = T^{\pm b}$$

so the theorem is true in this case.

If $c = 1$ then $ad - b = 1$ and

$$\begin{bmatrix} a & ad-1 \\ 1 & d \end{bmatrix} = T^a S T^d$$

so the theorem is true in this case too.

Now assume the theorem has been proved for all matrices in Γ with lower left entry $< c$ for some $c > 1$. Since $ad - bc = 1$, $\gcd(c, d) = 1$ and we have $d = cq + r$ with $0 < r < c$. Moreover,

$$\begin{bmatrix} a & b \\ c & d \end{bmatrix} T^{-q} S = \begin{bmatrix} -aq+b & -a \\ r & -c \end{bmatrix}$$

and hence the induction hypothesis gives us the result.

Corollary. Every element of Γ has the form $ST^pST^q \dots ST^z$ where p, q, …, z are integers. (Note that $T = ST^{-1}ST^{-1}S$.)

7.13 Dedekind Sums

Throughout this section we assume k is a positive integer, and h is an integer relatively prime to k.

The Dedekind sum is named after Richard Dedekind (1831–1916), the man who first defined an infinite set as one that can be put into one-to-one correspondence with a proper subset of itself. The Dedekind sum is defined as follows.

$$s(h,k) = \sum_{r=1}^{k-1} \frac{r}{k}\left(\frac{hr}{k} - \left[\frac{hr}{k}\right] - \frac{1}{2}\right)$$

And $s(0,1) = 0$.

To help derive the properties of this function, we use another function defined as follows.

$$((x)) = x - [x] - \frac{1}{2} \quad \text{if } x \text{ is not an integer}$$

and $((x)) = 0$ if x is an integer.

This is a periodic function with period 1. Note that it is an odd function, in the sense that $((-x)) = -((x))$. Moreover, $h_1 \equiv h_2 \pmod{k}$ implies $((h_1/k)) = ((h_2/k))$, since $((\ ))$ has period 1. The numbers h, $2h$, ..., $(k-1)h$, 0 are a complete set of residues mod k. If k is odd, another complete set of residues mod k is $-(k-1)/2$, $-(k-1)/2+1$, ..., -1, 0, 1, ..., $(k-1)/2$. Hence, if k is odd,

$$\sum_{r=1}^{k-1}((rh/k)) = 0$$

(since the $((\ ))$ function is odd). If k is even, we get an extra term,

$$((\ (k/2)/k\ )) = 0$$

but the result is the same. Hence

$$s(h,k) = \sum_{r=1}^{k-1} \frac{r}{k}((hr/k))$$

$$= \sum_{r=1}^{k-1}\left(\frac{r}{k} - \frac{1}{2}\right)((hr/k)) = \sum_{r=1}^{k-1}((r/k))((hr/k))$$

This shows that $s(-h,k) = -s(h,k)$ — since $((-hr/k)) = -((hr/k))$.

Theorem 7.13.1 *If h^{-1} is the inverse of h mod k, then*

$$s(h^{-1},k)=\sum_{r \bmod k}((r/k))((h^{-1}r/k))=\sum_{t=1}^{k-1}((ht/k))((t/k))=s(h,k)$$

Corollary. If $h^2+1\equiv 0 \pmod{k}$ then $s(h,k)=0$. (For if the congruence holds, $-s(h,k)=-s(h^{-1},k)=s(-h^{-1},k)=s(h,k)$.)

We conclude this section with the Reciprocity Law for Dedekind Sums. For this we need two lemmas.

Theorem 7.13.2

$$\sum_{r=1}^{k-1}[hr/k]([hr/k]+1)=2hs(k,h)+(h-1)(hk/3+k/3-h/2)$$

Proof: As r goes from 1 to $k-1$, $[hr/k]$ goes from 0 to $h-1$. Now if $1\le v\le h$, we have $[hr/k]=v-1$ just in case $k(v-1)/h<r<kv/h$ (with equality impossible). Hence the number of values of r for which $[hr/k]=v-1$ is $[kv/h]-[k(v-1)/h]$ — unless $v=h$, in which case it is $[kv/h]-[k(v-1)/h]-1$ (since $r=k$ is excluded). Hence

$$\sum_{r=1}^{k-1}[hr/k]([hr/k+1])$$

$$=\sum_{v=1}^{h-1}(v-1)v([kv/h]-[k(v-1)/h])+h(h-1)(k-1-[k(h-1)/h])$$

$$=-2\sum_{v=1}^{h-1}v[vk/h]+h(h-1)(k-1)$$

(telescoping).

Also

$$2hs(k,h)=2\sum_{v=1}^{h-1}v(kv/h-[kv/h]-1/2)$$

$$=-2\sum_{v=1}^{h-1}v[kv/h]+(2k/h)(h-1)h(2h-1)/6-(h-1)h/2$$

Hence the original left hand side summation equals

$$2hs(k,h) - (h-1)(2k(2h-1)/6 - h/2 - h(k-1))$$
$$= 2hs(k,h) - (h-1)(-hk/3 - k/3 + h/2)$$

Theorem 7.13.3

$$\sum_{r=1}^{k-1}((hr/k))^2 = (k-1)(1/12 - 1/6k)$$

Proof:

$$LHS = \sum_{r=1}^{k-1}((r/k))^2 = \sum_{r=1}^{k-1}(r/k - 1/2)^2 = RHS$$

We now give the Reciprocity Law for Dedekind Sums.

Theorem 7.13.4 *If $h > 0$ then*

$$12hk\ s(h,k) + 12kh\ s(k,h) = h^2 + k^2 - 3hk + 1$$

Proof:

$$\sum_{r=1}^{k-1}((hr/k))^2$$
$$= \sum_{r=1}^{k-1} h^2r^2/k^2 + [hr/k]^2 + 1/4 - hr/k + [hr/k] - 2(hr/k)[hr/k]$$
$$= 2h\sum_{r=1}^{k-1}(r/k)(hr/k - [hr/k] - 1/2) + \sum_{r=1}^{k-1}[hr/k]([hr/k]+1)$$
$$- \sum_{r=1}^{k-1} h^2r^2/k^2 + \sum_{r=1}^{k-1} 1/4$$
$$= 2hs(h,k) + 2hs(k,h) + (h-1)(hk/3 + k/3 - h/2)$$
$$-(h^2/k^2)(k-1)k(2k-1)/6 + (k-1)/4$$

Hence, using Theorem 7.13.3 and multiplying by $6k$,

$$(k-1)(k/2-1)$$

$$= 12hk\ s(h,k) + 12kh\ s(k,h) + (h-1)k(2hk + 2k - 3h)$$
$$-h^2(k-1)(2k-1) + 3(k-1)k/2$$

Thus

$$12hk\ s(h,k) + 12kh\ s(k,h)$$
$$= (k-1)(k/2 - 1 - 3k/2) + h^2(k-1)(2k-1) - (h-1)k(2hk + 2k - 3h)$$
$$= 1 - k^2 + 2h^2k^2 - 3kh^2 + h^2 - 2h^2k^2 - 2hk^2 + 3h^2k + 2hk^2 + 2k^2 - 3hk$$
$$= 1 + k^2 + h^2 - 3hk - 3kh^2 - 2hk^2 + 3h^2k + 2hk^2$$

and the result follows.

7.14 Eta Function

Dedekind's eta function is defined as follows. Where τ is in the upper half plane H,

$$\eta(\tau) = e^{\pi i\tau/12} \prod_{n=1}^{\infty} (1 - e^{2\pi i n\tau})$$

The meaning of the product is

$$e^{\sum_{n=1}^{\infty} \log(1 - e^{2\pi i n\tau})}$$

The eta product converges absolutely and uniformly (on any compact subset of H) — in the sense that the log series does (see the proof of Theorem 7.10.1).

The fact that the log series converges absolutely and uniformly implies that $\eta(\tau)$ is never 0, and it also implies that η is holomorphic (analytic) on H.

Note that $\eta(\tau + 1) = e^{\pi i/12}\eta(\tau)$. Hence the η function is periodic, with period 1.

The key result in this section is *Dedekind's Functional Equation,* namely, if a, b, c, d are integers such that $ad - bc = 1$, and $c > 0$, then (with $s(h,k)$ the Dedekind sum defined above)

$$\eta\left(\frac{a\tau + b}{c\tau + d}\right) = e^{\pi i(s(-d,c) + (a+d)/12c)} \sqrt{-i(c\tau + d)}\ \eta(\tau)$$

To prove the Functional Equation, we use an approach discovered by B. Gordon.

Theorem 7.14.1 *The Functional Equation holds if* $a = 0$, $b = -1$, $c = 1$ *and* $d = 0$.

Proof. In this case the equation reads

$$\eta(-1/\tau) = \sqrt{-i\tau}\,\eta(\tau)$$

The functions on the left and right of this equation are both analytic in the upper half plane H. Since a function that is analytic in a connected open set D is uniquely determined over D by its values along an arc interior to D, it suffices to show that the equation holds for numbers $\tau = iy$ with y a positive real. In that case the equation is equivalent to

$$\eta(i/y) = \sqrt{y}\,\eta(iy)$$

or

$$\log\eta(i/y) - \log\eta(iy) = \frac{1}{2}\log y$$

(We can take logs since the η function is never 0.)

Note that the graph of $\eta(iy)$ with $y > 0$ ascends from $(0,0)$ to a point near $(.5,.8)$ and then decreases, becoming concave up, to 0.

Now

$$\log\eta(iy) = -\frac{\pi y}{12} + \sum_{n=1}^{\infty}\log(1-e^{-2\pi ny}) = -\frac{\pi y}{12} - \sum_{n=1}^{\infty}\sum_{m=1}^{\infty}\frac{e^{-2\pi mny}}{m}$$

using the Taylor series. We can switch the order of the summation signs, and, by summing the GP's in n, we obtain

$$\log\eta(iy) = -\frac{\pi y}{12} + \sum_{m=1}^{\infty}\frac{1}{m(1-e^{2\pi my})}$$

We can replace y by $1/y$ in the above, and hence it suffices to show that

$$-\frac{\pi}{12y} + \sum_{m=1}^{\infty}\frac{1}{m(1-e^{2\pi m/y})} + \frac{\pi y}{12} - \sum_{m=1}^{\infty}\frac{1}{m(1-e^{2\pi my})} = \frac{1}{2}\log y \quad (*)$$

To prove this we use residues. For a fixed $y > 0$ and n any positive integer, let

$$F_n(z) = -\frac{1}{8z}\cot(\pi i(n+1/2)z)\cot\frac{\pi(n+1/2)z}{y}$$

Let C be the parallelogram joining the vertices y, i, $-y$, and $-i$, in that order. Inside C, the function F_n has simple poles at $z = ik/(n+1/2)$ and at $z = ky/(n+1/2)$ for $k = \pm 1, \ \pm 2, \ \pm 3, \ \ldots, \ \pm n$. There is also a triple pole at $z = 0$ with residue $i(y - 1/y)/24$. The residue at $z = ik/(n+1/2)$ is

$$\frac{\cot(\pi ik/y)}{8\pi k}$$

and the residue at $z = ky/N$ is

$$-\frac{\cot \pi iky}{8\pi k}$$

(To prove the above we use the following facts. The cot is cos / sin and sin has zeros precisely at integer multiples of π, while cos has zeros precisely at numbers $\pi/2$ greater than these numbers. The Laurent series for cot is

$$\cot z = \frac{1}{z} - \frac{z}{3} - \frac{z^3}{45} + \cdots$$

In general, if two functions p and q are analytic at z_0 and $p(z_0) \neq 0$, $q(z_0) = 0$, and $q'(z_0) \neq 0$, then z_0 is a simple pole of the quotient $p(z)/q(z)$, and the residue there is $p(z_0)/q'(z_0)$.)

Since

$$\frac{\cot(\pi ik/y)}{8\pi k}$$

is an even function of k, we have

$$\sum_{k=-n}^{n} \operatorname{Res} F_n(z) = i(y - 1/y)/24 + 2\sum_{k=1}^{n}\frac{\cot(\pi ik/y)}{8\pi k} - 2\sum_{k=1}^{n}\frac{\cot \pi iky}{8\pi k}$$

But

$$\cot iw = i\frac{e^{-w}+e^{w}}{e^{-w}-e^{w}} = -i\frac{e^{2w}+1}{e^{2w}-1} = \frac{1}{i}\left(1 - \frac{2}{1-e^{2w}}\right)$$

Hence the sum of the residues is

$$\frac{i(y-1/y)}{24}+\frac{1}{4\pi i}\sum_{k=1}^{n}\frac{1}{k}-\frac{1}{2\pi i}\sum_{k=1}^{n}\frac{1}{k(1-e^{2\pi k/y})}$$

$$+\frac{i}{4\pi}\sum_{k=1}^{n}\frac{1}{k}-\frac{i}{2\pi}\sum_{k=1}^{n}\frac{1}{k(1-e^{2\pi ky})}$$

Thus $2\pi i$ times the sum of all the residues of $F_n(z)$ inside C is an expression whose limit, as $n\to\infty$, is equal to the left member of (*). Therefore, by the Residue Theorem, it now suffices to show

$$\lim_{n\to\infty}\int_C F_n(z)dz=-\frac{1}{2}\log y$$

Now

$$F_n(z)=-\frac{1}{8z}\frac{1}{i}\left(1-\frac{2}{1-e^{2\pi(n+1/2)z}}\right)\frac{1}{i}\left(1-\frac{2}{1-e^{-2i\pi(n+1/2)z/y}}\right)$$

This function is bounded on each of the four sides of C, and in a way that is independent of n. For example, on the side joining y to i, we have $z(t)=(1-t)y+ti$ with $0\le t\le 1$. As t goes from 0 to $1-1/(4n+2)$,

$$|1-e^{2\pi(n+1/2)z}|>e^{\pi y/2}-1$$

As t goes from $1-1/(4n+2)$ to 1, $e^{2\pi(n+1/2)z}$ goes from $e^{\pi/2}i$ to -1 and without leaving the third quadrant. Thus, for these t's

$$|1-e^{2\pi(n+1/2)z}|>1$$

Thus, regardless of what n is,

$$\left|\frac{1}{i}\left(1-\frac{2}{1-e^{2\pi(n+1/2)z}}\right)\right|<1+\frac{2}{\min(1,\ e^{\pi y/2}-1)}$$

In other words, independent of n, there is a bound on $\cot \pi i(n+1/2)z$. Similarly, there is a bound, independent of n, on $\cot\pi(n+1/2)z/y$. Finally, there is a bound on $1/z$ (namely, $\sqrt{1+y^2}\,/y$). Hence, on the side of C joining y to i, we have a bound on $F_n(z)$ which is independent of n. And similarly, such a bound exists for the other sides of C. Hence, by the Lebesgue Dominated Convergence Theorem,

$$\lim_{n\to\infty}\int_C F_n(z)\,dz = \int_C \lim_{n\to\infty} F_n(z)\,dz$$

Now $\lim_{n\to\infty} zF_n(z) = 1/8$ on the edges of C connecting y, i, and $-y$, $-i$, but the limit is $-1/8$ on the other two edges. Hence

$$\lim_{n\to\infty}\int_C F_n(z)\,dz = \int_{-i}^{y}\frac{-1}{8z}\,dz + \int_{y}^{i}\frac{1}{8z}\,dz + \int_{i}^{-y}\frac{-1}{8z}\,dz + \int_{-y}^{-i}\frac{1}{8z}\,dz$$

and this equals

$$\frac{-\log y + \log(-i) + \log i - \log y - \log(-y) + \log i + \log(-i) - \log(-y)}{8}$$

For the segment from i to $-y$ we must take $\log(-y) = \log y + \pi i$ but for the segment from $-y$ to $-i$ we must take $\log(-y) = \log y - \pi i$. (With improper integrals we are taking a limit.) Hence this gives $-\frac{1}{2}\log y$ as required.

Corollary: The Dedekind Functional Equation now follows for $\begin{bmatrix} a & b \\ c & d \end{bmatrix}$ with $d = 0$ (using the fact that η is periodic, with period 1).

To prove Dedekind's Functional Equation in general, we use the following theorems.

Theorem 7.14.2 *If* $A = \begin{bmatrix} a & b \\ c & d \end{bmatrix} \in \Gamma$ *and* $c > 0$ *then, for every integer* m

$$\exp(\ \pi i((a+cm+d)/(12c) - s(cm+d,c))\ )$$
$$= \exp(\pi i m/12)\exp(\ \pi i((a+d)/(12c) - s(d,c))\ )$$

If we abbreviate the function $\exp(\pi i((a+d)/(12c) - s(d,c)))$ *as* $f(A)$, *then we can abbreviate the above equation as*

$$f(AT^m) = e^{\pi i m/12} f(A)$$

Proof. When discussing Dedekind sums in Section 7.13, we showed that $s(cm+d,c) = s(d,c)$. (This is true even if $d = 0$.)

Theorem 7.14.3 *If* $A = \begin{bmatrix} a & b \\ c & d \end{bmatrix} \epsilon\, \Gamma$ *and* $c > 0$ *then*
if $d > 0,\ f(AS) = e^{-\pi i/4} f(A)$, *and*
if $d < 0,\ f(-AS) = e^{\pi i/4} f(A)$.

Proof.

$$\begin{bmatrix} a & b \\ c & d \end{bmatrix} \begin{bmatrix} 0 & -1 \\ 1 & 0 \end{bmatrix} = \begin{bmatrix} b & -a \\ d & -c \end{bmatrix}$$

For $d > 0$, we have

$$f(AS) = \exp(\ \pi i((b-c)/(12d) - s(-c,d))\ )$$

$$= \exp(\ \pi i(b-c)/(12d) + s(c,d))\ )$$

from the properties of the Dedekind sum. Now the Reciprocity Law for Dedekind sums gives us

$$s(c,d) + s(d,c) = \frac{c}{12d} + \frac{d}{12c} - \frac{1}{4} + \frac{1}{12cd}$$

Since $ad - bc = 1$, we obtain

$$\frac{b-c}{12d} + s(c,d) = \frac{a+d}{12c} - s(d,c) - \frac{1}{4}$$

Substituting, we get

$$\begin{aligned} f(AS) &= \exp(\pi i \left(\frac{a+d}{12c} - s(d,c)\right)) \exp(\pi i(-1/4)) \\ &= e^{-\pi i/4} f(A) \end{aligned}$$

For $d < 0$, we have

$$f(-AS) = \exp(\ \pi i((-b+c)/(12(-d)) - s(c,-d))\ )$$

The Reciprocity Law gives

$$s(c,-d)+s(-d,c)=\frac{c}{-12d}-\frac{d}{12c}-\frac{1}{4}-\frac{ad-bc}{12cd}$$

$$\frac{-b+c}{-12d}-s(c,-d)=\frac{a+d}{12c}-s(d,c)+\frac{1}{4}$$

Substituting, we get

$$\begin{aligned} f(-AS) &= \exp(\,\pi i((a+d)/(12c)-s(d,c))\,)\exp(\pi i/4) \\ &= f(A)e^{\pi i/4} \end{aligned}$$

Theorem 7.14.4 *Suppose the Dedekind Functional Equation holds for some $A \in \Gamma$ (with $c>0$). Then (1) it is also satisfied for AT^m. (2) If $d>0$ then it is also satisfied for AS. (3) If $d<0$ then it is also satisfied for $-AS$.*

Proof. (1) Since the Dedekind Functional Equation holds for A, we have (taking $\tau=\tau+m$)

$$\eta\left(\frac{a\tau+am+b}{c\tau+cm+d}\right)=f(A)\sqrt{-i(c\tau+cm+d)}\;\eta(\tau+m)$$

$$=e^{\frac{\pi i m}{12}}f(A)\sqrt{-i(c\tau+cm+d)}\;\eta(\tau)=f(AT^m)\sqrt{-i(c\tau+cm+d)}\;\eta(\tau)$$

Hence

$$\eta(AT^m\tau)=f(AT^m)\sqrt{-i(c\tau+cm+d)}\;\eta(\tau)$$

(2) Suppose $d>0$. Since the Dedekind Functional Equation holds for A, we have (taking $\tau=-1/\tau$)

$$\eta(AS\tau)=f(A)\sqrt{-i(-c/\tau+d)}\;\eta(-1/\tau)$$

By the previous theorem,

$$\eta(-1/\tau)=\sqrt{-i\tau}\;\eta(\tau)$$

so that

$$\eta(AS\tau)=f(A)\sqrt{-i(-c\overline{\tau}/|\tau|^2+d)}\;\sqrt{-i\tau}\;\eta(\tau)$$

If τ is in Q I (the first quadrant), then $-i\tau$ is in Q IV, and the number $-i(-c\bar{\tau}/|\tau|^2 + d)$ is in Q I or IV (since $\bar{\tau}$ is in Q II, $-c\bar{\tau}/|\tau|^2$ is in Q III). Hence the Third Law of Exponents applies (we do not cross the negative real axis when we multiply the above two numbers). Similarly, it applies if τ is in Q II. Hence

$$\eta(AS\tau) = f(A)\sqrt{c - d\tau}\,\eta(\tau)$$

$$= f(AS)e^{\pi i/4}\sqrt{c - d\tau}\,\eta(\tau) = f(AS)\sqrt{-i(d\tau - c)}\,\eta(\tau)$$

(3) Suppose $d < 0$. Then, as above,

$$\eta(AS\tau) = f(A)\sqrt{c - d\tau}\,\eta(\tau)$$

$$= f(-AS)e^{-\pi i/4}\sqrt{c - d\tau}\,\eta(\tau) = f(-AS)\sqrt{-i(-d\tau + c)}\,\eta(\tau)$$

Theorem 7.14.5 *The Dedekind Functional Equation holds for any* $\begin{bmatrix} a & b \\ c & d \end{bmatrix}$ *with* $c > 0$.

Proof. It holds for S (Theorem 7.14.1). Now any $A \in \Gamma$ can be written

$$ST^aST^b \ldots ST^w$$

(see Theorem 7.12.1). So we can prove the theorem by induction. There are three cases. (1) If the Functional Equation holds for A then it holds for AT^m (Theorem 7.14.4). (2) If the Functional Equation holds for A with $d \neq 0$ then it holds for AS (Theorem 7.14.4). (3) If the Functional Equation holds for A with $d = 0$ then

$$A = \begin{bmatrix} a & -1 \\ 1 & 0 \end{bmatrix}$$

and $AS = T^a$. Now $AST^b = T^{a+b}$ and $AST^bS = \begin{bmatrix} a+b & -1 \\ 1 & 0 \end{bmatrix}$ and the Functional Equation holds in this case (since it holds when $d = 0$). And then it holds for AST^bST^c, where the d is no longer 0.

Thus the Functional Equation holds for all elements of Γ with $c > 0$.

For the Partition Formula Theorem, we need to adapt Dedekind's Functional Equation as follows.

Theorem 7.14.6 *Let* $F(t) = 1/\prod_{m=1}^{\infty}(1-t^m)$ *with* $|t| < 1$. *Let*

$$x = \exp\left(\frac{2\pi i h}{k} - \frac{2\pi z}{k^2}\right)$$

$$x' = \exp\left(\frac{2\pi i H}{k} - \frac{2\pi}{z}\right)$$

where $Re(z) > 0$, k *is a positive integer,* h *is an integer relatively prime to* k, *and* H *is an integer such that* $hH \equiv -1 \pmod{k}$. *(If* $k = 1$ *then* $h = 0$ *and we take* $H = 0$ *also.) Then*

$$F(x) = e^{\pi i s(h,k)}\sqrt{z/k}\,\exp\left(\frac{\pi}{12z} - \frac{\pi z}{12k^2}\right) F(x')$$

where $s(h,k)$ *is the Dedekind sum (see above).*

Proof. If $\begin{bmatrix} a & b \\ c & d \end{bmatrix} \in \Gamma$ with $c > 0$ then Dedekind's Functional Equation implies

$$\frac{1}{\eta(\tau)} = \frac{1}{\eta(\tau')}\sqrt{-i(c\tau+d)}\ \exp(\ \pi i((a+d)/12c + s(-d,c))\ )$$

where $\tau' = (a\tau+b)/(c\tau+d)$. Since

$$F(e^{2\pi i\tau}) = e^{\pi i\tau/12}/\eta(\tau)$$

this gives

$$F(e^{2\pi i\tau})$$
$$= F(e^{2\pi i\tau'})e^{\pi i(\tau-\tau')/12}\sqrt{-i(c\tau+d)}\exp(\pi i((a+d)/12c + s(-d,c)))$$

If $a = H$, $c = k$, $d = -h$, and $b = -(hH+1)/k$, and if $\tau = (iz+h)/k$ then $\tau' = (i/z+H)/k$ and we obtain

$$F\left(\exp\left(\frac{2\pi i h}{k} - \frac{2\pi z}{k}\right)\right)$$

$$= F\left(\exp\left(\frac{2\pi iH}{k} - \frac{2\pi}{kz}\right)\right) \sqrt{z}\, \exp\left(\frac{\pi}{12kz} - \frac{\pi z}{12k} + \pi i s(h,k)\right)$$

Replacing z by z/k, we obtain the result.

If $k = 1$, $h = 0$, the theorem reads

$$F(e^{-2\pi z}) = \sqrt{z}\, e^{\pi/12z - \pi z/12} F(e^{-2\pi/z})$$

7.15 Bessel Functions Avoided

At one point in Rademacher's proof he uses the following formulas from the theory of Bessel functions with purely imaginary argument:

$$\int_{c-\infty i}^{c+\infty i} t^{-5/2} e^{t+z^2/4t}\, dt = 2\pi i (z/2)^{-3/2} I_{3/2}(z)$$

(where $c > 0$ and z is any real number) and

$$I_{3/2}(z) = \sqrt{2z/\pi}((\sinh z)/z)'$$

(where the $'$ means 'differentiated'). In this section we prove a result that allows us to bypass these formulas, obtaining a less advanced proof of the partition function formula.

Theorem 7.15.1

$$\lim_{L\to\infty} e^{-L^2} \int_0^{\sqrt{L^2-(3/2)\ln L}} e^{t^2}\, dt = 0$$

Proof. The integral is bounded above by

$$\sqrt{L^2 - (3/2)\ln L}\; e^{L^2} e^{-(3/2)\ln L}$$

and

$$\lim_{L\to\infty} \frac{\sqrt{L^2 - (3/2)\ln L}}{L^{3/2}} = 0$$

Theorem 7.15.2

$$\lim_{L\to\infty} e^{-L^2} \int_{\sqrt{L^2-(3/2)\ln L}}^{L} e^{t^2}\, dt = 0$$

Proof. The integral is bounded by

$$e^{L^2}(L - \sqrt{L^2 - (3/2)\ln L})$$

and

$$\lim_{L\to\infty} (L - \sqrt{L^2 - (3/2)\ln L}) = 0$$

Theorem 7.15.3 *If c is any positive real,*

$$\lim_{L\to\infty} e^{-L^2} \int_{L}^{\sqrt{L^2+c}} e^{t^2}\, dt = 0$$

Proof. Since $L + c > \sqrt{L^2 + c}$, the integral is bounded by

$$e^{L^2} e^{c}(\sqrt{L^2 + c} - L)$$

and

$$\lim_{L\to\infty} (\sqrt{L^2 + c} - L) = 0$$

Theorem 7.15.4 *If c is any positive real,*

$$\lim_{L\to\infty} e^{-L^2} \int_{0}^{\sqrt{L^2+c}} e^{t^2}\, dt = 0$$

Proof. Use the first 3 theorems in this section.

Theorem 7.15.5 *Let L and c be positive real numbers. If C is the vertical line joining $\pm L$ to $\pm L + i\sqrt{L^2 + c}$ then*

$$\lim_{L\to\infty} \int_{C} e^{-t^2}\, dt = 0$$

Proof: This integral equals

$$\int_0^{\sqrt{L^2+c}} e^{-(L+ti)^2} i \, dt$$

and hence its absolute value is bounded by the integral in Theorem 7.15.4.

Theorem 7.15.6 *Let c be a positive real, and let C be the contour $v = \sqrt{u^2+c}$ with u going from $-\infty$ to ∞. Then*

$$\int_C e^{-t^2} \, dt = \sqrt{\pi}$$

Proof: Consider the contour D going from $-L$ to L along the real axis, then straight up to $L + i\sqrt{L^2+c}$, then to the left along the contour C to the point $-L + i\sqrt{L^2+c}$, and then, finally, straight back down to $-L$. Using Cauchy's Theorem,

$$0 = \int_D e^{-t^2} \, dt = \int_{-L}^{L} e^{-t^2} \, dt + \int_L^{L+i\sqrt{L^2+c}} e^{-t^2} \, dt$$

$$+ \int_{C,\ L+i\sqrt{L^2+c} \text{ back to } -L+i\sqrt{L^2+c}} e^{-t^2} \, dt + \int_{-L+i\sqrt{L^2+c}}^{-L} e^{-t^2} \, dt$$

As $L \to \infty$, the first integral in the sum of four integrals tends to $\sqrt{\pi}$ (from probability theory), while the second and fourth tend to 0 (by Theorem 7.15.5). The result now follows.

Theorem 7.15.7 *Let c be a positive real. Then*

$$\int_{c-\infty i}^{c+\infty i} \frac{e^t}{\sqrt{t}} \, dt = 2i\sqrt{\pi}$$

Proof: Let $s(t) = i\sqrt{t}$. Then as t goes vertically up from $c - \infty i$ to $c + \infty i$, s goes from right to left along the curve $v = \sqrt{u^2+c}$. For s takes $c + wi$ to

$$ie^{(1/2)(\ln\sqrt{c^2+w^2}+i\arctan(w/c))}$$

$$= -\sqrt[4]{c^2+w^2}\sin((1/2)\arctan(w/c)) + i\sqrt[4]{c^2+w^2}\cos((1/2)\arctan(w/c))$$

If this is $u + vi$ then

$$u^2 + v^2 = \sqrt{c^2+w^2}$$

$$uv = -w/2$$

— the latter since

$$\sin((1/2)\arctan(w/c))\cos((1/2)\arctan(w/c)) = (1/2)\sin\arctan(w/c)$$

Thus

$$u^4 + 2u^2v^2 + v^4 = c^2 + 4u^2v^2$$

$$(u^2 - v^2)^2 = c^2$$

with $s(c) = i\sqrt{c}$. Hence the image of the vertical line under $s(t)$ is $v = \sqrt{c+u^2}$. Call this curve E.

Substituting s for t in the given integral, we obtain

$$-2i\int_E e^{-s^2}\,ds$$

From Theorem 7.15.6 this equals $2i\sqrt{\pi}$, as required.

Theorem 7.15.8 *If c is any positive real, and k is a positive integer,*

$$\int_{c-\infty i}^{c+\infty i} \frac{e^t}{t^{(1/2)+k}}\,dt = \frac{1}{k-1/2}\int_{c-\infty i}^{c+\infty i} \frac{e^t}{t^{(1/2)+k-1}}\,dt$$

Proof: Use integration by parts and the fact that

$$\lim_{L\to\infty} \frac{1}{e^{(k-1/2)\ln\sqrt{c^2+L^2}}} = 0$$

Theorem 7.15.9 *If c is any positive real, and n is a positive integer,*

$$\int_{c-\infty i}^{c+\infty} \frac{e^t}{4^n n! t^{(5/2)+n}}\,dt = \frac{16(n+1)}{(2n+3)!}\sqrt{\pi}\,i$$

Proof: Use $k = n + 2$ in the previous theorem.

Theorem 7.15.10 *If n is a nonnegative integer, and z is a fixed complex number, let*

$$f_n(t) = \frac{e^t z^{2n}}{t^{(5/2)+n} 4^n n!}$$

Then if E is the set of complex numbers on the vertical line through $c > 0$, $\sum f_n(t)$ converges uniformly on E.

Proof: The n-th term of the series is bounded by

$$\frac{e^c}{c^{5/2}} \frac{(|z|^2/4c)^n}{n!}$$

on the vertical line. The result follows by the Weierstrass M-test.

Theorem 7.15.11 *Let c be a positive real, and z any complex number. Then*

$$\int_{c-\infty i}^{c+\infty i} \frac{e^t e^{z^2/4t}}{t^{5/2}}\, dt = \frac{8i\sqrt{\pi}}{z} \frac{d}{dz} \frac{\sinh z}{z}$$

Proof: Where $f_n(t)$ is defined in the previous theorem, the integrand is $\sum_{n=0}^{\infty} f_n(t)$. Because of its uniform convergence (Theorem 7.15.10), we can switch the integration and summation signs, obtaining

$$\sum_{n=0}^{\infty} z^{2n} \int_{c-\infty i}^{c+\infty i} \frac{e^t}{4^n n! t^{(5/2)+n}}\, dt$$

Using Theorem 7.15.9, we get

$$\sum_{n=0}^{\infty} z^{2n} \frac{16(n+1)}{(2n+3)!} \sqrt{\pi} i$$

and this equals

$$\frac{8\sqrt{\pi} i}{z} \left(\frac{2}{3!} z + \frac{4}{5!} z^3 + \frac{6}{7!} z^5 + \frac{8}{9!} z^7 + \cdots \right)$$

But

$$\frac{\sinh z}{z} = 1 + \frac{z^2}{3!} + \frac{z^4}{5!} + \frac{z^6}{7!} + \cdots$$

and, differentiating, we obtain the result.

7.16 Rademacher's Proof

Theorem 7.16.1 *If $n \geq 1$ the partition function $p(n)$ is represented by a convergent series:*

$$p(n) = \frac{1}{\pi\sqrt{2}} \sum_{k=1}^{\infty} A_k(n)\, \sqrt{k}\, \frac{d}{dn} \left(\frac{\sinh\left((\pi/k)\sqrt{(2/3)(n-1/24)}\,\right)}{\sqrt{n-1/24}} \right)$$

where

$$A_k(n) = \sum_{0 \leq h < k,\ (h,k)=1} e^{\pi i s(h,k) - 2\pi i n h/k}$$

Note that $A_1(n) = 1$.

Proof: By Euler's formula, if $0 < |x| < 1$,

$$\frac{F(x)}{x^{n+1}} = \sum_{k=0}^{\infty} \frac{p(k)x^k}{x^{n+1}}$$

for each nonnegative integer n. The series is the Laurent series of $F(x)/x^{n+1}$ in the punctured disk $0 < |x| < 1$. This function has a pole at $x = 0$ with residue $p(n)$. Hence, by Cauchy's residue theorem,

$$p(n) = \frac{1}{2\pi i} \int_C \frac{F(x)}{x^{n+1}}\, dx$$

where C is a counterclockwise circle with centre the origin and radius $e^{-2\pi}$. Let $\tau = (\ln x)/2\pi i$. Then corresponding to the unit circle in the x-plane we have an infinite rectangle in the τ plane, bounded by the real axis, and the lines $x = 0$, $x = 1$. The image of C in the τ plane is given by

$$\frac{\ln(e^{-2\pi}e^{i\theta})}{2\pi i} = \frac{-2\pi + i\theta}{2\pi i} = i + \theta/2\pi$$

with $0 \leq \theta < 2\pi$. In order words, the image contour is the straight line from i to $i+1$. For integration purposes, this path is equivalent to the Rademacher Ford circle path $P(N)$. (Note that the pre-image, in the x-plane, of $P(N)$ is a curve that loops over to each root of unity — the roots of unity are the pre-images of the rationals.) Thus

$$p(n) = \int_i^{i+1} \frac{F(e^{2\pi i\tau})}{e^{2\pi in\tau}}\, d\tau = \int_{P(N)} \frac{F(e^{2\pi i\tau})}{e^{2\pi in\tau}}\, d\tau$$

Where $\gamma(h,k)$ denotes the upper arc of the circle $C(h,k)$, this equals

$$\sum_{k=1}^{N} \sum_{0\leq h<k,\ (h,k)=1} \int_{\gamma(h,k)} \frac{F(e^{2\pi i\tau})}{e^{2\pi in\tau}}\, d\tau$$

For each of the integrals in this double sum, we now make the substitution discussed in Theorem 7.11.3 above:

$$z = -ik^2(\tau - h/k)$$

so that $\tau = h/k + iz/k^2$. Using the notation of Theorem 7.11.3, $p(n) =$

$$\sum_{k=1}^{N} \sum_{0\leq h<k,\ (h,k)=1} \int_{C'} F\left(\exp\left(\frac{2\pi ih}{k} - \frac{2\pi z}{k^2}\right)\right) \frac{i}{k^2} e^{-2\pi inh/k} e^{2n\pi z/k^2}\, dz$$

$$= \sum_{k=1}^{N} \sum_{0\leq h<k,\ (h,k)=1} \frac{i}{k^2}\, e^{-2\pi inh/k} \int_{z_1}^{z_2} e^{2n\pi z/k^2} F\left(\exp\left(\frac{2\pi ih}{k} - \frac{2\pi z}{k^2}\right)\right)\, dz$$

where C' is the arc on the circle with centre $1/2$ and radius $1/2$ going from $z_1(h,k)$ to $z_2(h,k)$.

We now use the Dedekind Functional Equation in the form of Theorem 7.14.6. Since $F(x') = 1 + (F(x') - 1)$ (where x' is as defined in Theorem 7.14.6), we obtain

$$p(n) = \sum_{k=1}^{\infty} \sum_{0\leq h<k,\ (h,k)=1} ik^{-5/2} e^{\pi is(h,k)} e^{-2\pi inh/k} (I_1(h,k) + I_2(h,k))$$

where

$$I_1(h,k) = \int_{z_1}^{z_2} \sqrt{z} e^{\pi/12z - \pi z/12k^2} e^{2n\pi z/k^2}\, dz$$

and

$$I_2(h,k) = \int_{z_1}^{z_2} \sqrt{z} e^{\pi/12z - \pi z/12k^2} e^{2n\pi z/k^2} \left(F(e^{2\pi i H/k - 2\pi/z}) - 1\right) \, dz$$

We next put a bound on I_2. First note that the disk bounded by the circle with centre 1/2 and radius 1/2 is mapped onto the half-plane $\mathrm{Re}(w) \geq 1$ by $w = 1/z$. If z is on the circumference of this circle, then $\mathrm{Re}(1/z) = 1$. Let $a = \mathrm{Re}(z)$ and $b = \mathrm{Re}(1/z)$. Using Theorem 7.10.6, we estimate the integrand of I_2 on the chord from z_1 to z_2:

$$|\sqrt{z} e^{\pi/12z - \pi z/12k^2} e^{2n\pi z/k^2} \left(F(e^{2\pi i H/k - 2\pi/z}) - 1\right)|$$

$$= \sqrt{|z|} \exp\left(\frac{\pi b}{12} - \frac{\pi a}{12k^2}\right) e^{2n\pi a/k^2} \left| \sum_{m=1}^{\infty} p(m) e^{2\pi i H m/k} e^{-2\pi m/z} \right|$$

$$\leq \sqrt{|z|} \exp\left(\frac{\pi b}{12}\right) e^{2n\pi/k^2} \sum_{m=1}^{\infty} p(m) e^{-2\pi m b}$$

since $a = \mathrm{Re}(z) \leq 1$. Thus the integrand of I_2 is

$$\leq \sqrt{|z|} e^{2n\pi} \sum_{m=1}^{\infty} p(m) e^{-2\pi(m-1/24)b}$$

$$\leq \sqrt{|z|} e^{2n\pi} e^{-2\pi(24m-1)/24}$$

since $b = \mathrm{Re}(1/z) \geq 1$. Thus the integrand is

$$\leq \sqrt{|z|} e^{2n\pi} \sum_{m=1}^{\infty} p(24m-1)(e^{-\pi/12})^{24m-1}$$

$$\leq \sqrt{|z|} e^{2n\pi} F(e^{-\pi/12}) \leq 110\sqrt{|z|} e^{2n\pi}$$

Since, in I_2, z is on the chord, $|z| < \sqrt{2}k/N$ (Theorem 7.11.4). The length of the path is less than $2\sqrt{2}k/N$ (Theorem 7.11.4), so

$$|I_2(h,k)| \leq 110(2)^{1/4}\sqrt{k/N} e^{2n/\pi} 2\sqrt{2}k/N \leq 370(k/N)^{3/2} e^{2n\pi}$$

Hence

$$\left| \sum_{k=1}^{N} \sum_{0 \leq h < k, \ (h,k)=1} i k^{5/2} e^{\pi i s(h,k)} e^{-2\pi i n h/k} I_2(h,k) \right|$$

$$\leq \sum_{k=1}^{N} \sum_{0\leq h<k,\ (h,k)=1} 370k^{-1}N^{-3/2}e^{2n\pi}$$

$$\leq 370N^{-3/2}e^{2n\pi}\sum_{k=1}^{N}\frac{\phi(k)}{k}$$

$$\leq 370N^{-3/2}e^{2n\pi}N = 370e^{2n\pi}/\sqrt{N}$$

In order to deal with I_1, we express it as follows (where $K(-)$ is the improper integral path from 0 once around the circle $(x-1/2)^2+y^2 = 1/4$ clockwise):

$$I_1(h,k) = \int_{K(-)} \sqrt{z}e^{\pi/12z-\pi z/12k^2}e^{2n\pi z/k^2}\,dz$$

$$-\int_0^{z_1} \sqrt{z}e^{\pi/12z-\pi z/12k^2}e^{2n\pi z/k^2}\,dz - \int_{z_2}^{0} \sqrt{z}e^{\pi/12z-\pi z/12k^2}e^{2n\pi z/k^2}\,dz$$

We call the last two integrals J_1 and J_2 respectively.

To estimate J_2, using Theorem 7.11.4, its pathlength is less than

$$\pi|z_2| < \pi\sqrt{2}k/N$$

Since, on the circumference $(x-1/2)^2+y^2 = 1/4$ we have $b = \mathrm{Re}(1/z) = 1$ and $0 < a = \mathrm{Re}(z) \leq 1$, the absolute value of the integrand of J_2 is bounded by

$$\sqrt{|z|}\exp\left(\frac{\pi b}{12} - \frac{\pi a}{12k^2}\right)e^{2n\pi a/k^2}$$

$$\leq 2^{1/4}\sqrt{k/N}e^{\pi/12}e^{2n\pi}$$

and hence $|J_2|$ is bounded by

$$2^{1/4}\sqrt{k/N}e^{\pi/12}e^{2n\pi}\pi\sqrt{2}k/N$$

$$\leq 7e^{2n\pi}(k/N)^{3/2}$$

and, similarly, $|J_1| \leq 7e^{2n\pi}(k/N)^{3/2}$.

From the above it follows that

$$\left|\sum_{k=1}^{N} \sum_{0\leq h<k,\ (h,k)=1} ik^{-5/2}e^{\pi is(h,k)}J_2\right|$$

$$\leq 7e^{2n\pi} N^{-3/2} \sum_{k=1}^{N} \sum_{0 \leq h < k, (h,k)=1} 1/k \leq 7e^{2n\pi} N^{-1/2}$$

Thus $p(n) =$

$$\sum_{k=1}^{N} \sum_{0 \leq h < k,\ (h,k)=1} ik^{-5/2} e^{\pi i s(h,k)} e^{-2\pi i n h/k} \int_{K(-)} \sqrt{z} e^{\pi/12z - \pi z/12k^2} e^{2n\pi z/k^2}\, dz$$

$$+S(N)$$

where $S(N)$ tends to 0 as $N \to \infty$. Letting N go to infinity, we obtain

$$p(n) = i \sum_{k=1}^{\infty} A_k(n)\ k^{-5/2} \int_{K(-)} \sqrt{z} e^{\pi/12z - \pi z/12k^2} e^{2n\pi z/k^2}\, dz$$

where

$$A_k(n) = \sum_{0 \leq h < k,\ (h,k)=1} e^{\pi i s(h,k) - 2\pi i n h/k}$$

taking care of the h's. Note that $A_1(n) = 1$.

We change the variable for the integral: $w = 1/z$. This changes the path to the straight line from $1 - \infty i$ to $1 + \infty i$. We obtain

$$p(n) = -i \sum_{k=1}^{\infty} A_k(n) k^{-5/2} \int_{1-\infty i}^{1+\infty i} w^{-5/2} \exp\left(\frac{\pi w}{12} + \frac{2\pi(n - 1/24)}{wk^2}\right) dw$$

Now substituting $t = \pi w/12$, we get

$$p(n) = -i(\pi/12)^{3/2} \sum_{k=1}^{\infty} A_k(n) k^{-5/2} \int_{\pi/12-\infty i}^{\pi/12+\infty i} t^{-5/2} \exp(t + z^2/4t)\, dt$$

if $z^2 = 4\pi^2(n - 1/24)/6k^2$.

By Theorem 7.15.11,

$$\int_{c-\infty i}^{c+\infty i} \frac{e^t e^{z^2/4t}}{t^{5/2}}\, dt = \frac{8\sqrt{\pi} i}{z} \frac{d}{dz} \frac{\sinh z}{z}$$

Hence

$$p(n) = -i(\pi/12)^{3/2} \sum_{k=1}^{\infty} A_k(n) k^{-5/2} \frac{8\sqrt{\pi} i}{z} \frac{d}{dz} \frac{\sinh z}{z}$$

with z as above. By the Chain Rule,

$$\frac{d}{dn}\frac{\sinh z(n)}{z(n)} = \frac{d}{dz}\sinh z \times \frac{dz}{dn}$$

and Rademacher's formula follows.

7.17 Numerical Calculations

In order to use the above formula to find $p(n)$ for a given n, we find a simpler way of writing $A_k(n)$, we take the derivative in the formula, and we establish an error bound for using N terms of the series.

Theorem 7.17.1

$$A_k(n) = \sum_{0\leq h\leq[k/2],\ (h,k)=1} 2\cos(\pi(s(h,k) - 2nh/k))$$

Proof: Use the fact that $s(k-h,k) = -s(h,k)$.

Theorem 7.17.2 *If $\alpha = n - 1/24$, the derivative in Rademacher's formula equals*

$$\frac{\frac{\pi}{\sqrt{6}\,k}\cosh\left(\frac{\pi}{k}\sqrt{\frac{2\alpha}{3}}\right) - \frac{1}{2\sqrt{\alpha}}\sinh\left(\frac{\pi}{k}\sqrt{\frac{2\alpha}{3}}\right)}{\alpha}$$

Theorem 7.17.3

$$\left|\frac{1}{\pi\sqrt{2}}\sum_{k=N+1}^{\infty} A_k(n)\sqrt{k}\,\frac{d}{dn}\left(\frac{\sinh\left((\pi/k)\sqrt{(2/3)(n-1/24)}\right)}{\sqrt{n-1/24}}\right)\right|$$

$$\leq \frac{44\pi^2}{225\sqrt{3N}} + \frac{\sqrt{2}\pi}{75\sqrt{n-1/24}}\sqrt{N}\sinh\left(\frac{\pi\sqrt{\frac{2}{3}(n-\frac{1}{24})}}{N}\right)$$

$$< \frac{1.12}{\sqrt{N}} + \frac{0.06\sqrt{N}}{\sqrt{n-1/24}} \sinh(2.57\sqrt{n}/N)$$

$$< \frac{1.12}{\sqrt{N}} + \frac{0.03\sqrt{N}}{\sqrt{n-1}} e^{2.57\sqrt{n}/N}$$

Proof: Because $|A_k(n)| \leq k$, the LHS above is bounded by

$$\frac{1}{\pi\sqrt{2}} \sum_{k=N+1}^{\infty} k^{3/2} \frac{d}{dn} \sum_{\nu=0}^{\infty} \frac{1}{(2\nu+1)!} \left(\frac{\pi}{k}\sqrt{\frac{2}{3}}\right)^{2\nu+1} (n-1/24)^{\nu}$$

$$= \frac{1}{\pi\sqrt{2}} \sum_{k=N+1}^{\infty} k^{3/2} \sum_{\nu=1}^{\infty} \frac{\nu}{(2\nu+1)!} \left(\frac{\pi}{k}\sqrt{\frac{2}{3}}\right)^{2\nu+1} (n-1/24)^{\nu-1}$$

Since the double sequence converges absolutely (all the numbers are positive anyway), we can switch the order of the summation signs. Since

$$\sum_{N=1}^{\infty} k^{-2\nu+1/2} < \int_{N}^{\infty} k^{-2\nu+1/2}\, dk = \frac{1}{(2\nu-3/2)N^{2\nu-3/2}}$$

it follows that the previous expression is bounded by

$$\frac{\pi\sqrt{2}}{3\sqrt{n-1/24}} \sum_{\nu=1}^{\infty} \frac{\nu}{(2\nu-3/2)(2\nu+1)!} \left(\pi\sqrt{(2/3)(n-1/24)}\right)^{2\nu-1} \frac{1}{N^{2\nu-3/2}}$$

$$= \frac{\pi\sqrt{2N}}{3\sqrt{n-1/24}} \left(\frac{\pi\sqrt{2(n-1/24)}}{3\sqrt{3}N} + \sum_{\nu=2}^{\infty} \frac{(\pi\sqrt{2/3}/N)^{2\nu-1}}{(2\nu-1)!(2\nu+1)(4\nu-3)}\right)$$

$$< \frac{\pi\sqrt{2N}}{3\sqrt{n-1/24}} \left(\frac{\pi\sqrt{2(n-1/24)}}{3\sqrt{3}N} + \frac{1}{25}\sum_{\nu=2}^{\infty} \frac{(\pi\sqrt{2/3}/N)^{2\nu-1}}{(2\nu-1)!}\right)$$

$$= \frac{\pi\sqrt{2N}}{3\sqrt{n-1/24}} \times$$

$$\left((1/3-1/25)\frac{\pi\sqrt{(2/3)(n-1/24)}}{N}+\frac{1}{25}\sinh\left(\frac{\pi\sqrt{(2/3)(n-1/24)}}{N}\right)\right)$$

$$=\frac{44\pi^2}{225\sqrt{3N}}+\frac{\sqrt{2}\pi}{75\sqrt{n-1/24}}\sqrt{N}\sinh\left(\frac{\pi\sqrt{\frac{2}{3}(n-\frac{1}{24})}}{N}\right)$$

Applying Theorem 7.17.3 to $n = 1000000$ and $N = 400$ we see that the error is at most 0.43. So with 400 terms of the series, we can calculate the exact value of $p(1000000)$.

Exercises 7.17

1. Using Rademacher's formula, calculate $p(5)$.
2. Calculate $p(100)$.

Bibliography

[1] Barlow, P. *An Elementary Investigation of the Theory of Numbers.* London: J. Johnson, 1811.

[2] Beiler, A. H. *Recreations in the Theory of Numbers.* New York: Dover, 1964.

[3] Chrystal, G. *Textbook of Algebra.* Vol II. New York: Dover, 1961.

[4] Davenport, H. *The Higher Arithmetic.* London: Hutchinson, 1968.

[5] Gauss, C. F. *Disquisitiones Arithmeticae.* Trans. Arthur A. Clarke. New Haven: Yale University Press, 1966.

[6] Grosswald, E. *Topics from the Theory of Numbers.* 2nd edn. Boston: Birkhäuser, 1984.

[7] Hardy, G. H., and E. M. Wright. *An Introduction to the Theory of Numbers.* 4th edn. New York: Oxford University Press, 1960.

[8] Niven, I., H. S. Zuckermann, and H. L. Mongomery. *An Introduction to the Theory of Numbers.* 5th edn. New York: John Wiley & Sons, 1991.

[9] Perron, O. *Die Lehre Von Den Kettenbrüchen.* New York: Chelsea Publications, 1929.

[10] Stark, H. M. *An Introduction to Number Theory.* Chicago: Markham Publishing Co., 1970.

Appendix A

Appendix: Answers to Selected Exercises

Answers for Exercises 1.1

14. Suppose there is an integer m (such as 1000) such that there is no largest member in the set of proper fractions which can be written as a sum of m or fewer distinct unit fractions. Indeed, let m be the least such. Then $m > 2$. Let L be the largest proper fraction which can be written as a sum of $m-1$ or fewer distinct unit fractions. Then L cannot be written as a sum of fewer than $m-1$ unit fractions. Let q be a positive integer such that $L + 1/q < 1$. From the definition of m, it follows that there is an infinite sequence of positive rationals $e_1, e_2, \ldots$ such that (1)

$$L + 1/q < L + e_1 < L + e_2 < \ldots < 1$$

and (2) for $j = 1, 2, \ldots,\ \ L + e_j$ can be written as a sum of m or fewer distinct unit fractions. However, if

$$L + e_j = 1/x_1 + \cdots + 1/x_k$$

with $k \leq m$ and $x_1 < \ldots < x_k$ then $L \geq 1/x_1 + \ldots + 1/x_{k-1}$ (from the definition of L), and hence $e_j \leq 1/x_k$, and so $x_k < q$. Hence there are only finitely many possibilities for the x's, whereas there are infinitely many numbers $L + e_j$. Contradiction. A further question is which is the smallest proper fraction which requires more than 999 unit fractions for its 'Egyptian fraction expression'.

We can also answer question 14 as follows. Let $P(n)$ be the statement 'there is a largest proper fraction which can be written as a sum of n distinct unit fractions'. Then $P(1)$ and $P(2)$ are true. Suppose $P(n)$ is true, and let L be the largest proper fraction which can be written as a sum of n distinct unit fractions. Let

$$e = \frac{1}{\left[\frac{1}{1-L} + 1\right]}$$

Then $f = L + e$ is a proper fraction which is a sum of $n + 1$ distinct unit fractions. Let

$$g = 1/x_1 + 1/x_2 + \cdots + 1/x_{n+1}$$

be any proper fraction with $x_1 < x_2 < \ldots < x_{n+1}$. If $g > f$ then $L + 1/x_{n+1} > L + e$, so that $1/e > x_{n+1}$. Hence only finitely many proper fractions which can be written as a sum of $n + 1$ distinct unit fractions are greater than f. Hence there is a largest proper fraction which can be written as a sum of $n + 1$ distinct unit fractions. That is, $P(n)$ implies $P(n + 1)$. Hence, by MI, for every n, there is a largest proper fraction that can be written as a sum of n distinct unit fractions. Thus, for any n, there are proper fractions which cannot be written as a sum of n (or fewer) distinct unit fractions.

Answers for Exercises 1.2

1. $B_8 = -1/30$, $B_{10} = 5/66$, $B_{12} = -691/2730$, $B_{14} = 7/6$.

Answers for Exercises 1.3

16. For all $n > 1$, $f(f(n) + n) = Qf(n) + f(n)$ and $Q = 0$, and hence $g(n) = f(f(n)+n) - f(n)$ has infinitely many zeros, which is impossible.

Answers for Exercises 1.4

1. 45 360 is the smallest natural number with exactly 100 divisors.
5. $\frac{1}{2}t(n)$ is the number of factorisations of n. Each factorisation $n = ab$ can be written

$$n = (a, b)^2 \frac{a}{(a, b)} \frac{b}{(a, b)}$$

and the last two factors are relatively prime.

Answers for Exercises 1.5

6. Let e be the largest exponent such that p^e divides one of 1, 2, ..., n. Then $p^e \leq n < p^{e+1}$, and hence

$$e \log p \leq \log n < (e + 1) \log p$$

and $e = [(\log n)/\log p]$.

Answers for Exercises 1.6

6. There are 8 solutions.
10. (24 012, 66 005, 70 237).

Answers for Exercises 1.7

1. There is no solution.
3. Using the Conic Transformation Theorem, we obtain

$$(2x - 2y - 1)(2x + 2y + 9) = 99$$

There are exactly 12 solutions.
4. There are exactly 4 solutions: $x = -7, -3, 5, 9$.
5. $x^4 + 6x^3 + 11x^2 + 6x + 1 = (x^2 + 3x + 1)^2$, so there are infinitely many solutions.

6. Here x is odd and $(2y)^2 + w^2 = z^2$, with $(2y, w) = 1$. By the Pythagorean Triangle Theorem,

$$2x^2 = z^2 + w^2 = (a^2 + b^2)^2 + (a^2 - b^2)^2 = 2a^4 + 2b^4$$

However, as Fermat showed, this is impossible for nonzero integers. Hence $y = 0$ gives the only solution.
7. This implies $(\frac{1}{2}(x^4 + 1))^2 = x^4 + y^4$. Hence $y = 0$.
8. Here $x = 1$ gives the only natural number solution to the equation.
Proof: Since $(2x^2 - 1, 2x^2 + 1) = 1$ and $(2x^2 - 1)(2x^2 + 1) = 3y^2$, either

$$\text{(i) } 2x^2 - 1 = 3z^2 \text{ and } 2x^2 + 1 = w^2$$

or

$$\text{(ii) } 2x^2 - 1 = z^2 \text{ and } 2x^2 + 1 = 3w^2.$$

Now $2x^2 - 1$ has the form $8a - 1$ or $8a + 1$, whereas $3z^2$ has the form $8a$, $8a + 3$, or $8a + 4$. Hence only (ii) is possible.

Here z is odd, and $(2x - z, 2x + z) = 1$. Since $(2x - z)(2x + z) = 3w^2$, one of these factors is a square, and the other is three times a square. In either case, $4x = 3u^2 + v^2$ and $2z = |3u^2 - v^2|$. Since $2x^2 - 1 = z^2$, we obtain

$$8(v^4 - 1) = 9(u^2 - v^2)^2.$$

Hence $v^4 - 1 = 2t^2$, and $t = 2t'$. This gives

$$t'^2 = \frac{1}{2}(v - 1)\frac{1}{2}(v + 1)\frac{1}{2}(v^2 + 1).$$

The three factors $\frac{1}{2}(v-1)$, $\frac{1}{2}(v+1)$, and $\frac{1}{2}(v^2+1)$ are pairwise relatively prime, and hence all squares. We have $v - 1 = 2m^2$, $v + 1 = 2n^2$ and hence

$$2m^2 + 1 = 2n^2 - 1,$$

so that $n^2 - m^2 = 1$ and hence $m = 0$. This means that $v^2 = 1$ and $u^2 = 1$. Hence $x = 1$ and $y = 1$ is the only solution in natural numbers.

10. T. Skolem used his 'p-adic' method to show that the only solutions are given by $x = 1$ and $x = 9$. However, the following much easier proof establishes the same thing.

To obtain a contradiction, suppose there is a solution in natural numbers with $y > 2$. Let x, y be the least such solution.

If x is even, then y is odd, and $5y^4+1$ has the form $4z+2$. However, if x is even, x^4 has the form $4z$. Hence x is odd, and $y = 2y'$ for some integer y' which is > 1.

Since x is odd, $x^2 - 1$ is divisible by 8. Since

$$\frac{1}{2}(x^2+1)\frac{1}{8}(x^2-1) = 5y'^2,$$

one of the natural numbers $\frac{1}{2}(x^2+1)$ and $\frac{1}{8}(x^2-1)$ is a fourth power. If $(x^2-1)/8 = w^4$ then $8w^4+1 = x^2$ and, by Theorem 1.7.4, $x = 1$ or 3. But then $y = 0$ or 2 — against the supposition that $y > 2$. Thus the other number is the fourth power.

We now have $\frac{1}{2}(x^2+1) = w^4$ and $\frac{1}{8}(x^2-1) = 5t^4$, and hence $1 = w^4 - 20t^4$, and so

$$5t^4 = \frac{1}{2}(w^2-1)\,\frac{1}{2}(w^2+1)$$

— where w is odd, since $w^4 = 20t^4+1$. From this it follows that one of the integers $\frac{1}{2}(w^2-1)$ and $\frac{1}{2}(w^2+1)$ is a fourth power. If $\frac{1}{2}(w^2-1) = a^4$ then $2a^4+1 = w^2$, and, by Theorem 1.7.2, $w^2 = 1$, and hence $x = 1$. But then $y = 0$ — against the supposition that $y > 2$.

Hence $\frac{1}{2}(w^2+1) = a^4$ and $\frac{1}{2}(w^2-1) = 5b^2$. From this we obtain $a^4 - 5b^4 = 1$.

Now $b^2 < w$ and $w^2 < x$ so $b^4 < x$ and

$$b^{16} < 5y^4+1 < 6y^4 < y^{16}$$

and hence $b < y$. From y's minimality, subject to the condition $y > 2$, we have $b \leq 2$. If $b = 0$ then $w^2 = 1$ and $x = 1$, $y = 0$ — against the condition that $y > 2$. If $b = 1$ then $5b^4+1$ is not a fourth power. If $b = 2$ then $w^2 - 1 = 160$, which is impossible. So $b > 2$. Contradiction.

11. Suppose that x, y, and z are positive, pairwise relatively prime integers such that $x^2 + y^4 = 2z^4$. Suppose $z > 1$. Then there are positive integers w, a, and s such that $s < z$ and $w^2 + a^4 = 2s^4$ and, for one of

$$t = \frac{a|as \pm w|}{2s^2 + a^2},$$

$z = s^2 + t^2$ and $y = |a^2 - 2(st/a)^2|$.

Proof: Note that x, y, and z are all odd. Let $m = \frac{1}{2}(y^2 + x)$ and $n = \frac{1}{2}(y^2 - x)$. Then $(m, n) = 1$ and $m^2 + n^2 = z^4$. Hence there are positive integers u and v such that $z^2 = u^2 + v^2$ and

$$y^2 = m + n = u^2 - v^2 \pm 2uv = (u \pm v)^2 - 2v^2$$

— with $(u, v) = 1$. Since $2v^2 = (u \pm v)^2 - y^2$, v is even — say, $v = 2v'$ — and

$$2v'^2 = \frac{1}{2}(u \pm v - y)\,\frac{1}{2}(u \pm v - y)$$

Since $y > |u \pm v|$, it follows that $u \pm v + y$ and $u \pm v - y$ are positive. Thus there are positive integers a and b such that

$$\begin{aligned} v &= 2ab \\ u \pm v &= a^2 + 2b^2 \\ u &= a^2 + 2b^2 \mp 2ab \end{aligned}$$

Since $u^2 + v^2 = z^2$ with $(u, v) = 1$, there are positive integers s and t with $z = s^2 + t^2$, $v = 2st$ and $u = s^2 - t^2$ (v having been shown to be even). Furthermore,

$$s^2 - t^2 = u = a^2 + 2b^2 \mp 2ab$$

so that

$$s^2 - t^2 = a^2 + 2(st/a)^2 \mp 2st$$

Solving this for t, we find that

$$t = \frac{\pm a^2 s \pm \sqrt{2s^4 - a^4}}{2s^2 + a^2}$$

(where there are four possibilities for the signs). Since t is an integer, $2s^4 - a^4$ is a square, say, w^2, where w is a positive integer. QED.

Thus each relatively prime solution can be derived from a smaller solution, descending until we reach the solution with $z = 1$. Conversely, each relatively prime solution has multiples which give rise to one or two greater relatively prime solutions. We can find all the solutions by starting with the one where $z = 1$ and working backwards through the above proof.

Corresponding to the solution with $z = 1$, there is a family of solutions $w = k^2$, $a = k$ and $s = k$. Here, $t = 0$ or $2k/3$. To have t an integer, we can take $k = 3$. Then $z = s^2 + t^2 = 13$, and $y = 1$, $x = 239$.

Corresponding to the family with $z = 13$, there is a family of solutions $w = 239k^2$, $a = k$ and $s = 13k$. Here, $t = 84k/113$ or $2k/3$. With $k = 113$, we $z = 2,165,017$ and $y = 2,372,159$. With $k = 3$, we get $z = 1525$, $y = 1343$ (and $x = 2,750,257$). Each of these two relatively prime solutions gives rise to others.

The 4 smallest relatively prime solutions are with $z = 1$, 13, 1525, and 2,165,017.

This problem is related to one proposed by Fermat in 1643. He asked Mersenne to find a Pythagorean triangle the sum of whose legs, and whose hypotenuse were both squares. If X, Y, and Z are the sides of the triangle, with $X < Y < Z$, then $X + Y = a^2$, $Z = s^2$ and $X^2 + Y^2 = Z^2$. Let $w = Y - X > 0$. Then $a^2 > w$ and

$$w^2 + a^4 = 2X^2 + 2Y^2 = 2Z^2 = 2s^4.$$

The smallest solution with $a^2 > w$ is the one with $s = 2,165,017$ and $a = 2,372,159$. We have

$$X = 1,061,652,293,520$$

$$Y = 4,565,486,027,761$$

Answers for Exercises 1.8

1. $1007 = 31^2 + 6^2 + 3^2 + 1^2$.

Answers for Exercises 1.9

5. Suppose $x^2 + 2^2 = y^3$. If x is odd then Theorem 1.9.6 applies and the only solution is $x = 11$, $y = 5$. If $x = 2x'$ then $y = 2y'$ and $x'^2 + 1 = 2y'^3$. Then $x' = 2m + 1$ and we have

$$m^2 + (m+1)^2 = y'^3$$

and, by Theorem 1.9.6, $m = u^3 - 3uv^2$, $m+1 = 3u^2v - v^3$. Subtracting, $-1 = (u+v)(u^2 - 4uv + v^2)$. Thus $u+v = \pm 1$ and $(u+v)^2 = 1$, which, by subtraction from $u^2 - 4uv + v^2$, gives $6uv = 0$ or 2. Hence u or v is 0, and $y' = u^2 + v^2 = 1$. Thus if x is even, the only solution is $x = 2$ and $y = 2$. The equation has exactly two solutions in natural numbers.

6. There are no solutions.

7. The only solution is with $x = 46$.

8. Suppose $x^3 + y^3 = 2z^3$ but $x \neq \pm y$. We may take it that $|xyz|$ is minimised under this condition. Hence $(x,y) = (x,z) = (y,z) = 1$, and thus x and y are both odd. Let $a = \frac{1}{2}(x+y)$ and $b = \frac{1}{2}(x-y)$ so that $x = a+b$, $y = a-b$ and $(a,b) = 1$. Then the equation gives $a(a^2 + 3b^2) = z^3$.

First suppose that a is not a multiple of 3. Then $(a, a^2 + 3b^2) = 1$ so that $a^2 + 3b^2 = t^3$ and $a = u^3$. By Theorem 1.9.6, we have $t = r^2 + 3s^2$, $a = r^3 - 9rs^2$ and $b = 3r^2s - 3s^3$. Since $(a,b) = 1$, it follows that $(r, 3s) = 1$, and r and s have different parity. Hence $(r+3s,\ r-3s) = 1$. Since

$$u^3 = a = r(r+3s)(r-3s)$$

it follows that $r+3s = k^3$, $r-3s = m^3$ and $r = n^3$ with $k^3 + m^3 = 2n^3$. Also

$$|kmn|^3 = |a| = \frac{1}{2}|x+y| < |xy| \leq |xyz| \leq |xyz|^3$$

Hence $k = \pm m$. If $k = m$ then $s = 0$, so that $b = 0$ and $x = y$. If $k = -m$ then $n = 0$ and $r = 0$, so that $a = 0$ and $x = -y$. Contradiction.

Thus we must suppose that $a = 3a'$, giving $9a'(3a'^2 + b^2) = z^3$. Since $(a,b) = 1$, it follows that $(9a', 3a'^2 + b^2) = 1$ and $3a'^2 + b^2 = t^3$, $9a' = u^3$. Hence, by Theorem 1.9.6, $t = r^2 + 3s^2$, $b = r^3 - 9rs^2$, and $a' = 3r^2s - 3s^3$. Also $(r,s) = 1$ and r and s have different parity. Hence $(r+s,\ r-s) = 1$ and, since $u^3 = 27s(r-s)(r+s)$, we have $r+s = k^3$, $r-s = m^3$, $s = n^3$ with $k^3 + (-m)^3 = 2n^3$. Moreover,

$$|kmn|^3 = \frac{|a'|}{3} = \frac{|a|}{9} = \frac{|x+y|}{18} < |xyz|^3$$

Hence $k = \pm m$, and we get a contradiction.

9. If x is even, $(x-1,\ x+1)=1$, so that $x-1$ and $x+1$ are both cubes. The only cubes whose difference is 2 are 1 and -1. Thus, if x is even, $x=0$.

Suppose now that $x=2m+1$. Then $m(m+1)=2y'^3$ where $y'=\frac{1}{2}y$. If m is even then $m=2a^3$ and $m+1=b^3$, giving $b^3+(-1)^3=2a^3$. By the previous exercise, $b=\pm 1$ and $x=-3$ or 1. If m is odd, $m=a^3$ and $m+1=2b^3$, giving $a^3+1^3=2b^3$. Hence, by the previous exercise, $a=\pm 1$, and $x=-1$ or 3. Thus there are exactly 3 natural number solutions.

10. If $\frac{1}{2}m(m+1)=y'^3$ then, as in the previous answer, $m=-2,\ -1,$ 0, or 1.

Answers for Exercises 1.10

2. The right triangle with sides 17/6, 24, and 145/6 has area 34.
3. The triangles with sides 20, 21, 29, and 12, 35, 37 each have area 210.

Answers for Exercises 2.1

5. $x=(1,x)$ so that $x^2=x+1$.

Answers for Exercises 2.2

3. (3, 7, 15, 1, 292, ...).

Answers for Exercises 2.3

3. If $a_2=1$ then the n-th convergent of $-r$ is $-(a_1,a_2,\ldots,a_{n+1})$.

Answers for Exercises 2.4

3. The Farey series F_{15} has 71 members.

Answers for Exercises 2.5

1. He bought 88 animals at the start.
2. The smallest possible number of maids is 292.

Answers for Exercises 2.6

1. The smallest positive integer solution is $x = 1,766,319,049$ and $y = 226,153,980$.
2. First suppose n is odd, so that $f_n/g_n \leq x$. If p/q is closer to x than f_n/g_n is, then $p/q > f_n/g_n$ and hence

$$\frac{f_{n-1}}{g_{n-1}} - \frac{f_n}{g_n} > \frac{f_{n-1}}{g_{n-1}} - \frac{p}{q}$$

so that, by Plato's Theorem, $q > (f_{n-1}q - pg_{n-1})g_n$. Since, by Theorem 2.6.1, $f_{n-1}/g_{n-1} - p/q > 0$, it follows that $f_{n-1}q - pg_{n-1}$ is a positive integer. Hence

$$q > (f_{n-1}q - pg_{n-1})g_n \geq g_n$$

When n is even, the proof is similar.

Answers for Exercises 2.7

3. Suppose Q_n is even and $2Q_n$ is a factor of $P_n^2 - R$. Then $(P_n^2 - R)/Q_n$ is an even integer, and

$$Q_{n+1} = \frac{R - P_{n+1}^2}{Q_n} = \frac{R - P_n^2}{Q_n} + 2a_nP_n - a_n^2Q_n$$

is even. Since $(R - P_{n+1}^2)/2Q_{n+1} = Q_n/2$, an integer, it follows that $2Q_{n+1}$ is a factor of $P_{n+1}^2 - R$.
4. Suppose all the Q_n's are even. Then $(R - P_2^2)/2Q_1 = Q_2/2$ is an integer. Since $P_2 = a_1Q_1 - P_1$, it follows that $2Q_1$ is a factor of $P_1^2 - R$ (using the fact that Q_1 is even). The converse follows from the previous exercise.
5. $(3, \underline{1}, 1, 1, 1, \underline{6})$.
6. The SCF expansions are $(\underline{6a+2}, 2, 1, 3a, 1, \underline{2})$ and $(\underline{6a+10}, 1, a, \underline{1})$.

Answers for Exercises 2.8

2. (1, 8, $\underline{1}$, 4, 3, 1, 1, 2, 2, 2, 2, 1, 7, 1, 4, 2, 306, 2, 4, 1, 7, 1, 2, 2, 2, 2, 1, 1, 3, 4, 1, 11, 1, 42, 1, $\underline{11}$).
5. If $P_{n+1} = P_n$ then $a_nQ_n - P_n = P_n$ so that $Q_n | 2P_n$. Conversely, if $Q_n | 2P_n$ then, since

$$1 + 2P_n/Q_n > (P_n + \sqrt{R})/Q_n > 2P_n/Q_n$$

(Theorem 2.8.5), it follows that $a_n = P_n/Q_n$. Hence $P_{n+1} = a_nQ_n - P_n = P_n$.
6. Let P be an integer such that $\sqrt{9n^2-2} > P > \sqrt{9n^2-2} - 3$ and 3 is a factor of $P^2 - (9n^2 - 2)$. Then $(P + \sqrt{9n^2-2})/3$ has a purely periodic SCF ending. One of the Q's in this ending is 3.

Furthermore, $3n - 1 + \sqrt{9n^2-2}$ has the following SCF ending.

P	$3n-1$	$3n-1$	$3n-2$	$3n-2$	$3n-1$
Q	1	$6n-3$	2	$6n-3$	1
	$6n-2$	1	$3n-2$	1	$6n-2$

Thus, unless $n = 1$, there are at least two SCF endings for $9n^2 - 2$.

Answers for Exercises 2.9

1. $a = 500,001$ and $b = 53,000$.
4. If $p = 4n+3$ then the equation has no solution. Suppose $p = 4n+1$

and (a, b) is the least positive solution of $x^2 - py^2 = 1$. Then a is odd, and b is even, and

$$\frac{1}{2}(a-1)\,\frac{1}{2}(a+1) = p(\frac{1}{2}b)^2$$

If $\frac{1}{2}(a-1) = pr^2$ and $\frac{1}{2}(a+1) = s^2$ then $s^2 - pr^2 = 1$, against the fact that (a, b) is the least positive solution of $x^2 - py^2 = 1$. Hence $\frac{1}{2}(a-1) = r^2$, and $\frac{1}{2}(a+1) = ps^2$, so that $r^2 - ps^2 = -1$.

(Similarly, $x^2 - py^2 = 2$ has a solution iff the prime p has the form $8n + 7$.)

5. By Theorem 2.9.1, $\pm Q_{n+1} = f_n^2 - Rg_n^2$. See the corollary to Theorem 2.8.3.

6. If the sides are $m-1$, m, and $m+1$, the semiperimeter is $3m/2$ and, by Heron's formula (actually known to Archimedes), the square of the area of the triangle is

$$n^2 = \left(\frac{3m}{2}\right)\left(\frac{m}{2}+1\right)\left(\frac{m}{2}-1\right)\left(\frac{m}{2}\right)$$

The right side of this equation is an integer only if m is even. Let $m = 2x$. Then $n^2 = 3x^2(x^2-1)$ so that $3x$ is a factor of n. Let $y = n/3x$. Then $x^2 - 3y^2 = 1$. Solving this equation, we find that the smallest sides of the 4 smallest triangles with consecutive integer sides are 3, 13, 51, and 193.

7. $x^2 + (x+1)^2 = y^2$ iff $(2y)^2 - 2(2x+1)^2 = 2$. Since no square has the form $8n+2$, there is no solution of $X^2 - 2Y^2 = 2$ with Y even. Thus to solve our problem, it suffices to solve $X^2 - 2Y^2 = 2$ and then set $x = \frac{1}{2}(Y-1)$, and $y = \frac{1}{2}X$. The triangles are (3, 4, 5), (20, 21, 29), (119, 120, 169), and (696, 697, 985).

8. $\frac{1}{2}x(x+1) = 23y^2$ iff $(2x+1)^2 - 46(2y)^2 = 1$. From the least nontrivial solution, we get 74,024,028 gold coins.

Answers to Exercises 2.10

2. In the SCF expansion of $\sqrt{1621}$, $P_{40} = 29$, $P_{41} = 10$, $Q_{40} = Q_{41} = 39$, and $a_{40} = a_{41} = 1$. Thus $g_{38} = G - G'$. By Theorem 2.9.2, $f_{40} =$

$10G + 39G'$ and $f_{39} = 39G - 10G'$. By Theorem 2.10.4, $s = 79$, and, by Theorem 2.1.16, $f_{79} = f_{40}g_{40} + f_{39}g_{39}$ and $g_{79} = g_{40}^2 + g_{39}^2$. Thus $A = f_{79}$ and $B = g_{79}$. By Theorem 2.9.5, $f_{2s} = A^2 + 1621B^2$, and the result follows by Theorems 2.9.2 and 2.9.4.

Answers for Exercises 3.1

1. $a_1 10^m + \cdots + a_m 10 + a_{m+1} \equiv a_1 + \cdots + a_m + a_{m+1} \pmod 9$.
2. Every even perfect number has the form $2^{n-1}(2^n - 1)$ where $2^n - 1$ is prime. If $2^n - 1$ is prime, and $n > 2$ then n is odd and $2^n \equiv 2 \pmod{10}$ or $2^n \equiv 8 \pmod{10}$. In the first case, $2^{n-1} \equiv 6 \pmod{10}$ and $2^n - 1 \equiv 1 \pmod{10}$. In the second case, $2^{n-1} \equiv 4 \pmod{10}$ and $2^n - 1 \equiv 7 \pmod{10}$.

Answers for Exercises 3.2

1. Consider $1/n, 2/n, \ldots, n/n$. Let d be a divisor of n. Among the n fractions are fractions equal to $1/d, 2/d, \ldots, d/d$. And $\phi(d)$ of these are in lowest terms. Thus $\phi(d)$ of the original fractions reduce to a fraction with denominator d.
2. If p is a prime factor of x, and $\phi(x) \leq N$ then $p - 1 \leq N$ (by Theorem 3.2.2). Thus the largest prime factor of x is less than or equal to $N + 1$. Hence $\phi(x) \leq N$ implies that

$$x \prod_{p|x} (1 - \frac{1}{p}) \leq N \quad \text{so that} \quad x \leq \frac{N}{\prod_{p \leq N+1} (1 - \frac{1}{p})}$$

3. $x = 2310$ is the largest of 37 solutions.
4. $(x - a)/10 + 10^m a = 4x$ leads to 102564 as the solution.
5. If $(n - 1)! \equiv -1 \pmod n$ then n has no nontrivial factors to divide into the factorial and make it congruent to 0.

Suppose $n = p$, a prime. Of the residues $1, 2, \ldots, p-1$, only ± 1 are their own inverses. All the other residues pair off, each with its unique

inverse. Hence $(p-1)! \equiv -1 \pmod p$.
6. The inverse of 19 mod $\phi(9991) = 96 \times 102$ is 4123. That is, $19 \times 4123 = \phi(9991)Q + 1$. Thus

$$(x^{19})^{4123} \equiv 4282^{4123} \pmod{9991}$$

implies that $x^{\phi(9991)Q+1} \equiv 7204 \pmod{9991}$, or $x \equiv 7204 \pmod{9991}$ (by Theorem 3.2.3).

Note that, to solve this problem, we need to know $\phi(9991)$ and hence the prime factorisation of 9991. If I knew this factorisation, but you were unable to find it, then I would easily obtain the 7204, but you would not.

Answers for Exercises 3.3

1. a^i generates the cyclic group a, a^2, ..., $a^m = 1$ iff $(i, m) = 1$.
2. 7 is the smallest primitive root of 71.
3. Let n be an even positive integer. If a is a positive integer $< n$ then $(a, n) = 1$ iff $(n-a, n) = 1$. Thus the sum S of the positive integers $< n$ and relatively prime to n is congruent to 0 mod n. Let g be a primitive root of p. Then the product of the primitive roots is g^S (mod p), if we take $n = p-1$. But S is a multiple of $n = p-1$. Hence $q \equiv g^S \equiv 1 \pmod p$.
4. This is true when $n = 1$ or 2. Suppose $n \geq 3$. By MI it follows that $5^{2^{n-3}}$ has the form $2^{n-1}h + 1$ where h is odd (*). Let d be the order of 5 in the multiplicative group of odd integers mod 2^n. Then d is a factor of the order of that group, namely, 2^{n-1}. From (*) it follows that $d = 2^{n-2}$. If $-5^j \equiv 5^k \pmod{2^n}$ then $-1 \equiv 1 \pmod 4$. Thus the negatives of the 2^{n-2} powers of 5 are distinct from those powers. Hence these $2 \times 2^{n-2}$ numbers together are the odd integers mod 2^n.

Answers for Exercises 3.4

2. If prime p has primitive root 10, then $10^{(p-1)/2} \not\equiv 1 \pmod p$. Since $10^{p-1} \equiv 1 \pmod p$, it follows that $10^{(p-1)/2} \equiv -1 \pmod p$. Thus

$10^{(p-1)/2}(1/p) + 1/p$ is an integer. Hence the fractional parts of the summands add up to $1 = .\overline{9}$.

Answers for Exercises 3.5

2. 52 is the answer to this Chinese Remainder Problem.
3. The pyramid has 201 steps.
4. The congruence has 16 solutions.

Answers for Exercises 3.7

2. If n is divisible by 4 or by a prime of the form $4m + 3$ then the answer is 0 — for $x^2 \equiv -1 \pmod{n}$ would have a solution. Otherwise, n can be written as a sum of two relatively prime squares in exactly 2^{k-1} ways — where k is the number of distinct odd primes dividing n.
3. $5 \times 13 \times 17 \times 29$ is the hypotenuse of exactly 8 primitive Pythagorean triangles.
5. His fee rose from \$49 to \$169 to \$289.
6. 160,225 soldiers.

Answers for Exercises 3.8

2. If p is a prime of the form $8m + 1$ then $x^2 \equiv -2 \pmod{p}$ has a solution, and hence, by Theorem 1.9.3, $p = a^2 + 2b^2$. Suppose we also have $p = c^2 + 2d^2$. Since

$$(bc - ad)(bc + ad) = (b^2 - d^2)p$$

it follows that p factors one of $bc \pm ad$. Since

$$(ac \mp 2bd)^2 + 2(bc \pm ad)^2 = p^2$$

it then follows that p factors one of $ac \mp 2bd$. We thus have

$$\left(\frac{ac \mp 2bd}{p}\right)^2 + 2\left(\frac{bc \pm ad}{p}\right)^2 = 1$$

with both summands nonnegative integers. Hence $bc \pm ad = 0$. Since $(a,b) = (c,d) = 1$, it follows that $a|c$ and $c|a$. Thus $a = \pm c$.

Answers for Exercises 3.11

1. There are 1994 answers: $x = 997 \times 991y$, where $y = 2 + 7m$ and $0 \leq y < 6979$.
2. ± 2189, ± 5668, ± 16823, ± 24680 (mod 128,331).
3. ± 2845, $\pm 2,094,307$ (mod 2^{22}).
4. $\pm 123,456,788$ (mod 3^{18}).
5. ± 2569 (mod 1,000,039).
6. ± 59999 (mod 1,000,033).
7. ± 128133, ± 201167, ± 298835, ± 371869 (mod 1,000,004).

Answers for Exercises 3.12

1. 12 pieces each.
2.

$$x = 1 + 14K, \quad y = -1 - 22K - 84K^2$$

$$x = 2 + 14K, \quad y = -3 - 34K - 84K^2$$

$$x = 8 + 14K, \quad y = -33 - 106K - 84K^2$$

$$x = 9 + 14K, \quad y = -41 - 118K - 84K^2$$

3.

$$x = 1 - 16K + 20K^2, \quad y = 2 - 46K + 60K^2$$

$$x = 14 - 36K + 20K^2, \quad y = 40 - 106K + 60K^2$$

4. Chrystal's equation has 8 disjoint families of solutions, as follows.

$$\begin{array}{llll}
x = & 2 + 4K - 96K^2 & y = & 3 - 6K - 144K^2 \\
x = & 1 - 20K - 96K^2 & y = & -42K - 144K^2 \\
x = & -28K - 96K^2 & y = & -2 - 54K - 144K^2 \\
x = & -3 + 44K - 96K^2 & y = & -2 + 54K - 144K^2 \\
x = & -5 - 52K - 96K^2 & y = & -11 - 90K - 144K^2 \\
x = & -10 + 68K - 96K^2 & y = & -11 + 90K - 144K^2 \\
x = & -13 + 76K - 98K^2 & y = & -15 + 102K - 144K^2 \\
x = & -24 + 100K - 96K^2 & y = & -30 + 138K - 144K^2
\end{array}$$

Answers for Exercises 4.1

2. $x = 77876$ gives one answer.
3. $r + 49 = x^2$ and $2r = y^2$ only if $x^2 - 2(y/2)^2 = 49$. The only answer in the right range is 392.
4. If p is a prime of the form $8m \pm 1$ then $z^2 \equiv 2 \pmod p$ has a solution. Since 2 is single hearted, and since the SCF ending has period of length 1, it follows by Siegfried's Sword that $x^2 - 2y^2 = p$ has a solution.

Answers for Exercises 4.2

4. $b_{kn} = a_n b_{(k-1)n} + a_{(k-1)n} b_n$, so, by MI, $b_n | b_{kn}$. Let $m = qn + r$ with $0 \leq r < n$. If $b_n | b_m$ then, since

$$b_m = a_{qn} b_r + a_r b_{qn}$$

the above implies that $b_n | b_r$. But $b_r < b_n$, so $b_r = 0$ and $r = 0$. Thus $n|m$.
5. $x(x+1)/2 = y^2$ iff $(2x+1)^2 - 2(2y)^2 = 1$. The n-th square triangular number is the integer nearest $\frac{(1+\sqrt{2})^{4n}}{32}$. The first 5 such numbers are 0, 1, 36, 1225, and 41616.
6. If $\frac{1}{2}(n-1)n\frac{1}{2}n(n+1)\frac{1}{2}(n+1)(n+2) = y^2$ and $x = 2n+1$ then the integer

$$\frac{x^2 - 9}{8} = \frac{(8y)^2}{(x-1)^2(x+1)^2} = t^2$$

Hence x is any positive integer such that, for some t, $x^2 - 8t^2 = 9$ and $y = (t^2 + 1)t$. For example, we can have $n = 25$, and the three consecutive triangular numbers are 300, 325, and 351.

Answers for Exercises 4.3

1. There are 17 such solutions.

Answers for Exercises 5.1

6. Suppose the larger circle has centre A and radius R, the smaller with centre B and radius r. With centre A and radius $R - r$, draw a circle. With AB as diameter, draw a circle to cut that circle in C. Join AC, and produce it to meet the larger circle in D. The perpendicular to AD at D is the required tangent.
7. Suppose the triangle is ABC, with $\angle A = \angle B$. Let AD bisect $\angle A$ and meet CB in D. Let M be in AB produced, so that $BM = BD$. Then triangles MBD and ADM are similar, so that $AD^2 = BD \times AM$. From a point P on a circle of diameter AB, draw a tangent $PQ = AB$. Join Q to the centre of the circle, meeting it at S. Then $QS = BD$, and the triangle ABD can be constructed.

Answers for Exercises 5.3

2. Fermat primes are 3, 5, 17, 257, and 65537.
4. No. For example, the irreducible polynomial

$$x^4 + 6x^2 - 60x + 36$$

has real roots

$$\frac{\sqrt{y-6} \pm \sqrt{\frac{120}{\sqrt{y-6}} - y - 6}}{2}$$

where y is the real root of the irreducible polynomial

$$y^3 - 6y^2 - 144y - 2736$$

If all algebraic numbers of degree 4 were constructible then the sum of the two roots of the original polynomial would be constructible, and hence y would be constructible. But this is not so.

Answers for Exercises 5.5

1. $2^{64} - 1 = 274,177 \times 67,280,421,310,721$.
2. There are 31 such polygons.

Answers for Exercises 5.6

1. Use the quadratic formula.
2. The sum of the areas of the lunes equals the area of the triangle.

Answers for Exercises 6.1

1. The reduced forms are $\begin{bmatrix} 1 & 1 & 10 \end{bmatrix}$, $\begin{bmatrix} 2 & \pm 1 & 5 \end{bmatrix}$, $\begin{bmatrix} 3 & 3 & 4 \end{bmatrix}$.
2. There is only one such form, namely, $\begin{bmatrix} 1 & 1 & 14 \end{bmatrix}$.

Answers for Exercises 6.5

1. When $D = -8003$, there are 26 equivalence classes of gaussian forms. The class represented by $\begin{bmatrix} 3 & 1 & 667 \end{bmatrix}$ represents 3, and $3(-2)^{-1} \equiv 4000 \equiv 1797^2 \pmod{8003}$, so we can take $h = 4000$ and $z = 1797$ (see Theorem 6.5.5). This leads to $N = 3594$, and $M = 7404$. The matrix R has $a = 6$, $b = 1$, $m = -6$, $c = 2 \times 667$, $n = -600$, and

$s = 275$. By generating an F sequence we can reduce this matrix to the identity. The matrix H such that HRH^T is the identity is

$$\begin{bmatrix} -45 & -22 & -49 \\ 27 & 13 & 29 \\ -64 & -31 & -69 \end{bmatrix}$$

Adding up the squares of its last column we do get 8003, resulting in the decomposition $1000 = 300 + 105 + 595$.

Answers for Exercises 6.6

1. By Theorem 1.8.1,

$$(1^2 + 1^2 + 0^2 + 0^2)(e^2 + f^2 + g^2 + h^2)$$

is a sum of 4 squares. Thus it suffices to prove the 4 square theorem for odd numbers. Since, for any odd k, there is an odd s between $\sqrt{3k-2}-1$ and $\sqrt{4k}$ inclusive, we have the result from Theorem 6.6.1.
7. Let

$$n - 169 = a^2 + b^2 + c^2 + d^2$$

If a, b, c, d are all nonzero we are done. If just one of them, say, $a = 0$, then we have

$$n = 5^2 + 12^2 + b^2 + c^2 + d^2$$

If two of them, say, a, $b = 0$, then we have

$$n = 12^2 + 4^2 + 3^2 + c^2 + d^2$$

If all but d are 0, we have

$$n = 10^2 + 8^2 + 2^2 + 1^2 + d^2$$

Answers for Exercises 7.1

1. Since $a^{-1}a \equiv 1 \pmod 4$, it follows that $(a^{-1} - 1)/2$ and $(a - 1)/2$ have the same parity. Since $\left(\frac{a^{-1}}{p}\right)\left(\frac{a}{p}\right)$ equals 1, the two Legendre symbols are either both 1 or both -1.

Answers for Exercises 7.3

1. If prime p divides n exactly t times then the sum contains $\ln p$ exactly t times. Hence if $n = \prod p^t$ then the sum is $\sum t \ln p = \ln n$. The second result follows from the Möbius Inversion Formula.

Answers for Exercises 7.6

1. $P(100) = Q(100) - Q(25) + Q(6\frac{1}{4}) = L(100) - L(25) = 6 - 2 = 4$. In fact the four triangles in question are $(3,4,5)$, $(5,12,13)$, $(8,15,17)$, and $7,24,25)$.
2. 5.3.
3. 3.9.
4. The lowest point on the bug's head corresponds to the leftmost point on a y versus x graph of $xy(x^2 - y^2) = t$. This point is where y starts to have more than one value for each x. The cubic discriminant for solving y in terms of x is $t^2/4x^2 - x^6/27$. When, and only when, this number is negative does y have more than one value for each x. Hence the leftpoint point occurs when $x = (27t^2/4)^{1/8}$ and $y = x/\sqrt{3}$. The lowest point on the bug's head is thus

$$((t^2/12)^{1/8}, (27t^2/4)^{1/8}) \approx (0.73t^{1/4}, 1.27t^{1/4})$$

Answers for Exercises 7.8

1. Let $c > 1$ (but close to 1). For sufficiently large n, $\ln p_n < p_n^{1/c}$, and, by the Prime Number Theorem, $p_n/(2\ln p_n) < \pi(p_n) = n$. Hence $p_n < 2np_n^{1/c}$ and thus $p_n < 4n^{c/(c-1)}$. Thus

$$\lim_{n\to\infty} \frac{\ln p_n}{\ln n} \le \frac{c}{c-1}$$

Since c is arbitrary, this limit actually equals 1. Hence

$$\lim_{n\to\infty} \frac{p_n}{n \ln n} = \lim_{n\to\infty} \frac{p_n}{n \ln p_n} \ln p_n \ln n = 1$$

using the Prime Number Theorem again.

Index

Kluwer Texts in the Mathematical Sciences

1. A.A. Harms and D.R. Wyman: *Mathematics and Physics of Neutron Radiography.* 1986 ISBN 90-277-2191-2
2. H.A. Mavromatis: *Exercises in Quantum Mechanics.* A Collection of Illustrative Problems and Their Solutions. 1987 ISBN 90-277-2288-9
3. V.I. Kukulin, V.M. Krasnopol'sky and J. Horácek: *Theory of Resonances.* Principles and Applications. 1989 ISBN 90-277-2364-8
4. M. Anderson and Todd Feil: *Lattice-Ordered Groups.* An Introduction. 1988 ISBN 90-277-2643-4
5. J. Avery: *Hyperspherical Harmonics.* Applications in Quantum Theory. 1989 ISBN 0-7923-0165-X
6. H.A. Mavromatis: *Exercises in Quantum Mechanics.* A Collection of Illustrative Problems and Their Solutions. Second Revised Edition. 1992 ISBN 0-7923-1557-X
7. G. Micula and P. Pavel: *Differential and Integral Equations through Practical Problems and Exercises.* 1992 ISBN 0-7923-1890-0
8. W.S. Anglin: *The Queen of Mathematics.* An Introduction to Number Theory. 1995 ISBN 0-7923-3287-3

KLUWER ACADEMIC PUBLISHERS – DORDRECHT / BOSTON / LONDON